Numbers and Narratives

Why have there been so few women mathematicians? This book does not seek an answer in absence but in the forces, ruptures, and intensities that shape the becoming of a *femme philosophe*—a mathematician, scientist, and philosopher—within the shifting assemblages of eighteenth- and nineteenth-century Europe. Moving beyond exclusion as mere negation, it traces the conditions of emergence, the differential speeds and slippages through which women entered, inhabited, and transformed the mathematical sciences.

Drawing on auto/biographical documents, literary and philosophical writings, and the materialities of the archive, this book approaches the digital turn not as a tool but as a plane of composition, where new trajectories of memory work unfold. Between historiography and fabulation, it maps a space where women's mathematical thought was not only possible but inevitable—if only in flashes, excesses, and détours.

This book will resonate with scholars in the sociology, history, and philosophy of science and mathematics, particularly those engaged with feminist thought, the politics of knowledge, and experimental archival methods.

Maria Tamboukou is Professor of Feminist Studies at the University of East London and Leverhulme Major Research Fellow (2022–5). She has held academic positions in a number of international institutions, including the 'Hannah Arendt' Centre for Political Studies at the University of Verona, Italy. Her research activity develops in the areas of philosophies and epistemologies in the social sciences, feminist theories, narrative analytics, and archival research. She has published in English, Greek, and French, and her work has been translated in Italian, Norwegian, Portuguese, Spanish, Welsh, and Greek. She is the author of nine monographs, two co-authored books, and four edited volumes on research methods, including her latest monograph, *Epistolary Narratives of Love, Gender and Agonistic Politics* in 2023. Writing histories of the present is the central focus of her work, currently configured as an assemblage of feminist genealogies.

Literary Methods in the Social Sciences

In the context of a rise of archival, auto/biographical and diary methods in the social sciences, this series addresses the transferability, consistency and validity of these methodological approaches. Presenting authored studies and edited collections, it welcomes volumes tackling the epistemological and methodological issues surrounding the use of 'literary methods' in the social sciences, whether through practical, 'how to' volumes or by way of studies that present and develop the deployment of such methods. Projects for the series are encouraged that address:

- Archival Research
- Auto/Biographical Methods
- Blogs and other Digital Platforms
- Diaries and Journals
- Documents Based Research
- Epistolary Research and the Use of Letters
- Narrative Research
- Social Media Platforms

Series Editor
Maria Tamboukou, University of East London, UK

Titles in this series

Numbers and Narratives
A Feminist Genealogy of Automathographies
Maria Tamboukou

For more information about this series, please visit:
https://www.routledge.com/Literary-Methods-in-the-Social-Sciences/book-series/LMSS

Numbers and Narratives

A Feminist Genealogy of Automathographies

Maria Tamboukou

Routledge
Taylor & Francis Group
LONDON AND NEW YORK

First published 2026
by Routledge
4 Park Square, Milton Park, Abingdon, Oxon OX14 4RN

and by Routledge
605 Third Avenue, New York, NY 10158

Routledge is an imprint of the Taylor & Francis Group, an informa business

British Library Cataloguing-in-Publication Data
A catalogue record for this book is available from the British Library

ISBN: 978-1-032-74325-7 (hbk)
ISBN: 978-1-032-87696-2 (pbk)
ISBN: 978-1-003-53399-3 (ebk)

DOI: 10.4324/9781003533993

Typeset in Times New Roman
by Deanta Global Publishing Services, Chennai, India

For Odiseas

Contents

Acknowledgements *viii*

Introduction: Becoming a Mathematician, Philosopher, and Scientist 1

1 Memory Machines and Archival Traces 30

2 Epistolary Entanglements: Between the Personal and the Scientific 82

3 The Dynamics and Politics of Translation 133

4 Exceptional Women? The *Agonistics* of Knowledge 159

5 Love and Mathematics: Existential Choices and Intellectual Freedom 194

6 Philosophical Ruminations and Process Epistemologies 223

7 Narrative Rhythmanalysis: Listening to the Echo of the Subject 269

Conclusion: Future Pasts: Charting Assemblages 294

Index *325*

Acknowledgements

I am profoundly grateful to the Leverhulme Trust for awarding me a Major Research Fellowship (MRF-2011-004), which allowed me to dedicate three uninterrupted years (2022–5) to this project. This support was an extraordinary and unexpected gift in my academic career.

This book draws on research conducted in Italy, France, Sweden, the UK, and the USA, in libraries and archives including the Veneranda Biblioteca Ambrosiana in Milan, the Moreniana Biblioteca in Florence, the Bibliothèque Nationale de France, the Library of the Mittag-Leffler Institute in Stockholm, the British Library in London, the Bodleian Libraries in Oxford, and the Morgan Library and Museum in New York. I extend my deepest thanks to the librarians and archivists who generously facilitated my work.

I am also indebted to the members of the International Advisory Board, whose insights and support have been invaluable throughout the research project that underpins the writing of this book: Prof Beatrice Acciaio, Dr Hessa Almotairi, Prof Penny Burke, Prof Anna Chronaki, Dr Albina Danilova, Prof Giulia Di Nunno, Dr Waltraud Ernst, Prof Xin Guo, Prof Ruth Hagengruber, Dr Eva Kaufholz-Soldat, Dr Sabrina Liccardo, Prof Julie McLeod, Dr Nicola Oswald, Prof Carrie Paechter, Prof Mithili Ramaswamy, Prof Ann Rigney, Prof Liz Stanley, Prof Caroline Trotot, and Dr Sophie van den Elzen. A special acknowledgement to Prof Joyce Goodman, whose passing is an irreplaceable loss to the community of historians in the UK and worldwide.

Above all, this research would not have been possible without the lifelong inspiration, countless conversations, and spirited debates shared with my partner, a mathematician whose deep immersion in the mathematical *modus vivendi* and thinking has been transformative. Her unwavering support extended beyond intellectual exchanges—she introduced me to the wider mathematical community across Europe and beyond and, on field trips across the continent, joined me in the search for the *lieux de mémoire* of women mathematicians. For all of this—and for the shared moments of discovery, frustration, and wonder—my gratitude is beyond words, but above all, it is deeply felt with love.

Introduction

Becoming a Mathematician, Philosopher, and Scientist

'It may be difficult for those removed from the mores of the scientific community to understand the enormous reticence with which anyone, especially a woman, would make public his or her personal impressions and experiences, particularly if they reflect negatively on the community', Evelyn Fox Keller wrote, drawing on her own auto/biographical experiences (1997, 90). In its multiple genres, auto/biographical writing violates the rules of disinterest and objectivity, which have been at the core of long-held discourses revolving around science and the figure of the scientist. If the latter also happens to be a woman, the challenges become even greater. Keller has further commented (90), pointing to the by-now-recognised argument that 'gender shapes the culture of science, just as it shapes any other culture' (Govoni 2014, 14).

Such restrictions and limitations were not only shaping the order of discourse in science, including the human and social sciences before the 1980s, but also what has now been configured as 'the narrative turns' (Hyvärinen 2010). In the wake of 'the narrative turns', however, auto/biographical approaches have emerged and developed in the wider field of science studies, as epistemological modalities for understanding not only the lives of scientists but also the conditions of possibility for the emergence of scientific problems, concepts, findings, and solutions.[1] More than illuminating the histories of science, scientific auto/biographies shed light on the social studies of science, including the moral and political economies of knowledge production and exchange (see Mazzotti 2014).

But, here it has to be noted that the narrative turn in the history of science was not taken easily. As Georgina Ferry has noted, 'for historians of science, biography was a troubled and troubling genre, rather than simply a good way to raise the profile of forgotten scientists and to make their science accessible to a general readership' (2014, 59). Moreover, Paola Govoni (2000) has discussed the dangerous liaisons between biographies and feminism, as underpinning the low epistemic status of the biographical approach to the histories of science. As she has critically observed: 'strong arguments against biography are to be sought in the 'female' character, which I suspect, is often tacitly attached to biography' (403). Finally, in the long history of the auto/biographical genre, that goes back to antiquity,[2] there was always a tension between a focus on heroic scientists and the image of the progress of human reason, framing and indeed constructing the persona of the scientist, as Massimo Mazzotti has pointed out (2014, 119), particularly commenting

DOI: 10.4324/9781003533993-1

on the fact that the biographical approach to science did not find a place either in the discursive spaces of social and cultural histories or in 'the new images of science that came to dominate the intellectual landscape of the mid-twentieth century' (120). Although scientific biography remained in the margins, its role has significantly changed in the last decades of the twentieth century, following Thomas Hankins' pathbreaking defence of biography in the history of science (1979). In the light of such difficulties regarding auto/biographical approaches to mathematical sciences, I now turn to Paul Halmos' influential intervention in the field.

'I like words more than numbers and I always did' (1985, 3), Halmos wrote in the very first pages of his 'automathography, a mathematical biography written by its subject' (3). In Halmos' view, an automathography should not be conflated with an autobiography, 'the story of my origins and my life' (3). But is such a divide or separation possible? Throughout his book, Halmos refers to childhood memories, desires, relations with significant others, and impressions of places and spaces as well as political and cultural events that shaped his desire to become a mathematician. Books and languages play a significant part in his automathography: 'I read a lot, write a lot, and love languages—and I suppose at bottom it all comes down to liking words' (8), he has emphatically written, highlighting the importance of culture in his automathography. As George Sarton has importantly argued, 'the history of mathematics should really be the kernel of the history of culture' (1936, 4).

Halmos is adamant that to become a mathematician, 'you must love mathematics more than anything else' (1985, 400). Pure love is not enough, of course, 'you must work at it hard and without stop, and you must never give up', Halmos notes (400). And yet, when it comes to the hierarchy of existential needs, desires, and strivings, the love of mathematics comes first: 'I am not saying that the love of mathematics is more important than the love of other things. What I am saying is that to the extent that one's loves can be ordered, the greatest love of a mathematician (the way I would like to use the term) is mathematics' (401).

Halmos' automathography is written from the perspective of a male mathematician who followed the networks and opportunities available to his gender in the long run of the twentieth century. This does not mean that he did not face the prejudices of being a Jewish immigrant and of carrying his Hungarian accent, despite the fact that he was educated in the USA: 'Then there was the accent. I was a foreigner, with or without pejorative adjectives, I felt like one, and I sounded like one', Halmos has noted (15). And yet, while reading his automathography, I often wondered how different things would be for a woman becoming a mathematician in the same period. In Chapter Six of his automathography, Halmos fondly remembers the Institute for Advanced Studies at Princeton University, where he came of age as a mathematician:

> The Institute was a small, cosy operation when I arrived in Princeton in 1939. The centre of the life of all the mathematicians in Princeton, both University and Institute, was Fine Hall (the old Fine Hall), which is still my Platonic ideal of a mathematics building. Dark corridors, leaded windows, heavy furniture, worn carpets; the common room always open, always in use; the

> library up to date, complete, run with an iron hand by Bunny Shields, tiny, white haired, probably born looking as if she were in her late 50's, severe, but always helpful (84).

Halmos has written that he 'virtually lived in the common room', where 'early morning or the middle of the night, there was always someone there, and some conversation about mathematics, the war, the quirks of the big shots, or the best nearby restaurant for oyster stew was always going on' (85). There were no such opportunities for knowledge and networking for women mathematicians, since Princeton only admitted women in its graduate courses in mathematics in the late 1960s (see Shafer 1987, x). Even when women were admitted, they would not always feel welcome and at home. As Alice Shafer has poignantly commented, 'at Harvard University, some professors refused to allow women to sit in their classrooms. They were however allowed, to sit in chairs placed just outside the classroom doors, so that they could hear the lectures' (x).

Despite such harsh gendered restrictions, which went on beyond the middle of the twentieth century, the love that Halmos has written about seems to surpass gender, time, and geographical boundaries. In her long engagement with Sofia Kovalevskaya's life and work, Russian mathematician Pelageya Kochina has chosen 'Love and Mathematics' as an encapsulating title for the scientific biography of the first woman to hold a chair in mathematics in modern Europe (Kochina 1985). Love is indeed amongst an array of intense feelings that have arisen from women's auto/biographical writings in the long process of engaging with mathematics, science, and philosophy, as I will further discuss in Chapter 5.

In addressing such affects, feelings, and existential quests, in the long process of becoming a woman mathematician in the early and late modern periods, I have thus configured the plane of my research that underpins this book, in the intersection or rather entanglement of mathematics with philosophy, literature, and poetry. In gendering Halmos' (1985) notion of automathographies', what I argue is that working auto/biographically with women mathematicians' life documents, as well as their literary and philosophical writings, is a way of better understanding not only the gendered micro-histories of the discipline but also and perhaps more importantly, the slow process of the long way to gender equality, which is still unfolding in our days. In this light, *automathographies* become an important tool in what, drawing on Michel Foucault (1986), I configure as 'counter-memory studies'—a turn to the past as a way of reconfiguring the present and prefiguring the future of women in mathematics, as I will further discuss in Chapter 1.

The inspiration for this book, as well as the research project that it draws on, came unexpectedly one evening in conversation with a mathematician friend who mentioned Kovalevskaya. 'Who is she?' I asked, and after reading one of her several biographical entries, I subsequently challenged myself: 'how many more women mathematicians in history are out there that I don't know about?' To my surprise, it was not only my ignorance that I had to address: in the course of my research, I found out that women mathematicians were largely invisible and unknown even when I presented my research in mathematical departments, let alone wider audiences in gender and literary studies, history of education, or philosophy.[3] What I was further amazed to find out in my initial literature research is that the six women who emerged from their archival hideouts were not only mathematicians, but also scientists and philosophers as well as creators of literature and poetry. What were then the conditions of possibility for these women to emerge as 'scholarly personae' (Niskanen and Barany 2021) in their times and geographies? It is some initial lines of the social, historical, cultural, and geographical contexts within which women as 'scholarly personae' were formed and challenged that I want to sketch in the next section, although the charting of these important contexts will continue in Chapter 3 as well as throughout the book.

Gendered histories and geographies of science and mathematics

As the dominant intellectual movement in Europe during the early modern period, humanism prepared the grounds for the eighteenth-century Enlightenment. Although led by males, it nevertheless created conditions of possibility for women's nature and social position to be re-examined. While the dominant stance among humanists and Enlightenment male authors was women's biological and intellectual inferiority in relation to men, a closer study of their ideas reveals complexities, ambivalences, gaps, and interstices that allowed the emergence of 'the other voice' (King and Rabil 2005).

Published around 1405, Christine de Pizan's *Book of the City of Ladies* (2000) is one of the earliest cultural expressions in the long process of re-evaluating the female sex and initiated a rich body of literature revolving around what is now known as the *querelle des femmes* or *the debate about women.* There were four main issues at the heart of this debate: 'the problem of chastity, the problem of power, the problem of speech and the problem of knowledge' (King and Rabil 2005, xxv). What is also important to note is that the debate about women's nature, as well as its possibilities and limitations, included philosophical and literary treatises and essays, but also an agonistic body of correspondence between opposing positions and views. As I will further discuss in Chapter 2, women mathematicians' letters and correspondences were important auto/biographical documents in my research, opening up new epistemological concerns and inquiries in epistolary analytics. Agonism here is understood not just as conflict or struggle, but also in its Arendtian sense as a positive force for achievement, creation, and recognition, a focus elaborated in Chapter 4.

In the wider context of the *agonal* spirit, then, while the majority of male authors from the Renaissance onwards held the view that it was useless for women to study natural philosophy and mathematics, there were others who encouraged science education for the women of the nobility. There were families who saw their daughters' education as a way to advance their prestige—as was the case with Maria Gaetana Agnesi—or simply allowed them to follow the scientific study and work of their fathers or other family members. The advent of printing changed the form of the humanists' pedagogical advice and enlarged its audiences. It was not letters written in Latin anymore, but rather treatises on education written in the vernacular and therefore facilitating their wider circulation and indeed translation, as I will further discuss in Chapter 3. This democratisation of knowledge also meant that daughters were not so much dependent on their families' decisions about their educational futures. As we will further see, Sophie Germain and Mary Fairfax Somerville pursued mathematics despite familial objections, relying on family and community libraries.

The socio-historical and cultural patterns of women's engagement with science in Europe from the Renaissance onwards have been well treated in the literature from a wide range of perspectives.[4] Patricia Phillips has particularly pointed to the emergence of 'the scientific lady' as a response to women's exclusion from the all-male field of classics: 'For many women science was a preferable alternative to the classics. It had two great advantages. It was a study that demanded serious attention and yet the student needed no more elaborate preparation than commitment, application and an independent mind' (1990, 28). Feminist historians and cultural theorists have also looked at the various ways women in the early modern period struggled to forge intellectual authority through textual and visual constructions of themselves as 'scholarly personae' (see Vanacker and van Deinsen 2022).

Here, it is important to acknowledge that during the seventeenth and early eighteenth centuries, science was increasingly practised by independent scholars outside the confines of institutions. Judith Zinsser has argued that during the period of this public turn, 'more women of the privileged classes had an opportunity to engage in the intellectual discourses of their day' (2005, 4). Karen Detlefsen, however, has taken issue with the narrative that 'when science belonged to exclusive, private institutions, it was dominated by men, and when science became more public, women became practitioners in greater numbers' (2002, 129). Without refuting the idea of the 'rise of public science', Detlefsen has posed some pertinent questions regarding the divide between 'the private' and the 'public' as well as the meanings that we ascribe to institutions (141). What Detlefsen has argued instead is that although there was 'a slight respite' of women's marginalisation within the world of scientific knowledge, ideas, and debates during the seventeenth century, their overall marginalisation was due to their 'exclusion from educational and scientific institutions' (142).

I concur with Detlefsen's argument, and throughout my work of writing feminist genealogies, I have persistently found that the history of women's education is a history of exclusions, or rather of 'marginalisations', as Ruth Watts has observed (2007, 1), notwithstanding the subtleties, nuances, and differentiations that my genealogical excavations have brought to the fore (see Tamboukou 2003, 2017). Over the years, scholars in the field of gender and education have worked tirelessly in recovering women's position in the history of education from different angles, perspectives, and disciplinary fields. In this context, there is a rich body of literature around the history of women's science education in the West.[5] Apart from different discourses around women's access to science education, which have been comprehensively treated in Leigh Whaley's (2003) important study on the different debates in the history of gender and science in the Western world, there were also important differences in women's opportunities for education even among neighbouring European states, given their social, cultural, political, and religious differences, as well as the diverse philosophical movements and scientific trends that developed within them, as I will further discuss in Chapter 3.

In this context of diversity, there was a significant difference in the educational opportunities for women in the Italian states during the seventeenth, eighteenth, and nineteenth centuries. While women faced hurdles across Europe and their education mostly happened within the private domain, it was only in the Italian states that some women were allowed to be connected with formal scientific institutions, such as academies and universities. Here, it is important to note that apart from the academies—both formal and informal—there was also a different tradition in the circulation of knowledge in the Italian states: the *conversazioni,* conversation clubs, where men and women gathered to talk mostly about literature, poetry, philosophy, and science, but also to listen to music and play word games. As Gabriella Berti Logan has shown, many official academies in Italy actually started as *conversazioni* (1999, 81). The Academy of Sciences of Bologna, for example, started from the Academy of the Inquieti, which was operating from the home of its founder, Eustachio Manfredi, with the involvement of his two sisters (94). It is not accidental then that it was the Bologna Academy that offered membership to Émilie Du Châtelet in 1746 and to Maria Gaetana Agnesi in 1748, in recognition of their important contributions to science and mathematics, as we will further see.

There were, of course, similarities between the *conversazioni* and the more well-known tradition of the *salons*, acting as social and intellectual platforms of women's active involvement in the political and cultural formations of modernity, as well as in their role in the dissemination of literary, philosophical, and scientific knowledge in France and throughout Europe (see Goodman 1994). Londa Schiebinger has emphasised that the Parisian *salon* was an 'institution of science [while] French *salons* of the seventeenth and eighteenth century competed with academies for the attention of the learned' (1989, 30). Although both forms of intellectual gatherings—*conversazioni* and *salons*—soon surpassed national and ethnic boundaries and became transnational and hybrid formations in the circulation of knowledge and culture, there was also a significant difference between them. While women's role in the *salons* was primarily to be promoters of literary and scientific knowledge, several women in the eighteenth-century Italian *conversazioni* made original contributions to the creation of knowledge. Marta Cavazza has pointed out that this was 'a unique historical case' in the diverse histories of the European Enlightenment.(2009, 279).

It is thus in this diverse field of socio-historical, political, and cultural conditions that the six women mathematicians, philosophers, and scientists of this study emerged. Here, it is important to note that these women were initially selected for their engagement with mathematics, in addition to their philosophical and scientific work. As I will discuss in the last section of this chapter and throughout the book, there were no clear disciplinary boundaries in the early modern period (see Smith 2009), but since the focus of my research was particularly on women mathematicians, I followed the bio-bibliographical source of *Women of Mathematics*

(Grinstein and Campbell 1987) as well as two online biographical sources, the Agnes Scott College project of *Biographies of Women Mathematicians* and the *MacTutor Index*.[6] As my research unfolded, a seventh woman mathematician, astronomer, and poet emerged, the Chinese Wang Zhenyi (1768–97), whose prolific scientific and literary work has not been translated in English or in any other European language. Thus, her archive remains inaccessible to Western audiences, although her posthumously published collected works, *Defengting chuji* [Collection of the Defeng Pavillion], has been digitised and is available online.[7] Since the short but valuable biographical entries around her (Leung 1998; Peterson 2000; Wang 2014) do not constitute sources for an in-depth study of her life and literary work, I decided to include her in the pen-portraits that I will sketch in the next section, thus marking a case in the project of widening the Western canon of women in mathematics, science, and philosophy (see Waithe and Dykeman 2023).

Émilie Du Châtelet (1706–49): science, happiness, and mathematics

'Be decided on who you want to be and on what you want to do, this is what most people lack; it is however the condition without which there is no happiness',[8] Du Châtelet wrote in presenting some of 'the great machines of happiness' in her philosophical treatise *Discours sur le bonheur* (1961, 16), which I will discuss at length in Chapter 6. Indeed, Du Châtelet's life is an actualisation of happiness as a lifelong endeavour. Born into an upper-class family of the old regime, she was expected to fulfil the role of a dutiful aristocratic daughter. However, her childhood took an unusual turn due to her perceived unattractiveness. Described by her father as unusually tall for her age, physically awkward, and possessing coarse skin and disproportionate features, she was deemed unlikely to marry despite her substantial dowry. As a result, a future in the convent seemed a likely path for her (see Edwards 1970, 4).

According to her biographers it was thus because of her grim future in the marriage trade that her father may have thought it a form of compensation to provide her with the best teachers.[9] This was not very difficult since Du Châtelet grew up alongside her younger brother Elisabeth-Théodore, who was given a full scholastic curriculum in his parents' home, as he was prepared for a church career (see Zinsser 2006, 28). Du Châtelet also had full access to her father's library, which occupied three rooms of the family home, and from an early age, she expressed her desire to become a scholar. By the age of ten, she had mastered Latin and knew long passages from Virgil, Lucretius, and Horace by heart. According to Voltaire's memoirs, it was her father who had taught her Latin (in Hamel 1910, 22). Likewise, she knew the philosophical works of Cicero, and by the time she was twelve, she was reading, writing, and was fluent in English, Italian, Spanish, and German, while she was also 'an accomplished musician' (23).

Du Châtelet's humanistic education extended to a passion for theatre, where she and her brother performed scenes from Corneille and Racine (Zinsser 2006, 29). This early interest grew into a lifelong love for theatre and opera, which she later reflected upon in her philosophical writings, as discussed in Chapter 6.

Alongside her engagement with the arts, Du Châtelet developed a serious interest in metaphysics and mathematics, possibly studying Euclid's theorems and proofs, as Zinsser suggests (28). She also explored Descartes' works, drawn to his integration of analytical geometry with meditations on the interplay of mind, matter, soul, and body (see Edwards 1970, 11). Her education was further enriched by her active participation in Parisian salons, including her father's Thursday gatherings, which brought together courtiers and members of the Académie française and Académie royale des sciences (see Zinsser 2006, 31). These salons offered a stimulating environment that shaped her social, intellectual, and cultural development.

Despite her remarkable humanistic, scientific, and cultural education, Du Châtelet's intellectual pursuits were paused when she reached marriageable age. No longer the 'ugly duckling', she was swept into the marriage trade, adhering to the norms of heteropatriarchal aristocracy. At nineteen, she married Florent-Claude, Marquis Du Châtelet-Lomont, and fulfilled her duties as a wife and mother, having three children: Gabrielle-Pauline, born in 1726; her first son, Florent-Louise, in 1727; and finally her second son, François-Victor, in 1733.

However, her seemingly conventional aristocratic life took a dramatic turn just months after the birth of her third child. In 1733, two significant *events* erupted in her life: she sought mathematical tutoring from the French academician Pierre-Louis Moreau de Maupertuis and began her famous relationship with Voltaire. These intertwined developments marked the start of her transformation into an influential scientist and a central figure of the French Enlightenment—an era now recognised as 'the Époque Émilienne' (Hagengruber 2022). This legacy, however, took nearly 300 years to gain acknowledgement. Further exploration of these *events* and their philosophical implications will follow in Chapters 1, 2, and 4.

Du Châtelet's tumultuous and impactful relationship with Voltaire, marked by love, collaboration, and enduring friendship, has been extensively documented.[10] They met in Paris in 1733, and shortly after, Voltaire faced persecution following the 1734 publication of his *Lettres philosophiques* [Letters Concerning the English Nation],[11] which lacked royal approval. Beyond this procedural violation, the book's critical essays on Voltaire's experiences in Britain were seen as an attack on the French political system, prompting an order for his arrest. This crisis led Voltaire to seek refuge at the Château de Cirey, Du Châtelet's family estate, where he stayed while she coordinated efforts with their influential circle to secure his pardon from the King (see Zinsser 2006, 90–1). Their partnership flourished during this period, blending intellectual collaboration with personal commitment.

Being refurbished with Voltaire's expenses, the Château de Cirey was turned into the lovers' scientific abode, where they spent many months 'Newtonising'—as they called their discussions of Newton's theories—reading, thinking, writing, and conducting experiments, while at the same time inviting friends and organising a range of social activities, including dinner parties, philosophical discussions over coffee, theatrical performances, masquerades, opera singing, and card games.[12] The Château eventually gained renown as a hub of scholarship, attracting mathematicians, writers, and philosophers from across Europe. It resembled a private academy more than the typical Enlightenment salon (see Zinsser 2006).

The Château de Cirey provided the essential space and conditions for Du Châtelet to pursue her groundbreaking scientific career. In an era when women faced significant barriers to participation in Enlightenment knowledge circles, her first intellectual project—a translation of Bernard Mandeville's *Fable of the Bees*—reflected a common strategy for women scientists to assert authority within the Republic of Letters, as I will discuss in Chapter 3.

Although this translation remains unpublished and largely unknown, Du Châtelet's later works gained her the recognition she sought. Her *Dissertation on the Nature and Propagation of* Fire (1739) marked the first essay by a woman published in the proceedings of the Royal Academy of Sciences, a milestone I will elaborate on in Chapter 4. This was followed by her influential *Institutions de Physique* (*Foundations of Physics*) in 1740 (Du Châtelet 1740). In her final years, she devoted herself to translating and commenting on Newton's *Principia*, a monumental achievement published posthumously in 1759 (Du Châtelet 1759) after her tragic death during childbirth on September 10, 1749. These pivotal works, which brought her both fulfilment and recognition, will be further explored in Chapters 3 and 6.

Du Châtelet's life ended amidst personal and intellectual intensities—her deep attachment to Newton's ideas and her passionate relationship with Jean François de Saint-Lambert. Aware of the risks of her unexpected pregnancy, she took deliberate steps to ensure her scientific legacy would endure, a theme I will return to in Chapter 1. Just a year after her untimely death, another woman, Maria Gaetana Agnesi, made history when she was awarded an honorary chair in mathematical analysis at the University of Bologna, following the success of her *Instituzioni Analitiche* (1748), a calculus textbook that secured her reputation as the first woman in the Western world to achieve prominence in mathematics—a legacy I will discuss next.

Maria Gaetana Agnesi (1718–99): performing the truth of mathematics

Maria Gaetana Agnesi was the eldest daughter of a rich silk merchant in Milan, who saw his daughter's linguistic talent, prodigious memory, and sharp intellectual abilities as a way to advance his social position in the Milanese nobility.[13] Agnesi's intellectual abilities emerged in her early childhood and as her first biographer wrote, 'she was so gifted with such a very singular talent, an impatient natural genius for scientific notions, and an admirable memory that she almost drank with her milk the first elements of the charming French language' (Frisi 1799, 8). Moreover, since there were several children in the Agnesi family, girls were educated alongside the boys by private tutors. While playing in the same room where her younger brother Giacomo was tutored in preparation for his college entrance, Agnesi displayed a surprising ease in retaining and repeating her brother's lessons in Latin just by listening to them, 'with admirable order and precision' (10). This unusual phenomenon was duly noted by the tutor Abbé Niccolò Gemelli, who informed her father and was subsequently asked to take over her tuition in Latin.

Agnesi's progress in Latin was fast and soon reached the heights of perfection, as testified by her first appearance in public at a soirée organised in the Palazzo gardens on 18 August 1727, where she defended women's ability to be educated in the liberal arts before a group of patricians (Agnesi 2005). Following this very successful public event, which I will further discuss in Chapter 4, Agnesi's overall education was carefully planned, and since money was not a problem, it encompassed some of the most prestigious intellectual figures in Milan, who were all ecclesiastic scholars with the exception of Count Carlo Belloni, her personal advisor, who coached her in disputation techniques and introduced her to the study of algebra and geometry as well as Newton's theories. As Agnesi's biographer Massimo Mazzotti (2007) has further shown, Agnesi became a mathematician in the context of what he configures and charts as Milan's version of 'the Catholic Enlightenment', a grid of intelligibility within which education in general and mathematical knowledge in particular become components of faith 'for a certain kind of enlightened Catholic, who perceived its philosophical certainty and clarity to be theologically illuminating' (Findlen 2011, 256). Moreover, Agnesi's oration was the first of many knowledge spectacles that she was asked to perform, usually in the company of her younger sister Maria Teresa, as we will further see in Chapters 2 and 4.

In her meticulous studies of gender and culture in eighteenth-century Italy, Cavazza has shown that the 'specularisation' of female knowledge was a widespread literary phenomenon and included many 'cases of young women celebrated for their knowledge in philosophical-scientific (and in two cases legal) subjects' (2009, 280). Moreover, specularisation paradoxically went hand in hand with the demand for 'modesty', the idea that while some 'exceptional' women could be celebrated for their intellectual and artistic skills, this should not deviate them from their traditional roles as dutiful daughters, mothers, and wives. As Cavazza has commented, the visibility of women philosophers and scientists in eighteenth-century Italy was difficult to reconcile 'with the image of modesty and 'seclusion (ritiratezza)' to which women were supposed to conform according to the behavioural codes and their associated values prevalent in Italy at the time' (286). While celebrating Agnesi's impressive intellectual skills and linguistic talents, Frisi has noted that the most wonderful thing about her was her 'natural reluctance to appear in public; reluctance that she had to overcome, with a virtuous fight, so as to obey her Father's wishes with total Filial submission' (1799, 11).

It seems, however, that Agnesi's 'filial submission' had its limits, and the duty to perform and display her knowledge publicly became a burden too heavy to bear, to the point that a 'mysterious illness' fell upon her, which disrupted her father's plans of exhibiting his talented daughter. All her biographers have speculated extensively on the seizures that tormented young Agnesi around the period that her early studies were concluded: she was in a short time taken by a convulsive sickness, whose strength forced her several times a day to start hopping all by herself, making it hard for her family aides to hold her down (18–19). Although she eventually recovered from this 'obstinate illness' (28), her own long-term resolution was to take the veil and retreat from the wild world of knowledge spectacle. Under the pressure of her father's insistence, however, a compromise was reached: she would continue living in the Palazzo under three conditions: 'to dress simply and modestly, to be able to go to church at her discretion, and to stop completely her attendance at dances, theatres and other profane amusements' (29). This compromise, however, was also the beginning of her serious engagement with the study of algebra and geometry, as 'the only provinces in the literary world where peace reigns' (30).

Agnesi withdrew from the public spectacle of knowledge to focus on philosophy and mathematics, producing significant works that highlighted women's educational potential and intellectual contributions. Her *Philosophical Propositions* (1738) articulated these views, while her *Analytical Institutions* (1748) earned her European acclaim and an honorary professorship at the University of Bologna for its clarity, precision, and synthesis—a legacy explored further in Chapters 4 and 6.

Agnesi's ideas about women's education, philosophy, science, and mathematics were gradually shaped through a long process of catalytic *events* that erupted in her life: her father's decision to use her linguistic and cognitive talent as a pathway to social advancement, the choice of her tutors, who significantly influenced her interest in scholarly work, and last but not least, the illustrious publications of her scientific work. But soon after her father's untimely death in 1752, she withdrew from the world of science and mathematics and devoted herself to the education and care of her immediate community. This unforeseen decision created the 'Agnesi enigma' (Anzoletti 1900) that is still debated in the literature revolving around her. Agnesi died penniless on 9 January 1799, just two years before Karl Friedrich Gauss published his *Disquisitiones Arithmeticae,* a monumental work in mathematical sciences that triggered his epistolary encounter with Sophie Germain, as we will see in the next section.

Sophie Germain (1776–1831): processual thought and the mathematics of the social

Sophie Germain's life has been told and retold from several angles and in different genres and media over the years. Almost all renditions of her life, however, draw on two historical sources, both written in the nineteenth century. The first biographical note came from her friend, Guglielmo Libri, an Italian mathematician and member of the French Academy of Sciences. Libri first wrote Germain's obituary in the *Journal Des Débats* on 18 May 1832, almost one year after her passing, and it was then included in the preliminaries of the first publication of her philosophical work, *Considérations Générales,* in 1833 (Germain 1833). The second was written by Jean-Léon-Hippolyte Stupuy, a poet, playwright, and literary figure. His study first appeared in the 1879 publication of her *Œuvres philosophiques* and was included again in its second edition in 1896 (Germain 1896). What I have found interesting in studying these first two biographical sources is the unacknowledged iterations that slip from the first to the second, eventually creating a biographical matrix within which all subsequent biographies are entangled.

Germain was born and raised in Paris, where political upheavals shaped her life from an early age. At thirteen, she witnessed the eruption of the French Revolution just outside her family home on Rue, St. Denis, in the heart of the city. More significantly, her father, Ambroise-François Germain, a wealthy merchant and member of the liberal bourgeoisie, was deeply involved in revolutionary politics. As Deputy of the Third Estate for Paris and later a member of the National Constituent Assembly, his political engagement and the debates within their home left a lasting impression on young Germain. This early exposure to the intellectual and cultural upheavals of revolutionary France profoundly influenced her worldview, later reflected in her unfinished philosophical treatise (Germain 1896), a theme I will explore further in Chapter 6.

Revolutions are as unsettling as they are transformative, especially for a young girl experiencing the turmoil of the radical phase of the French Revolution. For Germain, her father's library became a sanctuary, offering her a distraction from the pervasive fear. As her biographers note, she turned to study as a way to cope with the uncertainty and danger around her (see Libri 1832, 1).

This period was particularly precarious for intellectuals and scientists. The revolutionary turmoil of the early 1790s created an environment hostile to science, epitomised by the infamous declaration at Antoine Lavoisier's trial: 'The Republic has no need for geniuses' (see Jones 2016). Lavoisier, accused of tax collection, was executed, and foreign-born mathematicians such as Joseph-Louis Lagrange, who later became Germain's mentor and correspondent, lived under constant threat due to the 'law of suspects' (see Musielak 2020, 11–12). Such instability shaped Germain's early intellectual pursuits and her determination to seek refuge in the world of mathematics.

Given that, as a woman, Germain was also excluded from all formal educational institutions for higher studies—before as well as after the revolution—the library became the site of her self-education, and mathematics was chosen as her favourite discipline.[14] The details of her fascination with mathematics through the story of Archimedes' death as a result of his total immersion in the magic world of numbers, figures, and calculations have become a refrain in Germain's biographies: 'this geometrical science so endearing that nothing can distract from it, not even a threat of death, this science of which she hardly knows the name, this is the one that suits her; and, on the spot, she takes the heroic resolution to give herself completely to it' (Stupuy 1896, 6).

Sophie Germain's mathematical journey began at home but expanded significantly through her access to the lectures of the École Polytechnique via correspondence, introducing her to the Parisian scientific community, as I will discuss in Chapter 2. Her passion for number theory was ignited by Karl Friedrich Gauss's *Disquisitiones Arithmeticae* (1801), leading to a well-documented correspondence with Gauss during a turbulent era marked by the transition from the French Republic to the First Empire and the Napoleonic wars.

Amidst the uncertainty of the times, Germain also pursued her interest in the mathematical laws of acoustics, ultimately achieving an unprecedented milestone: winning the prestigious French Academy of Sciences prize in 1816 for her work on elasticity theory. This achievement, following two prior submissions in 1811 and 1813, cemented her reputation as a significant mathematical contributor, as noted by her biographers (see Libri 1832, 1).

After receiving her award, Sophie Germain oversaw the publication of her work (Germain 1821) and became more integrated into the Parisian scientific community, where she formed a friendship with Libri, primarily through correspondence. Their letters not only illuminate Germain's mathematical contributions but also offer insight into her personal thoughts and feelings, as I will explore in Chapter 2. In a letter to Libri dated 8 February 1830, Germain first mentioned her illness, expressing the pain and exhaustion that had hindered her ability to study (MBF/FPL/f432).[15] Her final letter to him, written on 18 April 1831, revealed the severity of her condition and the unbearable suffering it caused (BNF/NA4073).[16] Germain passed away two months later, on 27 June 1831.

An intriguing detail about Germain's life is that she never met Mary Fairfax Somerville, despite their overlapping timelines and shared mathematical interests. Somerville first visited Paris in 1817, one year after Germain's *Grand Prix de Mathématiques* victory, dining with several prominent French mathematicians who were familiar with Germain's work. While there is no evidence of direct contact, I have often wondered whether the two women knew of each other, given their proximity in age and overlapping networks. I will explore Somerville's life and contributions further in the next section.

Mary Fairfax Sommerville (1780–1872): automathographical desires

'My life has been domestic and quiet. I have no events to record that could interest the public. My only motive in writing it, is to show my country women that self-education is possible under the most unfavourable and even discouraging circumstances' (1872, 1).[17] In the very first lines of her unedited autobiography, Somerville makes a clear and brief statement about her decision to write a memoir, explicitly in defence of women's education and implicitly to show the strength of the female mind. Leaving aside the fact that this simple statement was not included in the published memoir—an effect of editorial interventions that I will discuss in Chapter 1—Somerville was indeed an autodidact, at least in the beginning of her engagement with mathematics. Her autobiography vividly recounts her first encounter with algebra in her hometown of Burntisland, Scotland. Invited by a friend to view her embroidery, Somerville's attention was drawn to a magazine featuring puzzles, charades, and a strange set of symbols involving letters and lines. Curious, she asked about it, only to learn it was 'a kind of arithmetic they call algebra' (32), which her friend could not explain further. Intrigued but unsupported, Somerville felt 'sad and forlorn' (32), with no guidance to satisfy her curiosity. Her breakthrough came unexpectedly through art. Her drawing lessons with landscape painter Alexander Nasmyth introduced her to Euclidean geometry. Nasmyth's suggestion to study *Euclid's Elements*—as the foundation of perspective, astronomy, and mechanical science—opened a new path for Somerville. This pivotal moment helped her begin exploring mathematics independently and marked the start of her journey into the 'magic world' of algebra (34).

Somerville's fascination with mathematics endured despite early obstacles. Unable to visit a bookshop herself, she obtained *Euclid's Elements* and Bonnycastle's *Algebra* through her brother's tutor, Mr Craw. With these resources, she began studying independently, eagerly immersing herself in the mystical world of mathematical symbols. Balancing her household duties as a dutiful daughter, she rose early to play piano, painted during daylight, and stayed up late at night reading Euclid by candlelight.

Somerville's mathematical pursuits, however, faced disruption in 1804 when she married Samuel Greig and moved to London. Her husband, dismissive of women's intellectual abilities, made little effort to support her passion, though he did not actively prevent her studies. In a small, poorly ventilated bachelor's house, Somerville's new life was far from ideal, particularly as she became a mother of two (54). Yet, despite her isolation and the demands of domestic life, she persevered, devoting her solitary days to her mathematical and intellectual pursuits with unwavering determination.

The untimely death of her husband in 1807 unexpectedly opened new doors for Somerville. As a young, financially independent widow, she returned to Scotland and immersed herself in mathematics, despite being largely self-taught. She diligently studied topics such as trigonometry, conic sections, and astronomy, though she struggled with Newton's *Principia* (57). Free from the disapproval of her extended family, Somerville divided her days between caring for her children, pursuing mathematics, and playing music.

Recognising the need for deeper engagement with mathematics, she found support from a small circle of Edinburgh intellectuals associated with the *Edinburgh Review*, who championed educational opportunities for women. With their guidance, she built a substantial library of mathematical works, primarily by French mathematicians, some of whom she would later meet during her travels. Encouraged by this group, she began contributing solutions to problems published in the *Mathematical Repository*, an endeavour further explored in Chapter 4.

Her second marriage to William Somerville was transformative, positioning her at the heart of the scientific and mathematical networks that would define her career. After relocating to London, William's involvement in scientific circles introduced Somerville to the French tradition of analytical mathematics, which was gaining prominence as British mathematics declined. Influenced by figures like Charles Babbage, she became deeply engaged with French methodologies, marking a turning point in her intellectual development.[18]

In 1817, her encounters with prominent French scientists such as Arago and Biot further cemented her connection to the European scientific community. Through social and professional gatherings in Paris, she gained direct access to leading thinkers and institutions, offering her a unique perspective on Parisian science. Her travels through France, Switzerland, and Italy expanded her network, allowing her to build lasting relationships with scientists across Europe and North America (see Somerville 1872, 72).

These connections culminated in the proposal to translate Laplace's *Mécanique Céleste* into English, a monumental recognition of her expertise. As Maria Edgeworth remarked, Somerville was regarded as the only woman capable of understanding Laplace's work—a testament to her intellectual prominence and the collaborative support of the scientific communities she engaged with throughout her life (Somerville 2001, 151).

By the time of Laplace's death in 1827, Somerville was widely recognised among English and French philosophers as a skilled interpreter of his work. As a token of esteem, she was even sent a lock of Laplace's hair—a gesture reflecting the high regard for her talents (Patterson 1983, 49). As I will further discuss in Chapter 3, the publication of her translation of *Mécanique Céleste* in 1831 marked the official start of her career in scientific writing.

In her autobiography, Somerville describes the challenges of balancing her scientific pursuits with her responsibilities as a wife, mother, and social hostess (1872, 62–3). Rising early to organise family matters, she carved out time for writing but often faced interruptions. Unlike men, who could prioritise work under the guise of business, Somerville had no such excuse. Visitors frequently disrupted her, sometimes arriving during moments of deep concentration, forcing her to pause and later resume her work as though marking a page in a book. Writing in the shared drawing-room during winter, she often hid her papers at the sound of the doorbell to protect her 'secret' (62–3) from discovery. Despite these obstacles, she developed the discipline to navigate these competing demands, laying the foundation for her prolific contributions to science.

Somerville's translation of Laplace's *Mécanique Céleste* was both her first and last book directly focused on mathematics. As Stenhouse notes (2021, iii), her subsequent works were shaped by a practical need to generate income, particularly as the Somerville family faced financial difficulties, leading to their permanent relocation to Italy in 1838.

While Somerville did not fully realise her aspiration to dedicate her life to mathematics, her influence inspired others, notably Ada Lovelace. In her autobiography, Somerville recalls her friendship with Lady Byron and Ada during her time in Chelsea, noting that Ada often stayed with her and studied mathematics on her advice (1872, 153). Their relationship continued through correspondence after Somerville moved to Italy, profoundly shaping Lovelace's turn to mathematics, as explored further in the next section.

Ada Lovelace (1815–52): mathematics as poetical science

At the age of twenty-six, while studying advanced mathematics with Augustus De Morgan, Ada Augusta Byron, Countess of Lovelace, began to delve into the nature of imagination. In a short essay dated 5 January 1841, titled *What Is Imagination?* (LB 175, ff. 199–200)[19], she explored this elusive faculty often associated with poets and artists, a theme that will be examined further in Chapter 6. During this period, Lovelace was also preparing the groundbreaking paper that would solidify her reputation: her annotated translation of Luigi Menabrea's *Notions sur la*

Machine Analytique de M. Charles Babbage (1843). But what had sparked her interest in the concept of imagination? The answer lies in her early education, shaped not only by her aristocratic upbringing but also by her inheritance as the daughter of the renowned poet, Lord Byron.

Born in London, Lovelace exhibited a remarkable aptitude for mathematics from an early age. Her mathematically inclined mother, Anne Isabella Noel Byron—whom Lord Byron nicknamed the 'Princess of Parallelograms'—sought to counterbalance any poetic sensibility or perceived instability Ada might inherit from her father (see Lewis 1995). As an educational reformer and philanthropist, Lady Byron ensured that her daughter's education was rigorously focused on science and mathematics. This was reinforced by a series of strictly supervised governesses and tutors, including Dr William King and Lady Byron's own childhood mentor, William Frend, a Unitarian minister (see Hollings, Martin, and Rice 2017, 6).

Lady Byron's adherence to Pestalozzi's principles of learning through action encouraged Lovelace's early fascination with practical skills, particularly her interest in flight. She dreamed of writing a book on 'flyology' and even designed a flying machine. In a letter to her mother dated 7 April 1828, she vividly described her invention: a horse-shaped device powered by a steam engine, equipped with large wings to lift a rider into the air (LB 41, ff.57–8). While her mother ultimately curtailed these experiments, deeming them too dangerous, Lovelace's letters reveal that her approach combined imaginative vision with serious study, laying the groundwork for her later intellectual achievements.

Recognising the importance of mathematics for channelling her creativity, Lovelace resolved to study it rigorously. In an 1834 letter to Dr William King, dated 9 March 1832, she expressed her determination to master Euclid, arithmetic, and algebra, confident in her ability to overcome any challenges (LB 172, f.126). While Dr King suggested a conventional British approach to mathematics (see Hollings, Martin, and Rice 2017, 9), her tutor William Frend recommended she connect with Somerville, renowned for her scientific achievements and personal warmth. Indeed, Somerville became both a mentor and lifelong friend to Lovelace, introducing her to the advanced world of continental mathematics and facilitating her connections with Babbage and her future husband, Lord William King. King, a friend of Somerville's son, Woronzow Greig, was seen as an ideal match for Lovelace, as suggested in Greig's unpublished memoir, which hints that he—and likely his mother—thought highly of the pairing (MSC, b206).

Lovelace's marriage to Lord William King, followed by the birth of three children in quick succession, temporarily interrupted her mathematical pursuits but never diminished her determination to continue her studies. During this period, she often turned to Somerville for guidance and reassurance, affirming both to her mentor and herself that her passion for mathematics remained undeterred. This unwavering commitment, as I will explore further in Chapter 2, became a defining feature of her life. By 1838, Somerville had relocated to Italy, and while their correspondence became less frequent, Lovelace's mathematical journey advanced under the tutelage of De Morgan, the first professor of mathematics at the University of

London and a strong advocate for women's higher education. Under his guidance, Lovelace made significant strides in mathematics, ultimately leading to the publication that secured her fame, as I will examine in Chapter 4.

Lovelace succumbed to uterine cancer on 27 November 1852. Honouring her final wishes, she was buried beside her father, whom she had never met, at the Church of St. Mary Magdalene in Hucknall, Nottinghamshire. Two years later, in 1854, a Russian aristocratic girl was born in Moscow, who would become the first woman professor of mathematics in modern Europe, as I will discuss next.

Sofia Kovalevskaya (1850–91): mathematician with the soul of a poet

Sofia Vasilevna Korvin-Krukovskaya, born on 15 January 1850, in Moscow, showed exceptional mathematical talent from an early age. Raised in a patriarchal aristocratic family, her intellectual abilities were nurtured when, at age eight, her father hired a mathematics tutor at their family estate in Palibino. Her tutor was impressed by her innovative approach to geometry, describing how she independently devised methods to understand key concepts. Later, at fifteen, she excelled in differential calculus under the guidance of Professor Alexander Strannolyubsky, who noted her remarkable grasp of advanced concepts.[20]

Kovalevskaya, despite her exceptional talent, could not pursue a university degree in nineteenth-century Russia due to gender restrictions. However, the 1860s brought a wave of intellectual freedom, with many young radicals supporting women's education (see Koblitz 1998). Through a 'white marriage' with progressive Vladimir Kovalevskii in 1868, Sofia was able to pursue her academic ambitions abroad, continuing her education in mathematics in Heidelberg. In 1870, Kovalevskaya moved to Berlin to work under Karl Weierstrass. Her academic journey was briefly interrupted by a trip to Paris in 1871, where she supported her sister involved in the Paris Commune. Returning to Berlin, she produced three groundbreaking papers, earning her doctorate from the University of Göttingen in 1874, becoming the first woman in Europe to do so. Following her academic achievement, Kovalevskaya returned to Russia, where her personal life took precedence. After years of emotional tension, her white marriage with Vladimir was consummated, and their daughter, Fufa[21], was born. However, Kovalevskaya and her husband faced professional challenges. She was denied academic positions due to her gender, while her husband's radical views further limited their opportunities. Tragically, her husband's financial ruin led to his suicide in 1883 while she was in Paris, trying to resume her academic career.

During this period, Kovalevskaya had received an invitation from Gösta Mittag-Leffler, a Swedish mathematician and disciple of Weierstrass, to join the newly established mathematics department at Stockholm University. Despite reservations about the financial terms, she accepted and arrived in Stockholm in November 1883. Her appointment was celebrated in the national press, with one article proclaiming her as 'the princess of science'; however, this position initially came without a salary, a fact she wryly noted in a letter to her brother-in-law: 'They would be better to assign me a salary. Well, yes, perhaps they will do that too' (Kovalevskaya 1951, 276). Her formal tenure would not come until six years later.

On 30 January 1884, Kovalevskaya delivered her first lecture at Stockholm University, a landmark event in the history of gender and mathematics. In addition to her twice-weekly lectures, Kovalevskaya actively participated in mathematical seminars hosted by Mittag-Leffler, where she both presented her work and guided students in their research. Mittag-Leffler praised her contributions in correspondence with university authorities, emphasising her profound knowledge and ability to inspire her students. These efforts were part of a broader campaign to secure her a more permanent academic position (in Björk 2002, 23). Her appointment, officially announced on 28 June 1884, was a significant achievement but came with notable financial disparities. Her annual salary of 4000 Swedish Kronor was considerably less than the standard 6000 Kronor earned by her male counterparts, and only half of it was funded by the university. The remaining amount was covered through private subscriptions.

In addition to her teaching duties, Kovalevskaya made significant contributions to Swedish mathematics through her involvement with *Acta Mathematica*, the journal Mittag-Leffler founded in 1882. In 1884, he invited her to join the editorial board and later assume the role of editor. She contributed by publishing her own research, reviewing submissions, and using her extensive networks in Russia, Germany, and France to draw prominent mathematicians and secure institutional subscriptions. Her efforts were instrumental in establishing *Acta Mathematica* as a leading international journal in the field.

Kovalevskaya travelled a lot during the academic holidays, mostly to keep in touch with the mathematical circles in France and Germany, but also to have some time for her research writing, as she was in the process of securing a permanent position, but also of establishing herself in the field not as a 'woman professor', but as a prominent mathematician. Being a member of the Paris Mathematical Society since 1882, she got to know some of the best French mathematicians of her time. It was in Paris that she lived the highlight of her career when, in December 1888, she received the *Prix Bordin* of the French Academy of Sciences for her paper 'On the Problem of the Rotation of a Solid Body About a Fixed Point' (Kovalevskaya 1889), as I will further discuss in Chapter 4.

Kovalevskaya spent the last year of her life in full recognition of her extraordinary achievements in mathematics, while her personal life was enriched with a new romance, Maxim Kovalevskii, a prominent sociologist and distant relative of her late husband. But while travelling from her partner's villa in the Riviera to Stockholm via Paris and Berlin, she caught pneumonia and died in her house in early February 1891. 'Too much happiness' became the discourse par excellence in the mythographies that have revolved around her life and which still keep obscuring a deeper understanding of her life (see Tamboukou 2022).

Throughout her life and work, Kovalevskaya emphasised the vital role of literary imagination in the mind and work of a mathematician, a theme I will explore further in Chapter 7. This poetic sensibility appears to be a shared trait among the six European women mathematicians whose pen-portraits I have presented as well as the

Chinese scholar Wang Zhenyi. Zhenyi's recognition as a mathematician and astronomer, though significant, has remained fragmented and overshadowed, a reflection of the colonial hierarchies that have long shaped the history of science. Including her portrait here is not simply an act of recovery or historical justice but an effort to deterritorialise the dominant narratives of mathematics, opening the field to its multiplicities and global becomings. Her life and work, entangled with the intellectual and cultural flows of the Qing dynasty, invite us to think beyond the Eurocentric frameworks that have constrained our understanding of women in mathematics.

Wang Zhenyi (1768–97): 'I love mathematics, I do not give up'[22]

'I am for extensive traveling and reading; even greater than man's is my ambition', wrote Wang Zhenyi in her poem *Praise of Manly Woman*. This line, part of her vast literary archive spanning thirteen volumes of poetry, prose, and essays, encapsulates her remarkable life as a scholar during the Qing dynasty—a life in which she defied traditional expectations for women by excelling in mathematics, astronomy, and literature (Peterson 2000, 342).

Born in Jiangning (present-day Nanjing), Zhenyi's intellectual foundation was shaped by her family. Her grandfather, who maintained an impressive library, nurtured her early education in astronomy and mathematics, while her grandmother introduced her to poetry. Her father, an educated man himself, further enriched her learning with lessons in geography, science, and medicine. Although primarily self-taught, Zhenyi's posthumously published work, *Collection of the Defeng Pavilion*, reveals her wide-ranging knowledge, drawn from both Chinese and Western sources (Leung 1998, 230).

Zhenyi's youth was marked by travel, which profoundly shaped her intellectual and creative development. At age eleven, she accompanied her family to Jiling for her grandfather's funeral, staying there for five years. During this period, she studied scientific subjects, learnt equestrian skills, practised martial arts, and mastered archery under the guidance of a Mongolian general's wife. Her poetry vividly reflects these travels, capturing scenes of intellectual and physical engagement. Unlike the traditional 'recumbent travel' of Chinese male literati, which sought artistic and political cultivation, Zhenyi's journeys were driven by her passion for scientific discovery and her desire to transcend the boundaries of the 'inner quarters' (Wang 2014, 24).

At sixteen, Zhenyi returned to the southern bank of the Yangzi River and later moved to the capital with her father, continuing her travels and documenting them in her poetry. Nearly a third of her poetic output was inspired by these experiences, offering a broader, more dynamic perspective on the world. However, her life became more sedentary after her marriage at twenty-five, when her duties as a wife limited her ability to travel. Despite this, it was during this period that she gained recognition for her work in poetry, astronomy, and mathematics, while also teaching both men and women (Leung 1998, 231).

Zhenyi's contributions to astronomy were groundbreaking. She calculated and explained complex celestial phenomena, including the movement of the equinoxes and the revolutions of the sun, moon, and planets. Her detailed description of a lunar eclipse demonstrated a meticulous, experimental approach to observation and calculation (Peterson 2000, 343–4). In mathematics, Zhenyi followed a similar path to her European contemporaries, striving to make complex ideas accessible. Her *Simple Principles of Calculation*, published at the age of twenty-four, simplified mathematical theories for broader understanding. In *Beyond the Study of Mathematics*, she wrote a rigorous and self-reflective account of her journey in mathematical sciences, predating Halmos' concept of automathography.

Although Zhenyi wrote six books on astronomy and mathematics, none have survived. However, three prefaces from her works are preserved in her *Collection of the Defeng Pavilion*, alongside her poetry and prose. These collected works, along with the *Jingling Series*, include nine volumes of prose, a single volume of poetry, and three volumes of *ci* poetry. To this day, her extant writings remain untranslated into English or other European languages, leaving her contributions largely inaccessible to global audiences.

Wang Zhenyi's life and work reflect her relentless pursuit of knowledge and her resistance to the constraints of her time. Her ambition and achievements position her not only as a key figure in the history of women in mathematics and science but also as a symbol of the potential for intellectual and creative transcendence across cultural and disciplinary boundaries.

Having drawn the pen-portraits of the six European women mathematicians, philosophers, and scientists that this study has focused on, as well as offering a preliminary sketch of a Chinese woman mathematician, astronomer, and poet, whose writings have yet to be made accessible to Anglophone and international audiences and readers, I now want to give an overview of the book, charting its epistemological background and approach.

The becoming of a transdisciplinary book: transpositions and diffractions

As Christina Hughes has aptly noted, 'there is a long history in feminist thought that has been concerned with the shortcomings of disciplinary knowledge', further adding that 'disciplines cut and chunk human, more-than-human and other-than-human experiences into separate and hierarchised knowledge fields' (2020, 1). Considering transdisciplinarity in the field of philosophy, Stella Sandford has pointed out that 'transdisciplinary theory and its concepts are not necessarily identifiable with any specific disciplinary fields, either in their origin or application' (2015, 160). In this light, Sandford considers philosophy 'the most tightly policed discipline in the humanities', (160) and its rejection of the transdisciplinary concepts, methods, and practices of feminist theory, a transdisciplinary area par excellence.

Looking back in the history of philosophy and science, we know, of course, that although distinctions between disciplines did exist, science and philosophy were interconnected, and the *savants* in the early modern period were active in a wide range of disciplines, including mathematics, physics, and the natural sciences as well as philosophy and literature (see Smith 2009). In the late modern period, however, disciplines had become much more bounded and specialised, while their strict hierarchies and boundaries have well reached our own days.

Transposing concepts, however, is not synonymous with transferring but rather includes radical changes or perhaps mutations, both in the concepts and the fields they are being transposed to. Rosi Braidotti has influentially theorised the importance of 'transpositions', searching its roots in music and genetics. The concept of transpositions 'indicates an intertextual, cross-boundary or transversal transfer, in the sense of a leap from one code, field or axis into another, not merely in the quantitative mode of plural multiplications, but rather in the qualitative sense of complex multiplicities', Braidotti has noted (2006, 5). As a musical term, transpositions can be taken as variations on a theme, non-linear but non-chaotic, while in genetics, transpositions refer to processes of mutation, neither random nor arbitrary: 'transposable moves appear to proceed by leaps and bounds, but are not deprived of their logic, or coherence' (5). In the field of epistemology, transpositions facilitate the emergence of other ways of knowing and offer 'a contemplative and creative stance that respects the visible and hidden complexities of the very phenomena it attempts to study', notes Braidotti (6).

Transpositions are indeed at the heart of this book, as it draws on a research project, which brings together insights, ideas, and concepts from philosophy, gender, feminist, and memory studies, the history of education, science, and mathematics as well as narrative, auto/biographical, and archival research methods. In navigating this transdisciplinary milieu, I follow lines of feminist affirmative approaches to theoretical texts, 'reading important insights through one another [and thus] reworking concepts that structure these insights in the traditions of thought from which they stem' (Barad 2003, 811). I will expand on the notion of diffractions in Chapter 1 and throughout the book, but what I want to do next is to draw a preliminary sketch of its chapters and themes.

Following this Introduction, Chapter 1 introduces the concept of *memory machines* to reimagine how we engage with the lives, contributions, and archives of six women mathematicians. *Memory machines* position memory not as passive retrieval or linear recounting but as an active, generative process—where fragments, absences, and connections are reassembled to create new understandings of the past. Through this concept, archives are no longer static repositories but dynamic sites of transformation, where traces such as letters, manuscripts, and reflections become catalysts for reshaping narratives and uncovering overlooked voices.

This perspective is particularly vital for constructing feminist genealogies, moving beyond recovering lost histories to reconfiguring how these histories resonate in the present. The lives and work of women mathematicians emerge not as static achievements but as forces that disrupt traditional narratives of mathematics and science, challenging hierarchical knowledge structures. By offering the concept of *memory machines*, this chapter sets the stage for the book's broader aim to re-examine the intellectual legacies of women mathematicians and propose new ways of thinking about history, memory, and the archive; It emphasises the past as an active presence, shaping the present and opening possibilities for transformation and connection.

Chapter 2 delves into the rich and complex world of women mathematicians' correspondences, proposing letters as vital sites of epistemological inquiry and cultural production. Drawing on the concept of *epistolary sensibility* (Tamboukou 2020), the chapter examines how women's letters intertwined the personal and the scientific, offering a space where mathematical ideas were exchanged, challenged, and refined. At the same time, these correspondences revealed the struggles of women excluded from formal institutions of knowledge, illuminating their efforts to position themselves within scientific networks. Through an analysis of key archival correspondences, the chapter explores the role of letters as assemblages of affects, intellect, and power dynamics. These epistolary entanglements reveal how women negotiated their new subject positions as mathematicians, navigated sociopolitical barriers, and contributed to the development and dissemination of mathematical and scientific knowledge.

By reframing letters not merely as historical artefacts but as active participants in the production of knowledge, this chapter advances the broader argument of the book that women mathematicians' lives and works must be understood as dynamic and interconnected processes of becoming, situated within the cultural, scientific, and gendered formations of their times. The chapter also highlights the enduring relevance of these correspondences, not only in reconstructing hidden feminist genealogies but also in reshaping how we think about science, collaboration, and intellectual legacy.

Chapter 3 examines the intertwined dynamics of gender, translation, and the production of scientific, mathematical, and philosophical knowledge, focusing on the contributions of women mathematicians. It explores their dual roles as translators of significant scientific texts and as original thinkers whose work has been translated across time and geographies. Translation is framed not simply as a tool for knowledge transfer but as a transformative, authorial practice central to intellectual and cultural exchange.

Structured thematically, the chapter first highlights how translation functioned as an original and vital means of disseminating knowledge. It then addresses historical and contemporary challenges in translating women's scientific and literary contributions, emphasising the erasure of their voices in twentieth-century scholarship and the recent revival of interest in their work. Finally, it explores the politics of translation, connecting it to gendered cultural and linguistic boundaries and arguing for the importance of transnational feminist approaches in reshaping narratives of scientific history.

Chapter 4 explores processes of becoming a woman mathematician, philosopher, and scientist through the lens of Hannah Arendt's (1998) concept of agonism, focusing on the affirmative contestation that shaped women's engagement with knowledge and learning. Drawing on archival materials such as autobiographies, scientific correspondences, and notebooks, it explores how knowledge production unfolded in contexts shaped by intellectual *agon*, exclusion, and systemic inequalities. By challenging the trope of the 'exceptional woman', the chapter proposes the concept of the *event* as a modality of understanding how women's contributions to science, mathematics, and philosophy were historically constituted. It highlights how their intellectual trajectories unfolded through dynamic processes of agonistic learning, creative experimentation, and negotiation within exclusionary structures.

The chapter argues that women mathematicians' achievements were not isolated acts of genius but outcomes of persistent effort, collaboration, and the determination to engage with a shared world of knowledge. Through this lens, it reimagines women's intellectual histories as vibrant and relational, contributing to feminist approaches to the history of mathematics, philosophy, and science and challenging enduring narratives of exclusion and exceptionality.

Chapter 5 looks into the intertwined themes of love and mathematics in the writings of women mathematicians, philosophers, and scientists, analysing how love has shaped their intellectual pursuits. It positions love not as a peripheral element but as a driving force, merging emotional intensity with intellectual passion to illuminate the deep connections between personal life and the history of mathematics. The chapter examines autobiographical, literary, and philosophical writings, exploring how love—romantic, intellectual, and existential—mobilised women's engagement with mathematics and science and reveals how love and mathematics are dynamically entangled in women's narratives. Through these analyses, the chapter argues that love operates not only as an emotional force but also as a transformative element in shaping knowledge creation, offering new perspectives on the relationship between intellectual life, emotional experience, and the pursuit of happiness in the history of science and mathematics.

Chapter 6 studies the philosophical writings of women mathematicians, analysing their contributions beyond mathematics and science to epistemology and the philosophy of knowledge. Examining texts such as prefaces, essays, and reflective writings, it highlights how these women navigated and challenged intellectual boundaries by synthesising mathematics, philosophy, and science.

The chapter argues that their philosophical reflections reveal a transdisciplinary approach, demonstrating innovative ways of thinking that resonate with contemporary discussions in the history of ideas. It examines recurring themes, including the nature of truth, creativity, and the intersections of metaphysics and scientific inquiry, showing how these thinkers critiqued and transformed the dominant discourses of their time. By foregrounding their philosophical contributions, this chapter emphasises the necessity of integrating their ideas into broader historical narratives, challenging the conventional separation of disciplines and offering new perspectives on the intertwined development of mathematics, science, and philosophy.

Chapter 7 deploys the concept of rhythm as a way of analysing Sofia Kovalevskaya's literary writings. Drawing on the approach of *narrative rhythmanalysis* (Tamboukou 2021), it examines how Kovalevskaya's works intertwine life, literature, poetry, and mathematics, forming dynamic assemblages that resist linear narratives or fixed interpretations. The chapter highlights how Kovalevskaya's subjectivity is expressed as a 'rhythmic knot', where intensities of life, thought, and creativity interweave and reconfigure across time and space. By tracing the pulsations and echoes in her texts, the chapter reveals how rhythms disrupt and reshape continuity, opening up multiplicities of becoming. Kovalevskaya's writings are thus read as spaces where poetic imagination and mathematical precision coexist, demonstrating the profound interplay of creative and intellectual forces.

In analysing these rhythmic entanglements, the chapter emphasises the importance of rhythmanalysis as both a method and a lens to understand the convergence of temporal, spatial, and emotional dimensions in Kovalevskaya's work, offering fresh perspectives on her legacy as both a mathematician and a writer.

Finally, the conclusion reflects on the broader themes and insights that have emerged throughout the book, emphasising the need for a genealogical approach to the lives and works of women mathematicians. It underscores the mechanisms of exclusion and erasure that have shaped the history of mathematics, revealing how women's contributions have often been marginalised or rendered invisible across different temporal and geographic contexts. By connecting these silences to the broader cultural, intellectual, and social assemblages in which these women lived and worked, the conclusion further explores the dynamic interactions that shaped their intellectual trajectories. It highlights the importance of disrupting dominant narratives to uncover the complex and often unexpected connections between these women's lives, their works, and the broader structures of power, knowledge, and history. Ultimately, the conclusion calls for a rethinking of how we approach the history of mathematics, proposing an *assemblage*-oriented perspective that not only recovers forgotten figures but also reimagines the possibilities for understanding the entanglements of mathematics, philosophy, and the literary arts.

Notes

1 For a comprehensive review of this body of literature, see Govoni 2014.
2 Diogenes Laertius' *Lives and Opinions of Eminent Philosophers*, was compiled sometime early in the 3rd century. See Warren 2007, for a recent discussion of this work.
3 For a list of seminars and presentations for this project see, https://sites.google.com/view/numbersandnarratives/dissemination [Accessed 9 February 2024].
4 See amongst others, Long 2010; Zinsser 2005; Fara 2004; Whaley 2003; Phillips 1990.
5 See, for example, Wills et al. (2023) for a recent overview of this field.
6 See, https://mathwomen.agnesscott.org/women/women.htm and https://mathshistory.st-andrews.ac.uk/
7 https://digital.library.mcgill.ca/mingqing/search/results-work.php?workID=76&language=ch [Accessed 2 March 2024]
8 Although there is an English translation of the *Discours* (Du Châtelet 2009), I have decided to follow the 1961 publication in French, as there were some subtleties in the text that I opted to translate myself.
9 See Hamel 1910; Edwards 1970.
10 See also Wade 1941, Mitford 1957, and Bodanis 2006.
11 See the Voltaire Foundation entry for more details on this influential publication: http://www.voltaire.ox.ac.uk/lettres-sur-les-anglais/ [Accessed 15 January 2023].
12 See Wade 1941, 1947, and Zinsser 2006, particularly Chapter 3.
13 For extended biographical details in English about Agnesi and her family, see amongst others, Mazzotti 2007 and Cupillari 2007.
14 See Fayolle 2012, for a discussion of women's education in the revolutionary and post-revolutionary period in France.

15 Moreniana Biblioteca di Firenze, fondo Palazzi-Libri, filza 432, thereafter and in the text (MBF/FPL), followed by folio number. Transcribed in Del Centina 2005, 68.
16 Bibliothèque Nationale de France, MS. Fr. Nouvelle Acq. 4073, thereafter and in-text (BnF), followed by manuscript and folio numbers. Also published in Delambre et al. 1879, translated in Bucciarelli and Dworsky, 1980, 121–2.
17 In this chapter, I am citing from the first draft of Somerville's autobiography, written in her own hand, at the 'Mary Somerville Collection' at the Bodleian Libraries in Oxford, Dep. c. 355, Folder MSAU-2, but I am following the pagination of the author's manuscript, not of the archives [fols. 72–363]. Thereafter and in text, MSC, followed by manuscript and folio numbers.
18 See Patterson 1983, 49 and Stenhouse 2021, iii.
19 Oxford Bodleian Libraries, Archive of the Noel Byron and Lovelace Families, Dep. Lovelace Byron. In the text and thereafter, LB followed by the folder and folio number. Unless otherwise indicated all letters cited in this chapter have been transcribed and published in Toole 1992.
20 See Tamboukou 2024 for more details on Kovalevskaya's early education in mathematics.
21 Nickname for Sofia Vladimirovna Kovaleskaya.
22 *Collection of the Defeng Pavilion,* Vol. 1, 20a, translated in Peterson 2000, 344.

References

Agnesi, Maria Gaetana. 1748. *Instituzioni analitiche ad uso della gioventù italiana*, 2 vols. Milan: Nella Regia-Ducal Corte.

Agnesi, Maria Gaetana. 1738. *Propositiones philosophicae quas crebris disputationibus domi habitus coram clarissimis viris explicabat extempore, et ab objectis vindicabat Maria Cajetana de Agnesiis mediolanensis.* [Philosophical Propositions which Maria Gaetana Agnesi of Milan Frequently Explained Extemporaneously and Defended against Objections in Discussions Held at Home Before the Most Illustrious Men]. Milan: Malatesta.

Agnesi, Maria Gaetana. 2005 [1707]. 'The Studies of the Liberal Arts by the Female Sex Are by No Means Inappropriate translated by Paula Findlen'. In *The Contest for Knowledge: Debates Over Women's Learning in Eighteenth-Century Italy*, edited by Rebecca Messbarger and Paula Findlen, 117–40. Chicago and London: The University of Chicago Press.

Anzoletti, Maria Luisa. 1900. *Maria Gaetana Agnesi*. Milan: Cogliati.

Arendt, Hannah. 1998 [1958]. *The Human Condition.* Chicago: University of Chicago Press.

Barad, Karem. 2003. 'Posthumanist Performativity: Toward an Understanding of How Matter Comes to Matter'. *Signs: Journal of Women in Culture and Society* 28 (3): 801–31.

Björk, Jan-Erik. 2002. 'Sonja Kovalevsky: Her Life and Professorship at Stockholm'. *Operator Theory: Advances and Applications* 132: 11–53.

Bodanis, David. 2006. *Passionate Minds: Émilie du Châtelet, Voltaire, and the Great Love Affair of the Enlightenment.* New York: Three Rivers Press.

Braidotti, Rosi. 2006. *Transpositions: On Nomadic Ethics*. Oxford: Polity Press.

Bucciarelli, Luis L., and Nancy Dworsky. 1980. *Sophie Germain: An Essay in the History of the Theory of Elasticity.* Boston, MA: Dordrecht.

Cavazza, Marta. 2009. 'Between Modesty and Spectacle: Women and Science in Eighteenth Century Italy'. In *Italy's Eighteenth Century: Gender and Culture in the Age of the Grand Tour,* edited by Paula Findlen, Catherine Sama, and Wendt Rowarth, 275–302. Stanford: Stanford University Press.

Cupillari, Antonella. 2007. *A Biography of Maria Gaetana Agnesi, an Eighteenth-Century Woman Mathematician: With Translations of Some of her Work from Italian into English.* New York: Edwin Mellen Press.

Del Centina, Andrea. 2005. 'Letters of Sophie Germain Preserved in Florence'. *Historia Mathematica* 32: 60–75.

Delambre, J. H. Choron, J. Fourier, Sophie Germain, J. L. de La Grange, J. Lalande, Guillaume Libri, and C. Henry. 1879. 'Les manuscrits de Sophie Germain et leur récent éditeur- documents nouveaux'. *Revue Philosophique de la France et de l'Étranger* 8: 619–41.

De Pizan, Christine. 2000 [c.1405]. *The Book of the City of Ladies*. Translated and edited by Rosalind Brown-Grant. London: Penguin Classics.

Detlefsen, Karen. 2022. 'The Rise of a Public Science? Women and Natural Philosophy in the Early Modern Period'. In *The Cambridge History of Philosophy of the Scientific Revolution*, edited by David Marshall and Dana Jalobeanu, 128–45. Cambridge: Cambridge University Press.

Du Châtelet, Émilie. 1739. 'Dissertation sur la Nature et la Propagation du Feu' [Dissertation on the Nature and Propagation of Fire]. In *Recueil des Pièces qui ont Remporté le Prix de l'Académie Royale des Sciences en 1738* [Collection of the Pieces that Won the Prize of the Royal Academy of Sciences in 1738], edited by Académie Royale des Sciences, 85–168. Paris: Imprimerie Royale.

Du Châtelet, Émilie. 1740. *Institutions de Physique* [Foundations of Physics]. Paris: Prault.

Du Châtelet, Émilie. 1759. *Principes Mathématiques de la Philosophie Naturelle*. Edited by Isaac Newton and E. Du Châtelet. 2 vols. Paris: Desaint & Saillant.

Du Châtelet, Émilie. 1961. *Discours sur le bonheur, édition critique et commente par Robert Mauzi*. Paris: Les Belles Lettres.

Du Châtelet, Émilie. 2009. *Selected Philosophical and Scientific Writings*. Edited and with an Introduction by Judith Zinsser, translated by Isabelle Bour and J. Zinsser. Chicago and London: The University of Chicago Press.

Edwards, Samuel. 1970. *The Divine Mistress: A Biography of Émilie du Châtelet, The Beloved of Voltaire*. New York: David Mckay.

Fara, Patricia. 2004. *Pandora's Breeches: Women, Science and Power in the Enlightenment*. London: Pimlico.

Fayolle, Caroline.2012. 'L'éducation est-elle un instrument de l'égalité? Les débats sur l'éducation des femmes à la période révolutionnaire et post-révolutionnaire'. In *Revisiter la 'querelle des femmes'. Discours sur l'égalité/inégalité des sexes, de 1750 aux lendemains de la Révolution*, edited by Éliane Viennot, 95–107. Saint-Étienne: Publications de l'Université de Saint-Etiennepp.

Ferry, Giorgina. 2014. 'Telling Stories or Making History? Two Lives in X-ray Crystallography'. In *Writing about Lives in Science: (Auto)Biography, Gender and Genre*, edited by Paola Govoni and Zelda Alice Franceschi, 55–63. Göttingen: V&R Unipress.

Findlen, Paula. 2011. 'Calculations of Faith: Mathematics, Philosophy, and Sanctity in 18th-century Italy (New Work on Maria Gaetana Agnesi)'. *Historia Mathematica* 38: 248–91.

Foucault, Michel. 1986. 'Nietzsche, Genealogy, History.' Translated by Donald F. Bouchard and Sherry Simon. In *The Foucault Reader*, edited by Paul Rabinow, 76–100. Harmondsworth: Peregrine.

Frisi, Antonio Francesco. 1799. *Elogio storico di D^A Maria Gaetana Agnesi, Milanese Dell'Academia Dell' Instituto Delle Scienze, e Lettrice Onoraria di Matematiche nella Università di Bologna*. Milan: Presso Giuseppe Galeazzi. *Reprint* with an introduction and additional notes by Arnaldo and Giuseppina Masotti, Milan: Scuola Tipografica del Pio Istituto pei Figli della Provvidenza, 1965. Translated by Antonella Cupillari in *A Biography of Maria Gaetana Agnesi, an Eighteenth-Century Woman Mathematician: With Translations of Some of her Work from Italian into English*, 33–106. New York: Edwin Mellen Press, 2007.

Germain, Sophie. 1821. *Recherches sur la théorie des surfaces élastiques*. Paris: Huzard-Courcier.
Gauss, Karl Friedrich. 1801. Disquisitiones Arithmeticae. Leipzig: Gerh. Fleischer
Germain, Sophie. 1833. *Considérations Generales Sur l'Etat Des Sciences Et Des Lettres Aux Différentes Époques de leur Culture*. Paris: Lachevardiere.
Germain, Sophie. 1896 [1879]. *Œuvres philosophiques de Sophie Germain, suivies de pensées et de lettres inédites. Et précédées d'une notice sur sa vie et ses œuvres par H[te] Stupuy (Nouvelle Édition)*. Paris: Firmin-Didot.
Goodman, Dena. 1994. *The Republic of Letters: A Cultural History of the French Enlightenment*. Ithaca, NY: Cornell University Press.
Govoni, Paola. 2000. 'Biography: A Critical Tool to Bridge the History of Science and the History of Women in Science'. *Nuncius* 1: 399–409.
Govoni, Paola. 2014. 'Crafting Scientific (Auto)Biographies'. In *Writing about Lives in Science: (Auto)Biography, Gender and Genre,* edited by Paola Govoni and Zelda Alice Franceschi, 7–30. Göttingen: V&R Unipress.
Grinstein, Louise, S., and Paul J. Campbell, eds. 1987. *Women of Mathematics: A Biobibliographic Sourcebook*. New York: Greenwood Press.
Hagengruber, Ruth, ed. 2022. *Époque Émilienne*. Cham: Springer.
Halmos, Paul. 1985. *I Want to become a Mathematician: An Automathography*. New York: Springer and Narratives.
Hamel, Frank. 1910. *An Eighteenth-Century Marquise: A Study of Émilie du Châtelet and Her Times. New York*: James Pot & Co.
Hollings, Christopher, Ursula Martin, and Adrian Rice. 2017. 'The Early Mathematical Education of Ada Lovelace'. *BSHM Bulletin: Journal of the British Society for the History of Mathematics* 32 (3): 221–34.
Hughes, Christina. 2020. 'Introduction'. In *Transdisciplinary Feminist Research: Innovations in Theory, Method and Practice,* edited by Carol Taylor, Christina Hughes, and Jasmine, B. Ulmer, 1–3. London: Routledge.
Hyvärinen, Matti. 2010. 'Revisiting the Narrative Turns'. *Life Writing* 7 (1), 69–82.
Jones, Steve. 2016. *No Need for Geniuses: Revolutionary Science in the Age of the Guillotine*. London: Little Brown.
Keller, Evelyn, Fox. 1997. 'The Anomaly of a Woman in Physics'. In *Working it Out: 23 Women Writers, Artists, Scientists, and Scholars Talk About Their Lives and Work*, edited by S. Ruddick and P. Daniels, 71–91. New York: Pantheon.
King, Margaret L., and Albert Rabil Jr. 2005. 'The Other Voice in Early Modern Europe: Introduction to the Series'. In *The Contest for Knowledge*, edited and translated by Rebecca Messbarger and Paula Findlen, xi–xxxi. Chicago and London: The University of Chicago Press.
Koblitz, Ann Hibner. 1998. 'Science, Women, and the Russian Intelligentsia: The Generation of the 1860s'. *Isis* 79 (2): 208–26.
Kochina, Pelageya. 1985. *Love and Mathematics: Sofia Kovalevskaya*. Translated by Michael Burov. Moscow: Mir Publishers.
Kovalevskaya, Sofia. 1951. *Memories and Letters* [Коваевской, Софье, В. Воспоминанця u nucьма]. Moscow: AN SSSR.
Kovalevskaya, Sofia. 1889. 'Sur le probleme de la rotation d'un corps solide autour d'un point fixe'. *Acta Mathematica* 12 (H.2): 177–232.
Leung, Angela, Ki Che. 1998. 'Wang Zhenyi', translated by W. Zang. In *Biographical Dictionary of Chinese Women: V. 1: The Qing Period, 1644–1911*, edited by Lily XiaoHong Lee, Clara Law, and A. D. Stefanowska, 230–32. London and New York: Routledge.
Lewis, Judith, S. 1995. 'Princess of Parallelograms and Her Daughter. Math and Gender in the Nineteenth Century English Aristocracy'. *Women's Studies International Forum* 18 (4): 387–94.

Libri, Guglielmo. 1832. 'Sciences Mathématiques-Mlle Germain'. *Journal Des Débats*, 18 May 1832, pp. 1–2.
Logan, Gabriella, Berti. 1999. 'Italian Women in Science from the Renaissance to the Nineteenth Century'. PhD diss., University of Ottawa. Accessed 24 October 2022. https://ruor.uottawa.ca/handle/10393/8789.
Long, Kathleen P., ed. 2010. *Gender and Scientific Discourse in Early Modern Culture.* Farnham: Ashgate.
Lovelace, Ada. 1843. 'Sketch of the Analytical Engine invented by Charles Babbage Esq. By L.F. Menabrea, of Turin, Officer of the Military Engineers, with Notes Upon the Memoir by the Translator'. *Taylor's Scientific Memoirs* 3: 666–731.
Mazzotti, Massimo. 2007. *The World of Maria Gaetana Agnesi, Mathematician of God.* Baltimore: Johns Hopkins University Press.
Mazzotti, Massimo. 2014. 'Rethinking Scientific Biography: The Enlightenement of Maria Gaetana Agnesi. In *Writing about Lives in Science: (Auto)Biography, Gender and Genre*, edited by Paola Govoni and Zelda, Alice Franceschi, 117–37. Göttingen: V&R Unipre
Mitford, Nancy. 1957. *Voltaire in Love.* London: Hamish Hamilton.
Musielak, Dora. 2020. *Sophie Germain: Revolutionary Mathematician.* 2nd ed. Cham: Springer.
Niskanen, Kirsti, and Michael, J. Barany. 2021. *Gender, Embodiment, and the History of the Scholarly Persona.* Cham: Palgrave Macmillan.
Patterson, Elizabeth Chambers. 1983. *Mary Somerville and the Cultivation of Science 1815–1840.* The Hague: Martinus Nijhoff.
Peterson, Barbara Bennett. 2000. 'Wang Zhenyi'. In *Notable Women of China: Shang Dynasty to the Early Twentieth Century*, edited by Barbara Bennett Peterson with He Hongfei, Han Tie, Wang Ziyu, and Zhang Guangyu, 341–46. New York: M. E. Sharpe.
Phillips, Patricia. 1990. *The Scientific Lady: A Social History of Women's Scientific Interests, 1520–1918.* London: Weidenfeld and Nicolson.
Sandford, Stella. 2015. 'Contradiction of Terms: Feminist Theory, Philosophy and Transdisciplinarity'. *Theory, Culture and Society* 32 (5–6): 159–82.
Sarton, George. 1936. *The Study of the History of Mathematics and the Study of the History of Science.* New York: Dover Publications.
Schiebinger, Londa L. 1989. *The Mind Has no Sex? Women in the Origins of Modern Science.* Cambridge, MA: Harvard University.
Shafer Alice. 1987. 'Foreword'. In *Women of Mathematics: A Biobibliographic Sourcebook*, edited by Louise S. Grinstein and J. Campbell Paul, ix–xii. New York: Greenwood Press.
Smith, Pamela. 2009. 'Science on the Move: Recent Trends in the History of Early Modern Science'. *Renaissance Quarterly* 62 (2): 345–75.
Somerville, Mary. 1872. Oxford: Bodleian Libraries, Dep. c. 355, MSAU-2, fols. 72–363.
Somerville, Mary. 2001 [1873]. *Queen of Science: Personal Recollections of Mary Somerville.* Edited by Dorothy McMillan. Edinburgh: Canongate.
Stenhouse, Brigitte. 2021. 'Mary Somerville: Being and Becoming a Mathematician'. PhD thesis The Open University. https://doi.org/10.21954/ou.ro.0001346f
Stupuy, Jean-Léon-Hippolyte. 1896 [1879]. 'Étude sur la Vie et les œuvres de Sophie Germain'. In *Œuvres philosophiques de Sophie Germain, suivies de pensées et de lettres inédites*, 1–73. Paris: Firmin-Didot.
Tamboukou Maria. 2003. *Women, Education, the Self: A Foucauldian Perspective.* Basingstoke: Palgrave, Macmillan.
Tamboukou Maria. 2017. *Women Workers' Education, Life Narratives and Politics: Geographies, Histories, Pedagogies.* Basingstoke: Palgrave.
Tamboukou, Maria. 2020. 'Epistolary Lives: Fragments, Sensibility, Assemblages in Auto/biographical Research'. In *The Palgrave Handbook of Auto/Biography*, edited by Anne Chapell and Julie M. Parsons, 157–64. Basingstoke: Palgrave.

Tamboukou, Maria. 2021. 'Narrative Rhythmanalysis: The Art and Politics of Listening to Women's Narratives of Forced Displacement'. *International Journal of Social Research Methodology* 24 (2): 149–62.
Tamboukou, Maria. 2022. 'Traces in the Archive: Re-imagining Sofia Kovalevskaya'. *Life Writing* 19 (3): 341–56.
Tamboukou Maria. 2024. 'A Princess of Science? Becoming the First Woman Professor in Mathematics in Modern Europe'. *Nordic Journal of Educational History* 11 (2): 57–79. https://doi.org/10.36368/njedh.v11i2.1059
Vanacker Beatrijs, and Lieke van Deinsen, eds. 2022. *Portraits and Poses: Female Intellectual Authority, Agency and Authorship in Early Modern Europe.* Leuven: Leuven University Press.
Wade, Ira O. 1941. *Voltaire and Madame du Châtelet: An Essay on the Intellectual Activity at Cirey.* Princeton: Princeton University Press.
Wade, Ira O. 1947 *Voltaire and Madame du Châtelet and Studies on Voltaire with Some Unpublished Papers of Mme Du Châtelet.* Princeton: Princeton University Press.
Waithe, Mary Ellen, and Boos Therese Dykeman, eds. 2023. *Women Philosophers from Non-Western Traditions: The First Four Thousand Years.* Cham: Springer.
Wang Yanning. 2014. *Reverie and Reality: Poetry on Travel by Late Imperial Chinese Women.* New York: Lexington.
Watts, Ruth. 2007. *Women in Science: A Social and Cultural History.* London: Routledge.
Whaley, Leigh. 2003. *Women's History as Scientists: A Guide to the Debates.* Santa Barbara: ABC-CLIO.
Wills, Hannah, Sadie Harrison, Erika Jones, Rebecca Martin, and Farrah Lawrence-Mackey, eds. 2023. *Women in the History of Science: A Sourcebook.* London: UCL Press.
Zinsser, Judith P. 2005. 'Introduction'. In *Men, Women and the Birthing of Modern Science*, edited by P. Zinsser Judith, 3–9. DeKalb, IL: Northern Illinois University Press.
Zinsser, Judith, P. 2006. *Daring Genius of the Enlightenment, Émilie du Châtelet.* New York: Penguin.

1 Memory Machines and Archival Traces

In looking back at the historiography of encyclopaedic and biographical references of Italian women scientists in early modern Europe, Paula Findlen (2014) has pointed to the absences, silences, and deselections in the work of early biographers and, therefore, in the current state of the archives. She has therefore argued that 'we cannot understand the visible centre of how the woman of science emerged in the eighteenth century without exploring all of these other genealogies that do not become the subject of biography but belong to a very different kind of history' (114).

In responding to Findlen's argument about the need for 'a different kind of history', I have thus configured the plane of my research as a feminist genealogy, working with Michel Foucault's notion of *counter-memories* (1986). As a theoretical and methodological approach, genealogy in Foucault's philosophical work (1986) investigates the processes, procedures, and apparatuses whereby truth and knowledge are produced. Genealogy writes the history of the present; it problematises the multiple, complex, and non-linear configurations of the socio-political and cultural formations of modernity. In the context of the research project that this book draws on: what were the conditions of possibility for women to be excluded from the world of mathematical sciences, how were they historically constituted as subjects of scientific knowledge, and why are they still on the margins of this discipline as students, teachers, and/or researchers?

Genealogy as a method of analysis looks into the archive to excavate forgotten stories and documents, which might shed light on the practices and discourses that have excluded and marginalised women from the field of mathematical sciences. But instead of seeing history as a continuous development of an ideal schema, genealogy searches in the maze of dispersed events to trace discontinuities, but also unexpected continuities. Women's ambivalent position in the world of mathematical sciences is a paradigmatic case of uneven historical developments, particularly if we consider that Laura Bassi was the first woman to hold a Chair in Physics in 1732, and it then took more than 150 years for Sofia Kovalevskaya to be appointed Professor of Mathematics at Stockholm University in 1884 (see Audin 2011). What happened in between, and why are women's positions in mathematics still in a sore state in the twenty-first century?

The archive is at the heart of this research, which has been contextualised in the wake of the 'archival turn', that is, the vast surge of interest in archives, memory,

DOI: 10.4324/9781003533993-2

and traces of the past that has occurred among both popular and academic audiences over the last few decades. I have written extensively on archival methods in the social sciences, particularly from the perspective of Foucauldian genealogies (see Tamboukou 2020). Elsewhere in my work with colleagues in the UK (Moore et al. 2016), we have put forward the notion of 'archival sensibility' as an overall epistemological and methodological plane for doing archival research, and later on in this chapter, I will expand on how 'archival sensibility' has informed the excavation, presentation, and analysis of women mathematicians' auto/biographical, epistolary, and literary documents.

The genealogical project of excavating *counter-memories* is thus of particular importance in this research. By revealing discontinuities in the supposed continuous development of history, Foucault's genealogical project also implies a discontinuity in present socio-cultural formations. Genealogy attempts to go further by tracing possible ways of thinking differently instead of accepting and legitimating what are already the 'truths' of our world. The aim is to create conditions of possibility for *counter-memories* to emerge that will help subjects re-imagine their present, specifically women's position in mathematics today in our case. This is the future to which genealogies aspire: opening possibilities for life by separating us from 'the contingency that has made us what we are, the possibility of no longer being, doing, or thinking what we are, do, or think', as Michael Mahon has succinctly put it (1992, 122).

If we also consider that the majority of women mathematicians' philosophical and literary writings, as well as their letters, have not been translated in English, we understand why *counter-memory* becomes so important for this research. Memory work is not so much directed to the past but actually reconfigures the present in terms of connecting disparate narratives and stories into 'a design that has a meaning', as feminist philosopher Adrianna Cavarero has aptly put it (2000, 1). This reconfiguration of the present can also become a prefiguration of the future in terms of empowering young women students in mathematics, as well as women mathematicians, to make sense of their abilities and of their future becomings. It is thus the importance of 'memory work' that I will discuss in the next section.

Counter-memories **and 'the work of memory'**

As already noted in the previous section, Foucault's notion of *counter-memory* is of particular importance in this study. But how is *counter-memory* to be understood? In considering the notion of historical sense in Nietzsche, Foucault rethinks the three Nietzschean modalities of traditional history, namely the monumental, antiquarian, and critical history (Mahon 1992, 97), to which genealogy returns not to follow but to metamorphose them. This return rests upon three anti-Platonic themes that Foucault deploys in detail.

First is the parodic, anti-realist theme as opposed to a 'remembering history' that offers subjects identities borrowed from the glorious past. Genealogy reveals the masquerade of past identities and addresses them in a playful manner as if they were figures in a carnival party, so that 'the veneration of monuments becomes parody' (Foucault 1986, 97). Past figurations of the scholarly personae, whether 'great [men] scientists' or 'exceptional women in science and mathematics' can thus be deconstructed in the wider context of the science carnival, revealing instead that such figures were the effects rather than the causes of long processes in the history of science and mathematics (see Keller 1987).

Second is the dissociative theme, disrupting supposed historical continuity and dissolving coherent identities, whose fragments were kept together within the masks: 'the respect for ancient continuities becomes systematic dissociation' (Foucault 1986, 97). The debates around women's position in the history and philosophy of science are important interventions in deconstructing discourses on the plane of the Foucauldian dissociative theme (see Schiebinger 1987).

Third is the sacrificial theme as directed against the knowing character of history and contesting the supposed neutrality and objectivity of the subject of knowledge: 'the critique of the injustices of the past by a truth held by [men] in the present becomes the destruction of the [man] who maintains knowledge by the injustice proper to the will to knowledge' (Foucault 1986, 97). In this light, Eviatar Zerubavel has argued (2003, 1-2) that we need 'a *sociomental* topography of the past' to understand how individuals and societies construct, structure, and organise memories, highlighting the importance of engaging with the ways we remember history, rather than its 'actual' events, which can never be pinned down anyway, as they are always mediated and interpreted.

As Foucault sees it, these three anti-Platonic themes, which guide the return of genealogy to history, are constructive elements of *counter-memory*, a totally new dimension that transforms traditional historiography into an 'effective history' (Dean 1994). By disturbing the legends of the past, genealogy as 'effective history' opens up paths for its subjects to set out for new, improbable identities. In this context, for Foucault, *counter-memory* is a modality of rethinking time and history, and it is not accidental that scholars who have engaged with the dynamics of memory have drawn on Foucauldian insights to rethink the mnemonic processes through which 'sites of memory' (Nora 1989) are continuously invested.

Drawing on Foucault's 'scarcity principle', Ann Rigney has considered the role of media 'in generating shared memories through processes of selection, convergence, recursivity and transfer' (2005, 11). In challenging the image of memory as 'an original storehouse', always in danger of being lost or eroded, and thus in need of preservation, as in Maurice Halbwachs' (1997 [1950]) model of 'lived memories', Rigney is also sceptical about the discourse of counter-memory: 'memory has tended in practice to become synonymous with "counter-memory", defined in opposition to hegemonic views of the past and associated with groups who have been "left out", as it were, of mainstream history' (2005, 12), she has argued. In Rigney's critique, such an approach to memory still draws on the image of the 'original storehouse', as well as the conviction that it is possible to 'recover' memories that have been 'lost' or 'hidden'.

What Rigney offers instead is 'a social-constructivist model that takes as its starting point the idea that memories of a shared past are collectively constructed and reconstructed in the present rather than resurrected from the past' (14). In doing so, Rigney draws on the concept of 'cultural memory' (Assmann 2012), which highlights 'the extent to which shared memories of the past are the product of mediation, textualization and acts of communication' (Rigney 2005, 14). In this sense, memory studies should focus on what Rigney identifies as *mnemonic technologies*, a term that points directly to Foucault's oeuvre and his notion of 'technologies' (Foucault 1988), the different ways in our culture that human beings have deployed to understand themselves and their world.

Foucault has actually mapped four domains of such technologies: first, the *technologies of production*, which 'permit us to produce, transform, or manipulate things' (18); second, the *technologies of sign systems*, 'which permit us to use signs, meanings, symbols, or significations' (18); third, the *technologies of power* 'which determine the conduct of individuals and submit them to certain ends or domination, an objectivizing of the subject' (18); and fourth, the *technologies of the self*, through which individuals attempt 'to transform themselves' (18). It is easy to see how 'technologies of memory' or *mnemonic technologies*, as Rigney calls them (2005, 16), can be deployed on all axes of Foucault's four-partite system of 'technologies', but Rigney specifically draws on Foucault's notion of the 'principle of scarcity' that comes from his earlier work on *The Archaeology of Knowledge* (1969). According to this principle, very few things about the world are actually said, written, or remembered and what in the end 'remains' either in texts, archives, and/or memory traces cannot represent 'the essence' of the real and therefore does not have any absolute value:

> The scarcity of utterances, the scrappy and incomplete character of the discursive field, the fact that in the end few things can be said, explains why utterances are not infinitely transparent, like the air we breathe; instead, they are transmitted and preserved; they are invested with value and people try to appropriate them; they are repeated, reproduced, transformed, and replicated, not just through copying and translation, but also through interpretation, commentary, and an internal proliferation of meaning (Foucault 1969, 156–7, cited in Rigney 2005, 16).

By transposing Foucault's 'scarcity principle' in the field of memory studies, Rigney charts five modalities in the work of cultural memory: 'the selectivity of recall, the convergence of memories, the recursivity in remembrance, the recycling of models of remembrance and memory transfers' (16). I will expand on these five modalities by giving some examples from my research with the women mathematicians' archival documents, which I will consider in more detail in the next section of this chapter.

Selection is important as a *mnemonic technology*, Rigney argues particularly drawing on Aleida Assmann's distinction between 'archival memory' and 'working memory' (Assmann 1999, cited in Rigney 2005, 17). What is the difference between 'archival' and 'working memory'? In Assmann's configuration, whatever its provenance, 'archival memory' is an assemblage of memories—effects of prior selections themselves—that are available but may or maynot be used in the actual act of remembrance. 'Working memory', on the other hand, connotes the effects of *mnemonic technologies* that are enacted and performed, including amongst others, narrating, reading, and writing as well as participating in commemorative events or engaging with digital media platforms in our days. Later in the chapter, I will consider, among other selection processes, the gaps between the intense memory work around Ada Lovelace's debated contribution to computer science and 'the archival memory' of her papers in the Bodleian Libraries in Oxford, which have given rise to contrasting mythologies around her life and work (see Hollings, Martin, and Rice 2017).

In the context of the work of memory then, convergence points to the assembling and reassembling of memories around Pierre Nora's famous concept of 'sites of memory' [*lieux de mémoire*], places literal or symbolic where 'memory crystallises and secretes itself' (Nora 1989, 7). In deploying the notion of convergence, Rigney makes connections between Nora's description of the 'sites of memory' providing 'a maximum amount of meaning in a minimum number of signs' and Foucault's reference to 'an internal proliferation of meaning', thus coining the notion of memory work as 'a self-perpetuating vortex of symbolic investment' (Rigney 2005, 18). In this light, 'sites of memory' function as 'cultural frameworks' of memory work, Rigney argues, in a transposition of Maurice Halbwachs's notion of a 'social framework' (Halbwachs 1994, cited in Rigney 2005, 18). In working with Sofia Kovalevskaya's meta-archive, I have mapped a dense network of such convergences and indeed conflations of memory and have let myself be whirled in Rigney's metaphor of the vortex (see Tamboukou 2022).

In the same way that utterances 'are repeated, reproduced, transformed and replicated' through various textual practices, including copying, translation, and commentary in Foucault's scarcity principle (cited in Rigney 2005, 20), so are memories, and particularly their textual and visual versions, Rigney notes, in considering 'recursivity' in memory work (20). It is through repetition that mnemonic practices become solidified in 'sites of memory' (Nora 1989), be they monuments, films, theatrical plays, popular literature, life-history writings, and even scientific texts. Just a glimpse into Émilie Du Châtelet's multi-modal archive creates a good exemplar of the work of 'recursivity' in memory work around her, a theme that will be developed more in the consideration of the adventures of her papers in general and her epistolary archive in particular.

But in the same way that Foucault's utterances are in a continuous process of discursive formation and transformation, modes of remembrance also enter an incessant cycle of modelling and remodelling. As Rigney argues this process depends on the social, cultural, and political frameworks within which acts of remembrance unfold, bearing in mind that 'the desire to recall, the availability of information, and the availability of suitable models of remembrance do not always coincide' (2005, 22). To put it simply, memory work is an ongoing process, a project in the making; it is never final and never finished. *Memory work is never done,* to paraphrase the famous feminist slogan. The most recent body of literature on Maria Gaetana Agnesi's legacy, whose focus has shifted from the woman to her world (see Mazzotti 2007), is a good reference point for the always unfinished project of memory work and the new models and forms it can take up.

In finally considering transferability and translation as a modality of remembrance, Rigney returns to the notion of *counter-memory*, highlighting the role of mnemonic practices in the symbolic struggles of marginalised groups in 'celebrating and re-enforcing their sense of a common past' (2005, 23) and thus enacting *counter-memories* of their historical constitution as subjects. Mary Somerville's multiple and edited memoirs are an exemplary case of transferability and translation in counter-memory work. As already argued at the beginning of this section however, *counter-memory* in its Foucauldian rendition is not inscribed in the dialectical opposition between dominant and marginalised memories 'from below', which Rigney's (2005) work has problematised. In tracing a critical genealogy of *counter-memory,* very much focusing on its Foucauldian take, Verónica Tello has proposed 'an alternative concept of counter-memory, one that is post-dialectical, not bound to the formulas of either/or, us/them or self/other, but which is instead committed to the endless accumulation and proximities of things – the and–and' (2022, 390).

In this backdrop, Rigney's (2005) five modalities of *mnemonic technologies* show that notwithstanding their role in social movements and identity politics, *counter-memories* themselves are in a continuous process of being culturally constructed and reconstructed, not only on a national but also—perhaps more importantly—on a transnational level, an argument that has given rise to a vibrant body of literature on transnational memory work.[1] My research with women mathematicians' archives is thus mapped in this wider web of transnational memory work, which according to Sophie van den Elzen, 'does not just make the past present, but also textures the present into a meaningful narrative, connecting disparate contemporary events into shared history' (2021, 24).

But in entering mnemonic laboratories, my approach has gone beyond the discursive to consider 'how matter matters' (Barad 2007) in memory work. In my previous work, I have focused on the material turn in philosophical treatments of memory, particularly looking at the strands of embodied and emplaced memories, while excavating memory traces in women workers' auto/biographical and political writings. In further following entanglements between memory, forgetting, and imagination, I have also examined how gender inflects the field of memory studies (see Tamboukou 2015, 2016c).

In my current research of writing a feminist genealogy of automathographies, my interest has shifted, however, from the materiality of mnemonic inscriptions in women's auto/biographical documents to the materiality and complexity of the work of memory itself. Drawing on my previous work of theorising the archive as a living organism (Tamboukou 2014), I am thus interested in unravelling discursive, material, spatial, and embodied entanglements in memory work. Given the genealogical framework of my research, I have considered Rigney's (2005) influential scholarship and its connections with the Foucauldian 'scarcity principle', as already discussed above, but I have also been intrigued by Laura Basu's notion of the 'memory dispositif' (2009). As I have discussed extensively in my work, the notion of the *dispositif* is central in Foucault's genealogical project, making important connections with Deleuze and Guattari's notion of the *machinic assemblage* (see Tamboukou 2003, 2008).

Here, it is important to remember that Deleuze and Guattari have used different terms for the notion of the *assemblage* [*agencement*] when referring to the composition of heterogeneous multiplicities, such as communities, organisations, or archives, in our case, which can be '*a machinic assemblage* of bodies, an intermingling of bodies reacting to one another', but also 'a *collective assemblage of enunciation*, of acts and statements, of incorporeal transformations, attributed to bodies' (1988, 88, emphasis in the original). In my previous work, I have used the notion of *the machinic assemblage* to emphasise materiality and embodiment in the *assemblages* of women, art, and education (Tamboukou 2008), and in following this line of thought, I will now turn to the consideration of what I will call *memory machines* as a way of engaging with space/time/matter entanglements in memory work.

Building memories

When engaging with the notions of *dispositifs* and *machines* in the work of Foucault, Deleuze, and Guattari, it is crucial to bear in mind that these thinkers have persistently refused to work within closed theoretical and methodological systems. As Rosi Braidotti has suggested, they have rather followed 'the problematic' model of philosophy (1991, 3), locating a problem that relates to our actuality and then mapping the problem within a matrix of polyvalent discourses and practices, attempting to trace the conditions of its possibility. Braidotti has used the metaphor of a building site to describe it: 'Theoretical work, especially philosophy, is rather like a building site: the selection of elements, the distribution of tasks, and the overall plan of the project are the key to what is called the "materiality" of ideas' (3). Paraphrasing Braidotti, memory work could also be taken as 'a building site': the selection of documents, the distribution of tasks, and the overall plan of the project become constitutive elements of the material turn in memory studies and research.

In the wider context of 'building memories', then, my philosophical toolbox includes the *dispositif,* which was first used as a concept by Foucault in *Discipline and Punish* (1977) but was further elaborated in the *History of Sexuality* (1978). The *dispositif* in Foucault's thought connotes a system of relations that can be established between heterogeneous elements, discursive and non-discursive practices, 'the said as well as the unsaid' (1980, 194). A *dispositif* can include 'discourses, institutions, architectural arrangements, regulations, laws, administrative measures, scientific statements, philosophic propositions, morality, philanthropy, etc.' (194). There is always 'an urgent need' (195) to which the functioning of the *dispositif* responds: 'the *dispositif* is essentially of a strategic nature, which means assuming that it is a matter of a certain manipulation of relations of forces, either developing in a particular direction, blocking them, stabilising them, utilising them, etc' (196). Being inscribed in a play of power, the *dispositif* also relates to certain types of knowledge which derive from it, but also condition it. In Foucault's view, therefore, the analyst has a *dispositif* when they have succeeded in isolating a cluster of power relations, sustaining, and being sustained by certain types of knowledge (196).

Drawing on this configuration, Basu (2009) has therefore coined the notion of the 'memory dispositif' in her treatment of the Ned Kelly *lieu de mémoire*. In examining how the memory site of the most well-known bush ranger in Australia 'may form and evolve over time, and in particular how it may produce, organise and transmit meaning, in terms of the mediatizations that are so essential to its existence' (140), Basu deploys the Foucauldian notion into the field of memory studies and tests the analytical effects emerging from this transposition. In doing so, she also draws on Rigney's metaphor of the 'memory vortex' (2005, 18), as already discussed in the previous section. What she argues is that the study of 'memory sites' as *dispositifs* reveals their complex constitution, highlights the heterogeneity of their components, maps power/knowledge relations transversing them at a given time, and also creates conditions of possibility for their continuous evolvement.

In her exposition of the Foucauldian notion, Basu also points to Deleuze's commentary on the Foucauldian *dispositif*, particularly highlighting his description as 'a tangle, a multilinear ensemble' (Deleuze 1992, 159, cited in Basu 2009, 141). Indeed, Deleuze has underlined two important consequences arising from a philosophy of *dispositifs*: the rejection of universals and a drive away from the Eternal and towards the new. As he has written: 'The new is the current. The current is not what we are but rather what we are in the process of becoming—that is the Other, our becoming other' (1992, 163–4). He therefore concludes that in each *dispositif* it is necessary to distinguish the historical part, what we are (what we are already no longer), and the current part, what we are in the process of becoming. In this context, *memory dispositifs* function along these two territories; on the one hand, they map the past—what we are already no longer—and on the other hand, they re-imagine the future—what we are in the process of becoming.

Further to his description of the *dispositif as* 'a tangle, a multilinear ensemble' (159), Deleuze has also written that the *dispositif* is composed of lines and zones that are difficult to determine and localise. These lines are usually deployed in unforeseen directions, while it is amidst crises that new lines are created and new directions open. As Deleuze sees it, in each *dispositif*, the analysis has 'to untangle the lines of the recent past and those of the near future: "that which belongs to the archive and that which belongs to the present"' (164). In the genealogical practice of untangling, there are lines of light, lines of force, and lines of subjectification emerging from each *dispositif* (160). According to the Deleuzian description, the lines of light form variable shapes inseparable from the *dispositif* itself, while each dispositif has its own system of managing light lines and producing light effects, 'distributing the visible and the invisible, giving birth to objects which are dependent on it for their existence and causing them to disappear' (162). As Deleuze notes, any *dispositif* 'can be defined from the point of view of the visible and from the point of view of that which can be enunciated' (162)

Following Deleuze's reading of the Foucauldian *dispositif*, memory work in the archive can indeed be charted as a 'multilinear ensemble' (159) on the grounds of which 'unknown landscapes' (159) emerge, while thinking evolves along moving lines. The archive of women mathematicians' personal, literary, and philosophical documents has indeed created regimes of exemplary visibility as well as regimes of shadow and darkness. Consider, for example, how the increasing visibility of women in the salons and *conversazioni* of the early modern period was sustained by the carefully kept invisibility of their bodies, their exclusion from the formal institutions of knowledge, and the restrictions imposed upon their mobility. Moreover, while most of these women were celebrated during their lifetime, they were forgotten after their death, and there have been ebbs and flows in the way they have been treated in acts and practices of memory work throughout the nineteenth and twentieth centuries. This play between darkness and light, visibility and invisibility, carries on to our own days through a cluster of discursive distortions and material practices by which women are still marginalised in science and mathematics.

In this regime of light and darkness, what I suggest is that, taken as a *dispositif*, memory work in the archives functions as an impressionist painting, blurring lines, colours, and shadows. Moreover, the artist/designer/researcher of the *dispositif* runs the risk of being themselves dispersed in the extravaganza of the light lines they allow to emerge or be hidden behind their shadows. To continue with untangling the 'multilinear ensemble' (159), the lines of forces move across and between points and lines within the cartography of the *dispositif* and make connections between disparate zones and layers.

As already pointed out at the beginning of this section, the notion of the *dispositif* is a critical area where connections between Foucault and Deleuze and Guattari's modes of thought can be made. Deleuze and Guattari have not used the notion of *dispositif*, however, even though Deleuze (1992) has theorised it in a most influential way, as we have already seen. One of the key concepts that permeates Deleuze and Guattari's work, and which runs in parallel with the notion of the *dispositif*, is that of the *machine*. Unlike closed organisms and fixed identities, *machines* are *assemblages* without any organising centre, which can only function as they connect with other machines in a constant process of becoming: 'a machine may be defined as a *system of interruptions* or breaks [...] Every machine, in the first place, is related to a continual material flow (*hylè*) that it cuts into' (Deleuze and Guattari 1984, 36, emphasis in the text).

The machine has no ground or foundation: 'it is nothing more than the connections and productions it makes; it is what it does; it therefore has no home; it is a constant process of deterritorialization, or becoming other than itself' (Colebrook 2002, 56). Claire Colebrook explains that 'there is no aspect of life that is not machinic; all life only works and is in so far as it connects with some other machine [...] so life is a proliferation of machinic connections' (56). The concept of the machine allows for the possibility of open configurations, continuous connections, and intense relations, incessantly transforming life: 'everywhere there are breaks-flows out of which desire wells up, thereby constituting its productivity and continually grafting the process of production onto the product' (Deleuze and Guattari 1984, 37). It is in this context that 'the machine' is used to denote the intermingling of bodies in Deleuze and Guattari's various terms of the *assemblage*, as already noted in the previous section.

To return to the field of memory studies, Red Chidgey has deployed the notion of the *memory assemblage* to denote 'a set of historically contingent vitalities that circulate across different scales and sites to compose the intelligibility of memorial entities' (2018, 42). In studying how feminist memories travel across mediated cultures, Chidgey has shown that feminist memories taken as *assemblages* are being formed by heterogeneous components and forces including, among others, 'people, objects, technologies, images and practices that contingently coalesce to give certain ideas or behaviours prominence at a particular time and place' (42). Chidgey has further explored what *memory assemblages* do by unpacking their capacities as a model as well as their possibilities for intervention (46–55). In the same vein, van den Elzen has considered anti-slavery *memory assemblages* as the collective effects of memory work in the transnational women's movement (2021). In this context, van den Elzen has pithily observed that 'memory work is not performed on the past, but on the material traces and mediations circulating in the present, and a productive way to analyse it is to observe how agents intervene in the assemblages on which shared memories rely' (24).

We have thus seen how both notions, the *dispositif* and the *assemblage*, have been deployed independently in the field of memory studies. What I suggest, however, is that given their polyvalent form as *assemblages* of heterogeneous elements as well as sites for the interplay of intense forces, the *dispositif* and the *assemblage* create an epistemological plane where Foucault's and Deleuze and Guattari's analytics can make connections. Considering these connections, what I therefore argue is that memory work in the archive and beyond should be seen within the conceptual framework of the *memory machine*, a plane for multiplicities to make connections, and a contested site for power relations and flows of desire to be enacted. In deploying Deleuze and Guattari's important notion of the *machine*, I want to highlight the dynamic process of production, transformation, and organisation of memories and show how *the machine of memory* encompasses a wide range of *dispositifs* and *assemblages*, including the socio-technical and epistemological effects of the digital revolution, as I will further discuss in the next section.

In offering the concept of the *memory machine* as a tool that brings together the analytical capacities of the *dispositif* and the *assemblage* in the wake of the digital revolution, I further highlight the differences not in what they are, but in what they can do. Recall that the 'strategic nature' of the *dispositif* is for Foucault 'a matter of a certain manipulation of relations of forces' (1980, 196). Deleuze and Guattari have also pointed out that it is not enough to chart the components of an *assemblage* but also to map its capacities (1984, 257). In this context, we should consider how flows of affects in *assemblages* saturate bodies and produce realities: 'we know nothing about a body until we know what it can do, in other words, what its affects are, how they can or cannot enter into composition with other affects', Deleuze and Guattari have argued (275).

Here, it is important to consider how Foucault (1980) considers power in relation to the *dispositif*, while Deleuze and Guattari (1984) refer to affects when thinking about the capacities of the *assemblage*. To put it simply, power is at the heart of the *dispositif*, and there are indeed important power relations to be mapped in the *dispositif* of memory work with women mathematicians' archives. What the *assemblage* brings to the fore, however, is the creative force of desire that is sidelined and downplayed in Foucault's philosophy in general and his configuration of the *dispositif* in particular, as I have written elsewhere in my work (see Tamboukou 2003, 2010, 2015).

Commenting on Deleuze and Guattari's work in relation to Foucault's, Paul Patton has seen them as 'engaged in parallel but complementary projects' (2000, 53), particularly in their analysis of power and desire. In Deleuze's and Guattari's analyses, desire is regarded as an autonomous and productive force shaping the social rather than being determined by it: 'desire is always constitutive of a social field' (Deleuze and Guattari 1984, 348). In this light, they have theorised multiple ways in the flowing of desire through *assemblages*. In the same vein, power is for Foucault a material entity of relations at play. As Deleuze stresses, power for Foucault 'isn't just the relation of a force to a being or an object but corresponds to the relation of a force to the other forces it affects, or even to forces that affect it' (Deleuze 1990, 117). According to Patton, this affective dimension of power in Foucault forms rhizomes with the Deleuzian notion of desire and becoming (2000, 53). What I therefore want to suggest is that memory work in the archive should be taken as a *memory machine*, a conceptual framework enabling the study of how embodied and embedded power relations and forces of desire are entangled not only in the contents of the archival documents but also in the ways they have been assembled, reassembled, studied, interpreted, and effectually rewritten in the historiography of gender, science, philosophy, and mathematics.

In the fold between power and desire in memory work: mapping different archives

As already discussed in the previous section, the archive is at the heart of my work. However, in engaging with the six mathematicians' auto/biographical, philosophical, and literary writings, I have worked with different archives as well as different forms of archival documents that I now want to present and discuss in this section through the lens of the *memory machine*. As I have written elsewhere in my work, apart from librarians, curators, and archivists who create and organise archives, researchers also create 'archival *assemblages*' when they bring together documents from diverse archives and sources around the world (see Tamboukou 2017). Indeed, all research projects create archival *assemblages*, whether they be documents, oral interviews, transcriptions, or other research data. But researchers often hide the archival messiness of their research through their immersion in the power relations of knowledge production that Foucault (1969) has influentially theorised in *The Archaeology of Knowledge*. It is, therefore, my situated position in the archive, as well as in the knowledge field of my research, that I want to map as part of the overall epistemological stance of 'situated knowledges' (Haraway 1988). In doing so, I will first look at the archival sites, histories, and practices of all six women mathematicians, philosophers, and scientists. In the final section, I will consider them in their interrelationships and the connections they have made in the *machines* of memory work.

Émilie Du Châtelet: archives in turbulent times

Shortly before she died, Du Châtelet deposited the manuscripts of her scientific publications with Claude Salier, the royal librarian: 'I use the liberty, you have given me to leave in your hands the manuscripts, which I have great interest in securing that they will remain after me' (in Kölving and Brown 2018, vol.2, 478), she wrote in one of her last extant letters. While wishing that all would go well with the birth, despite her fears, she nevertheless asked the librarian 'to put a number in the manuscripts and to register them, so that they are not lost' (478). In doing so, Du Châtelet deployed the mnemonic technology of 'archiving herself', a recurrent practice that I have traced in writing feminist genealogies throughout the eighteenth, nineteenth, and twentieth centuries (see Tamboukou 2016c, 2020). To put it simply, educated women like Du Châtelet were quite conscious of the fact that their work would be forgotten after their death and so they put in motion *memory machines* to preserve whatever they could for the archives of the future. As an effect of Du Châtelet's memory work, these manuscripts were indeed not lost, but many of her other philosophical, literary, and scientific papers, as well as several of her letters, were either destroyed or dispersed in different archives, and they are still emerging from their hideouts. As Andrew Brown has boldly put it, the history of Du Châtelet's manuscripts is and will remain obscure,[2] despite the intense memory work that is still going on around her, as I will further discuss.

The greatest loss in Du Châtelet's archive is, of course, her correspondence with Voltaire. There are only four letters in Du Châtelet's extant correspondence: two letter poems and a short note from Voltaire, written in 1736, 1740, and 1746, as well as a very short undated letter, probably written around 1743, from Du Châtelet asking Voltaire for some financial assistance.[3] One can only assume that their letters were deliberately destroyed either by Voltaire himself, following Du Châtelet's untimely death, or by somebody else after Voltaire's death. The disappearance of this correspondence, which is a great loss for the history of ideas, given the importance of both Du Châtelet and Voltaire in Enlightenment thinking, was most probably the effect of intense forces of desire in the *memory machines* of their time.

Power relations were also at work, particularly if we consider that Voltaire's papers were bought by Catherine the Great after his death in 1778. Voltaire was in correspondence with the Russian empress for over fifteen years, which might have been the reason why she was willing to pay a huge amount of money—30,000 rubles in gold, besides furs, jewels, and diamonds—to his niece and heir Marie-Louise Denis, so that she could obtain his papers that included her letters.[4] However, historians, librarians, and archivists now think that the reason for this acquisition had more to do with Catherine's memory work and cultural politics, namely her desire to forge herself as an Enlightened monarch as well as her political intention to mark the Russian empire as a civilised European country that knew how to appreciate and honour great minds.[5] At the same time, the empress was also castigating France for the way Voltaire, his ideas, and his work were persecuted during his lifetime as well as after his death.[6] We can see how *memory machines* were in full swing then and now if we consider the intense memory work that followed the initial acquisition.

Voltaire's library, comprising 70,000 volumes of books as well as thirteen bound volumes of manuscripts—which included Du Châtelet's papers in Volume 9—was initially archived in the *Hermitage*, the empress's residence at the time. This initial collection was actually organised by Voltaire's secretary Jean-Louis Wagnière, who was commissioned by the empress not only to carry the library to St. Petersburg and install it in the *Hermitage*, following the order of books in Voltaire's Ferney residence in France, but also to arrange and bind the manuscripts. Wagnière also created two inventories, one for the manuscripts and one for Voltaire's handwritten notes in the margins of various works.[7] Voltaire's Collection was subsequently transferred to the National [then Imperial] Library of Russia in St. Petersburg in 1861.[8]

In 1940, Ira O. Wade transcribed several manuscripts written by Du Châtelet and published them in 1947 in his *Studies on Voltaire*: 'Part Two—Unpublished Papers of Madame Du Châtelet'. These unpublished papers, which were all marked by Voltaire's handwriting as 'written by Madame Du Châtelet', included her translation of Mandeville's *The Fable of the Bees,* chapter 4 on the formation of colours of her essay on optics [*L'Essai sur l'optique,* Chapitre 4: *De la formation des couleurs*] and chapters 6–8 of her *Reasoned Grammar* [*La Grammaire raisonnée*] (Wade 1947).

In 2017, the Centre for the History of Women Philosophers and Scientists (HWPS) and the National Library of Russia, St. Petersburg (NLR) agreed upon a collaborative digital edition of the Émilie Du Châtelet manuscripts preserved in the Voltaire Collection of that library, and in 2020, they celebrated the release of the first digital and the first historical-critical edition of this important corpus of Du Châtelet's early works, a project which is still in progress.[9] This new digital and historical-critical edition also includes the essay *On Freedom* [*De la liberté*], which was probably intended to be part of her *Foundations of Physics* (see Jorati 2019, 256), Du Châtelet's extensive annotations to the St. Petersburg copy of Voltaire's *Traité de métaphysique* [Treatise on Metaphysics]—which might in the future be reattributed to Du Châtelet's partnership with Voltaire, according to Brown[10] as well as an early (undated) version of the *Préface du traducteur* [Translator's Preface] to the Mandeville translation.[11]

But while Voltaire's library and manuscripts, including some of Du Châtelet's papers, did find a house, albeit outside France, the main corpus of Du Châtelet's scientific manuscripts went through wild adventures. When she died in 1749, her manuscripts were moved to the Château de Cirey, where they stayed for several decades. After her husband's death, the Château was inherited by her son, but he and his wife were guillotined during the French Revolution in 1794. Since they had no children, all their possessions went to their niece, Diane Adélaïde de Simiane (née Damas), who was their designated heiress, including the Château's library and Du Châtelet's manuscripts. Although the Château was declared national property, Mme de Simiane managed to reclaim it and prevent its planned demolition, but her heirs sold it in 1892.

During this long period, Du Châtelet's manuscripts remained stored in the Château de Cirey for an extended period before being relocated to a nearby house in Rosnay-l'Hôpital (Aube) when the Château was sold. There, they were kept in ten wooden boxes in the attic, largely untouched—except by rodents, which caused irreparable damage to one box. In 2010, the manuscripts were 'discovered' and later auctioned in 2012, despite efforts by 1,400 researchers worldwide to have them designated as a national treasure. They were purchased by an investment company that later collapsed amid fraud allegations.[12] In 2016, the manuscripts were acquired in a second auction by a Swedish buyer described by Andrew Brown, who had long worked with Du Châtelet's manuscripts, as 'a friend of ours'.[13] Amid these political and financial upheavals, many of Du Châtelet's literary manuscripts were lost or scattered throughout the nineteenth century. Additionally, most of her anti-clerical writings were likely destroyed by the Damas family, either because they were deemed 'too shocking', as Brown suggests, or out of fear of persecution.[14]

Du Châtelet's letters have had an equally interesting editorial history, which began in 1746 while she was still alive with the publication of her election at the Academy of Sciences in Bologna and reaches our own days with the publication of her complete correspondence, edited by Ulla Kölving and Andrew Brown in 2018. Du Châtelet was indeed a voluminous correspondent, and the forms and types of her letters are polyvalent and multi-functional. Her letters are important auto/biographical documents, seizing moments of the everyday, dealing with worldly issues and business affairs, and carrying marks of socio-economic, historical, and political experiences, formations, and circumstances. Apart from the letters of the everyday, there are letters to friends and confidants, where existential questions are raised and confessions are made. There are also philosophical letters, evoking subjects of a cultural nature in the traditional style of literary correspondences, and there are her love letters, where discursive expressions of intense passions and feelings are etched, but also struggles with the beloved are staged, dreams unfold, and plans are made. Last but not least, there is a rich body of scientific correspondence with eminent mathematicians and scientists of her time, as I will further explore in Chapter 2. Taken together as an epistolarium (Stanley 2004), her letters utterly chart a map where the process of becoming a woman mathematician, scientist, and philosopher can be traced.

As I have shown so far, Du Châtelet's letters -and indeed her entire archive- are multi-modal, comprising, digital, and published documents, originally housed in several European countries including Austria, Belgium, France, Germany, Italy, Poland, Switzerland, and the UK as well as Russia and the USA.[15] The French National Library (BnF) has digitised the entirety of Du Châtelet's documents that are housed in its books and manuscripts departments, including her mathematical, scientific, and philosophical publications as well as her published and autograph letters.[16] Her letters to Maupertuis are included in Gallica—the digital library of the BnF—and it is the only digitised collection of autograph letters in the entirety of Du Châtelet's papers, which offers a sense not only of her handwriting but also of the form of her epistolary discourse.[17]

The second substantial collection of Du Châtelet's autograph letters that I was able to engage with is her correspondence with her lover, Jean-François de Saint Lambert. This collection of ninety-six autograph letters is housed at the Morgan Library and Museum in New York, and although it has not been digitised and is not available online, I was able to obtain digital images of the entire collection through a research grant from the University of East London, for which I am grateful. The study of the autograph letters alongside the three published versions of this correspondence (Besterman 1958; Soprani 1997; Kölving et Brown 2018), with all their differences in terms of the epistolary dates, as well as their annotations, has offered diverse epistemological insights that I will further discuss later in this chapter as well as in Chapter 5.

My virtual encounters with Du Châtelet's archive have foregrounded the discursive aspects of memory work but have also offered a panorama of power relations and forces of desire in the working of her *memory machine*. Although I have not engaged with analogue documents of Du Châtelet's papers, the physical and corporeal aspects of memory work were mostly enacted through my visits to Du Châtelet's house at No. 12 La Place des Vosges, at the heart of the Parisian district of Le Marais. In considering the spatiality and materiality of memory work, earlier in this chapter, I have referred to Nora's famous concept of 'sites of memory' [*lieux de mémoire*], and the connections that Rigney (2005) has made with Foucault's 'scarcity principle' (1969) in deploying the notion of convergence as a mnemonic technology. My repeated visits to Du Châtelet's house at La Place des Vosges have put in motion Rigney's 'memory vortex' (2005,18), but I was, at the same time, puzzled by the fact that there is no memorial plaque marking the historical significance of this location. As I have already observed above, 'memory work is never done', and as I have shown in my work with the Parisian seamstresses in the past,[18] women's *lieux de mémoire* have yet to be created, well into the twenty-first century, whatever their social, cultural, and/or scientific status.

What has been definitely preserved, however, is the Château de Cirey, where Du Châtelet lived with Voltaire and where she found her voice as a writer and established herself as a scientist, philosopher, and mathematician. Although a private property, the Château has been classified as a Historical Monument, but its website still indicates that it has been labelled as a 'Maison des Illustres', since it is the place 'where Voltaire resided from 1734 to 1749, invited by the Marquise du Châtelet'.[19] Du Châtelet's memory is also marked in the Château de Lunéville, in Lorraine,[20] where she had been invited with Voltaire by Stanislas, former king of Poland, to join his court, and where she died.[21] As *lieux de mémoire*, Du Châtelet's residences are still marked by Voltaire, and yet their architecture, spatial arrangements, furniture, and decoration, as well as their gardens, carry the sense and smell of intense intellectual and emotional histories and make Du Châtelet's *memory machine* work tirelessly to our own days.[22]

What should also be added to the clogs of this *memory machine* is a series of portraits, as stark etches of embodied memories. Beatrijs Vanacker and Lieke van Deinsen have pointed out that the relation between gender and the representation of intellectual authority is deeply rooted in European history, while women in the early modern period struggled to forge intellectual authority through textual and visual constructions of themselves as scholarly personae (2022, 7). Du Châtelet's portraits carry such embodied mnemonic traces in that they are carefully staged to foreground her status as an illustrious woman philosopher, mathematician, and scientist. While she is dressed as an aristocratic woman in all of them, the books, sheets of geometric figures, and dividers that she holds in her hands are strong signs of her devotion to mathematics.[23]

While memory work with Du Châtelet has mostly unfolded on a plane of published, digital, and digitised textual and visual documents, against the backdrop of very troubled archival histories, Maria Gaetana Agnesi's archive, housed at the Veneranda Biblioteca Ambrosiana in Milan, would initially appear as a classic archival collection. And yet, an in-depth engagement with Agnesi's archive and the history of its creation reveals many more layers of complexity, as I will further discuss in the next section.

Maria Gaetana Agnesi: archival rhythms and affects

My visit to the Veneranda Biblioteca Ambrosiana in Milan in November 2022 was a very important event, not only in the development of the research project that underpins this book, but also for me personally, as it was my first physical visit to archives after the lockdown, and I felt very emotional about it. But it was much more than the first post-pandemic archival visit. The Ambrosiana Biblioteca is a historic library in Milan, which was founded in 1607 by Cardinal Federico Borromeo and was inaugurated in 1609. When Borromeo donated his art collection to the Biblioteca, the Pinacoteca Ambrosiana was created on the same premises in 1618. Having been established as one of the earliest libraries to grant access to anyone who could read and write, the library and the art gallery grew and flourished over the years to become a renowned cultural institution in Europe and worldwide.[24]

Since Agnesi was born, grew up, lived, worked, and died in Milan, it is not surprising that her papers are housed in the Ambrosiana Biblioteca, while a reproduction of Agnesi's marble bust, created by the sculptor Giuseppe Franchi in 1781 is also part of the Pinacoteca. Agnesi's manuscripts were deposited in the Ambrosiana Biblioteca in 1831 by Lady Luigia Verri, widow of Confalonieri and heir of the Agnesi (see Masotti 1940, 92). They were collected in twenty-five codices of different formats and sizes, 0.180–0.204, and this organisation of the manuscripts still holds today.[25]

There is a period of thirty-one years between Agnesi's death in 1799 and the deposition of her papers in 1831. If we consider that Agnesi's impressive library at the Palazzo Agnesi was dispersed and sold, alongside other valuable objects and furniture, due to financial difficulties that the family went through following her father's death in 1752, we can only imagine that the deposited papers are only a very small fraction of Agnesi's books and manuscripts that comprised her library and overall intellectual heritage at the palazzo. Since the palazzo itself was bombed to the ground during World War II, it is only thanks to the detailed inventory that was created soon after her father's death that we know that Agnesi's working library included 400 volumes and was 'similar in size to a college lecturer's personal library' (see Mazzotti 2007, 93).

The publication dates of most of the volumes in this library indicate that it was created by Agnesi herself to serve her studies and scientific interests, and thus the inventory offers an overview of the cultural resources underpinning her scientific writings. What is deposited in the Ambrosiana Biblioteca, however, is only a handful of Agnesi's school notebooks and some letters with her tutors, as I will further discuss in Chapter 2. Since the notebooks of a young girl were not deemed valuable, they were not sold. They must have been kept in her sister Anna Confalonieri's house as family memorabilia, and they have eventually become Agnesi's only textual traces.

Mazzotti has argued that it is because of this scarcity of traces about her life that 'Agnesi has suffered a peculiar historiographical maltreatment' (2007, xii), in the sense that she has become an *assemblage* of highly contradictory images and personas: 'a symbol of the Enlightenment, a heroine of counterrevolution and political restoration, a prophet of the national unification of Italy, a Catholic modernist, a proto-feminist, the ideal fascist woman, and even a potential saint of the Church' (xii). We have here a forceful enactment of Foucault's 'scarcity principle' in the sense that Agnesi's very few traces were 'invested with value [...] reproduced, transformed, and replicated, not just through copying and translation, but also through interpretation, commentary, and an internal proliferation of meaning', as Rigney has commented in transposing the Foucauldian principle in memory work (2005, 16).

Indeed, the *memory machine* around Agnesi started its work while she was still alive, when Franchi, an eminent sculptor, who was also teaching at the Brera Academy in Milan, created a marble bust of her. Copies of this bust were later ordered by several ecclesiastic and secular members of the Milanese nobility and found a place even in the Ambrosiana, as we have seen above (see Anzoletti 1900, 386). Since Agnesi had flatly refused to sit for any portraits, Franchi decided to shadow her by following her daily activities at the *Pio Albergo Trivulzio*, a charitable institution in Milan for the care of the urban poor and invalid, where she had become the director of the women's section (see Mazzotti 2007, 147). When she was presented with the marble bust that Franchi had created for her, while visiting her sister Anna Confalonieri, she was both surprised and pleased. This is what she wrote to the artist in response to his apologies for having 'stolen' her portrait from her:

> I have received with perfect gratitude the gift that you have deigned to bestow upon me, and I cannot cease to admire the skill with which you have so aptly imitated me, albeit surreptitiously. You are right to say that the theft committed against me is no trifling matter; therefore, while I am far from despising it, I cannot approve of it, knowing full well how undeserving I am of the great honour bestowed upon me. I hope for some opportunity to demonstrate my gratitude, and I remain with utmost esteem, your devoted and obliged servant (in Anzoletti 1900, 386).

As already noted in the previous section, artistic representations have historically played an important role in the work of memory, particularly so in the early modern period, but what is intriguing in the case of Agnesi's marble bust is that not only did she not commission it, but she also opposed its creation. Agnesi was evidently working hard to eliminate the unbearable visibility of her life as a learned woman-spectacle in which her father's ambition had enwrapped her. Patriarchal power and filial desire were entangled forces in Agnesi's *memory machine*, creating intense regimes of light and darkness, visibility and invisibility, memory and forgetting.

But Agnesi's desire to become imperceptible was only partially realised. While she died penniless and was buried with fifteen other women in an unmarked grave, (see Mazzotti 2007, 149), her first biography, *Elogio Storico di Donna Maria Gaetana Agnesi*, written by Antonio Francesco Frisi (1799) and published only four months after her death, decisively inscribed her in historiography. Frisi, a close friend of the Agnesi family, had easy access to family documents and particularly letters. Being an established eighteenth-century historian himself, he knew how to interweave memories, life stories, and histories. Despite its hagiographic approach, the *Elogio* remains the main biographical source, a symbolic 'site of memory' par excellence, that has informed all life-writing projects about and around Agnesi and has been repeatedly used and abused in her *memory machines*. Recall the role of 'recursivity in memory work' and the way Rigney (2005, 20) has highlighted repetition as a mnemonic practice in life-history writing.

In translating it into English for the first time in 2007, Antonella Cupillari has written that the *Elogio* 'offers a fresh portrait of Agnesi through the eyes of a contemporary' (2007, 1). Luisa Anzoletti, the author of Agnesi's extended documentary biography, published in Milan in 1900, describes the *Elogio* as 'the port of safety which prevented the shipwreck of every confirmed memory of Maria Gaetana's life' (1900, 43). Given its importance not only as Agnesi's main biographical source but also in the historiography of mathematical sciences, the *Elogio* was revised and published in 1965 by Arnaldo and Giuseppina Masotti, who were historians and mathematicians and thus offered lengthy annotations and comments on her mathematical work. In her own biographical approach, Giovanna Tilche has notably argued that Frisi's memoir was the diary that Agnesi never wrote, a document of life that put Agnesi's *memory machine* in motion (1984, 32–3). While drawing on Frisi's memoir, as well as Anzoletti's documentary sources, Massimo Mazzotti has excavated the legal archives to unveil details about Agnesi's world that would have been lost forever. By studying 'thick files of handwritten sheets produced by notaries after Pietro's [Agnesi's father] death', Mazzotti offers a lively description of the Palazzo Agnesi:

> The large room in which Gaetana studied and wrote was far from the site of the conversazione and sheltered by a series of quiet rooms that were used by members of the family for prayer and rest. In one of these rooms, facing an ivory crucifix, was a walnut prie-dieu. Next to it, a door opened onto Gaetana's studio, a room notable for its sparse decoration and lack of precious textiles. The visitor's attention was instead captured by four large

> globes, three armillary spheres, and 'various instruments [ordigni] for the study of mathematics,' which were on display in a window case and on a desk. Another desk held piles of books and maps. Gaetana used two small Chinese style tables and a few padded chairs for conversation with family members and tutors. Against the wall was her growing library, the higher shelves accessible only by means of a ladder (Mazzotti 2007, 44).

Mazzotti's careful work in rewriting the notaries' detailed inventory has brought life to Agnesi's archive and has made an important contribution to the spatiality and materiality of memory work around and about her. It is through Mazzotti's recreation of Agnesi's world that we have a sense of her daily routine in relation to her studies, as I will further discuss in Chapter 4. As I was preparing for my visit to Agnesi's archive at the Ambrosiana Biblioteca, I was able to visualise the world her papers had emerged from, but I was at the same time apprehensive of my visit there, particularly in relation to my ability to read and understand the manuscripts.

As I have written elsewhere at length, an archive visit needs careful preparation and communication in advance (see Tamboukou 2016a), particularly so when the time and the budget of the visit are limited. And prepared I was. In the preparatory phase of the research project, I had already been in touch with the archivists, and I had a detailed inventory of Agnesi's manuscripts that I was planning to study. I had booked my visit there, following the advice of the archivists, but also the general rules of the manuscripts division: up to four pieces per day, although they could make exceptions for foreign visitors like me. It turned out that I did not need the exceptions, as my experience has always been: take your time with the archive (see Tamboukou 2014).

During the preparatory phase, I also got access to some digitised documents of the inventory to have an overall idea of Agnesi's handwriting, but nothing could have prepared me for the affective experience of sitting in the *Sala Lettura*, looking at and touching Agnesi's notebooks and letters. Archival sensibility (Moore et al. 2016) is not an idea; it is a material space/time experience, as my being in the *Sala Lettura* forcefully reminded me.

Just because of the cataloguing order of the manuscripts, the first notebook I was given was, '0183 sup.: Opuscolo mitologico d'incerto autore tradotto in greco da Maria Gaetana Agnesi e scritto di sua mano per propria istruzione' [Mythological pamphlet by an uncertain author translated into Greek by Maria Gaetana Agnesi and written in her own hand for her own instruction] (VBA.0.183. SUP).[26] Quite simply, it was a pamphlet on mythology written by an uncertain author, probably one of her tutors, translated into Greek by Agnesi and written by her hand for her own instruction. The 'archive fever' (Derrida 1995) reached its heights when Ancient Greek, the language I was taught at school, was in front of my eyes in a young girl's careful and calligraphic handwriting. What could I have desired more? The handwritten trace of language surfaced again in the second manuscript I was given: '0184: I due libri di supplemento a Quinto Curzio colla traduzione in italiano, francese, tedesco e greco di Maria Gaetana Agnesi e scritti di sua mano'

[The two supplementary books to Quinto Curzio with translation into Italian, French, German, and Greek, handwritten by Maria Gaetana Agnesi] (VBA.0.184. SUP).[27] In examining this document, I had the opportunity to follow her learning process in three languages I could read and understand: Latin, French, and Ancient Greek. Translation has come up as a strong theme in this project, as I will further discuss in Chapter 3.

My visit to the Ambrosiana Library in Milan signalled my return to the world of analogue documents, reading rooms, and majestic buildings, an experience that seemed very dim during the grey period of the pandemic. As it turned out, it was in Italy again that I had my second experience of 'physical' archival research, but not for an Italian mathematician this time. How on earth were Sophie Germain's papers found in the Moreniana Biblioteca in Florence? It is such archival mysteries that I will explore in the next section.

Sophie Germain: machinic mnemonics and archival crimes

'Events and political discussions have prevented us from drawing public attention to the loss which some time ago, the mathematical sciences suffered in the person of Mlle Sophie Germain' , Guglielmo Libri (1832, 1) wrote at the very beginning of his obituary, which eventually became her first biography, as we have seen in the Introduction. Libri was indeed puzzled by the fact that Germain's death went unnoticed, and he was the first to break the silence that enveloped her after her death. Perhaps he did not know at the time that her death certificate registered her as a 'rentiere-annuitant' [property holder] and not as a mathematician, scientist, or philosopher (see Mozans et al. 2022, 156). In 1835, only four years after her death, her philosophical work was highly commented on by Auguste Comte, the founder of positive philosophy, as I will further discuss in Chapter 6.

One would have expected that Comte's praise would have secured Germain's position in the French Pantheon. And yet, when in 1876, 'the literary and philosophical France forgot to celebrate the centenary of her death',[28] the scholar Hugo Gœring 'presented a study on the genius woman that her country seemed to overlook'[29] in the proceedings of a philosophical association in Berlin. Echoing his sentiment of disappointment, this is what the positivist journalist Jean-Léon-Hippolyte Stupuy wrote three years later, in the introduction of the publication of her philosophical work, which would become her second biography:

> You must painfully admit it. While so many women have found fame in frivolous writings, the only French woman who has succeeded in serious work, esteemed by mathematicians, from whom a whole aspect of her genius escapes, is hardly known to the public (Stupuy 1879, 7).

And yet, it seems that neither the publication of her philosophical work was enough to restore Germain's memory. When the Eiffel Tower was constructed in 1889, Germain's name was not included in the list of seventy-two distinguished scientists and mathematicians displayed on it, despite her important contributions to the theory of the elasticity of metals, a point of utmost importance in the engineering work for this monument (see Mozans et al. 2022, 156). 'Sophie Germain was forgotten', Stupuy wrote again in the Preface of the second edition of her philosophical work, published in 1896, as a move against forgetting. This second publication of her work not only included some of her unedited correspondence with important mathematicians of her time, as in the 1879 edition, but also a series of articles from scholars in France and overseas, as testaments to the importance of her philosophical work. Moreover, this second edition was just a cog in the *memory machine* that also included a series of commemoration events:

> Sophie Germain has taken on this second life made of the memory of the living: her work and biography have secured their place in the memory of humanity; her neglected tomb has been found and restored; the Municipal Council of Paris, always mindful of Parisian glories, has given the name of Sophie Germain to one of the streets of the Capital, and to one of its higher schools for girls, and her bust, reconstructed according to the phrenological head that exists in the Museum of Natural History, adorns the main courtyard of this school. Moreover, a commemorative plaque has been placed on the house where she died (Stupuy 1896, v–vi).

But the veil of silence around Germain and her work continued into the twentieth century: there was no centenary celebration for her death in 1931, as Harlor (1953) poignantly commented in his article on her life and work, published in the *Revue des Deux Mondes* in September 1953. Germain's persistent forgetting seems to enact the Foucauldian sacrificial theme, which contests the supposed neutrality and objectivity of the subject of knowledge, as we discussed in the first section. Following Zerubavel (2003, 2), we do need 'a *sociomental* topography of the past' so that we can understand how the French establishment constructed, structured, and organised memories around women, science, and mathematics, in general, and Germain, in particular, throughout the nineteenth and twentieth centuries. It also seems that the counter-memory discourses and practices were entangled with the adventures of her archive, as I will discuss next.

Germain died on 27 June 1831, during a month when fierce riots had erupted in the streets of Paris—as a rebellion of the *Misérables*. Libri himself was involved in these events, having taken an active part in the July 1830 uprising that overthrew the Bourbons, while also being involved in the unification movement in Italy.[30] While at the end of his obituary for Germain, he wrote that he was a 'stranger to her country, but not to her affections and to the objects of her studies' (Libri 1832, 2), he became a French citizen and got heavily involved in the cultural politics of the July Monarchy.[31] But despite his abilities as a mathematician and the fact that some of his theorems in number theory are still quoted (see Del Centina 2005, 61), Libri's passion turned out to be documenting and writing the history of mathematics and through this interest he came to own one of the largest private libraries in Europe, often under dubious circumstances, which eventually led to the accusation that he was stealing books and manuscripts to satisfy his passion for collection.[32]

Libri had to flee Paris after the demise of the 1848 Revolution,[33] and a large part of his archive, which included some of Germaine's letters, was confiscated. Although they are now housed at the *Bibliothèque Nationale* and have been digitised,[34] some of the manuscripts and letters that Libri took with him were later sold at an auction he held in London in 1859. In the same year, the Paris Academy received another part of Germaine's archive as a donation from her sister and nephew.[35] When Libri died ten years later, his archive and therefore Germaine's papers, manuscripts, and letters were dispersed, while half of his archive, including his correspondence with Germaine, has been lost for good, or they might appear someday, unexpectedly (see Del Centina and Fiocca 2004). Thus, while Libri became instrumental in triggering the intense memory work that followed Germaine's death, his nomadic lifestyle, his political adventures, and his illegal activities have had an important impact on Germaine's archive, which is still open with several unanswered questions demanding further research, as Dora Musielak has aptly remarked.[36]

In the backdrop of such archival troubles, Andrea Del Centina and Alessandra Fiocca (2018) have further discussed how Germain's letters were dispersed after her death, while the process of their rediscovery, which started in the late 1870s, is still ongoing.[37] Moreover, the publication of some of Germain's letters that were dispersed in different archives and public libraries in Europe has been equally adventurous and has included archivists, mathematicians, historians of mathematics, and publishing editors from the Paris, Göttingen, and Turin Academies of Sciences.[38] This interest was mostly because of her correspondence with Gauss and the intensive archival and editorial work that surrounded the celebrations of the first centenary of his birthday in 1877. In thus considering Germain's letters, we are dealing with a very complex meta-archive that includes analogue and digitised documents, published unedited letters, as well as transcriptions and translations in both historical and contemporary publications.

There is, of course, a burgeoning body of literature around digital and digitised archives and the way they have changed the epistemological nature of archival research in general, and the way we approach, understand, and analyse archival documents in particular.[39] Contextual loss is among the most important epistemological issues in the understanding and analysis of archival documents in general and letters in particular. To put it simply, transcriptions and publications of selected letters, as well as the more recent digitisation practices, often involve the extraction of individual documents from their original context, which may include the arrangement within a collection, annotations as well as relationships to other documents. The problem of the loss of context becomes even more acute if we consider the fact that women did not get the opportunity to have collections of their papers or *fonds* created, in the archival tradition, whether they were literary theorists, artists, activists, or scientists, as I have written extensively in my work (see Tamboukou 2020). Their papers, if preserved at all, are usually included in the *fonds* of famous men, dispersed in different archives around the world; they have often been taken out of their original context and have moved around different archival locations, mostly due to a lack of funding.

Regarding the problem of contextual loss, Germain's letters were dispersed, lost, and partially found in different collections all over Europe. Additionally, some of them were only found as drafts, since not all her recipients kept her letters (see Tamboukou 2024). In further considering Germain's meta-archive, I was particularly interested not only in the importance of different transcriptions and translations but also in different citations. As I have already mentioned in the Introduction, Germain's mathematical and philosophical work has not been translated in English, including her letters, with the exception of her correspondence with Gauss (see Musielak 2020) as well as some selected letters or fragments of them that have been included in her two main scientific biographies.[40] It is here that full and accurate citations can direct researchers of Germain's work in tracing 'archival trails' (Nygaard 2024) of her extant correspondence and can also enable the analysis of her epistolary discourse and the transferability of her ideas. Throughout this book, I have thus given full references not only to the archival sources of women mathematicians' letters but also to their transcriptions, publications, and existing translations, including my own. Here, I have followed Sara Ahmed's important remark that 'citation is feminist memory'; it is 'how we acknowledge our debt to those who have come before; those who have helped us find our way when the way was obscured because we deviated from the paths we were told to follow' (2017, 15–16).

So far, this section has revealed the layered complexities of Germain's memory work, shaped by forces of friendship, desire, politics, power dynamics surrounding her archive, and ongoing struggles between remembrance and forgetting. These tensions persist today, as I experienced during a visit to Père-Lachaise Cemetery in January 2025. My goal was to locate Germain's grave, but what unfolded was a confrontation with the silences and exclusions surrounding women's contributions to knowledge.

Germain's grave is absent from the cemetery's public map, an omission echoing earlier encounters, such as my search for Rosa Bonheur's grave fifteen years ago (see Tamboukou 2010). Despite Germain's historical significance, the cemetery staff initially failed to recognise her name or locate her resting place. Only after presenting documentation, including photographs and references, was I able to confirm its location in the 16th Division, plot 153. Finding the grave proved equally challenging. The plans provided were minimal, and the grave was obscured by a tree, making the search non-linear and reflective of the broader, hidden nature of Germain's legacy. The experience underscored how memory, space, and recognition—in short, the *lieux de mémoire*—are stratified, with Germain's grave serving as a metaphor for the exclusion and resistance she faced in life. Her contributions, often overshadowed, continue to demand recognition amid the rhythms of historical neglect. Standing before her grave, I was thus reminded of the persistence required to confront these silences. Germain's resting place is not just a site of remembrance but a *lieu de mémoire* par excellence, a call to re-examine how we structure memory and acknowledge whose voices are preserved and celebrated. Cemeteries, too, are *assemblages* of memory, where exclusions become visible and can be resisted, inviting us to rethink the flows of history, memory, and forgetting.

Finding the house where Germain spent her final years, at 13 Rue de Savoie in the 6th arrondissement, was surprisingly effortless. A simple plaque on its façade provided a clear, minimal trace—no need for deep excavation or revelations from the layers of time. The house's modesty stood in stark contrast to the grandeur of Du Châtelet's aristocratic mansion at Place des Vosges, which, as noted earlier, remains unmarked. This juxtaposition—two pivotal mathematicians, yet shaped by vastly different intersections of social class, remembrance, and erasure—encapsulates the complexities of memory explored in this chapter.

The stark differences between Germain's quiet trace and Du Châtelet's unmarked opulence reveal how power, class, and time shape their legacies, a theme I will return to in the conclusion. These women, remembered in unpredictable and often contradictory ways, remind us that history's rhythms do not flow in a straight line. Instead, they fold back on themselves, intertwining forgotten silences with the echoes of those who have been exalted, creating a tapestry of remembrance shaped as much by exclusion as by recognition.[41]

In light of the above, all five modalities that Rigney (2005) had identified in the work of cultural memory—namely selection, convergence, recursivity, and modelling as well as translation and transfer—have been seen at work in Germain's *memory machine*. What could also be added on the plane of her *machinic mnemonics* is her own intervention in the making of her legacy. Like Agnesi, Germain appeared indifferent to her visual representation, leaving no portraits, aside from a reconstructed bust by Zechariah Astrue, commissioned by the City Council of Paris.[42] Despite this, Germain was acutely aware of her intellectual identity as a woman mathematician in the Parisian scientific community. She was determined to establish herself as an independent mathematician in dialogue and collaboration with her contemporaries. In this context, she actively oversaw the publication of her mathematical works, asserting her presence in the Parisian academic circles. This was no small feat at a time when publishing outside formal academic networks posed significant challenges for women, as Jenny Boucard has highlighted, observing that Germain appears to have been 'the only woman who published on number theory in the first half of the nineteenth century' (2020, 191).

In tracing 'memorial practices' around Germain, Boucard has also looked at how Germain's *machinic mnemonics* were deployed in discourses around women's education in nineteenth-century France. These practices illustrate how Germain's work became part of a broader debate on the movement for women's higher education in Europe. This leads us to the concept of *memory machines* and their role in these discussions, with Mary Somerville's intellectual legacy forming the focus of the next section.

Mary Somerville: editing memories

In the final pages of her *Recollections*, Mary Somerville reflected on her advanced age, acknowledging the physical and cognitive challenges she faced, including severe hearing loss and a fading memory for everyday events and names. Despite these limitations, she took pride in her enduring mathematical abilities, highlighting her continued capacity to read, understand, and solve complex problems in higher algebra (2001, 303). Her determination remained unwavering, as she persisted in revisiting difficulties until they were resolved. Additionally, her passion for learning remained strong, as she enthusiastically engaged with new scientific discoveries and theories across various disciplines, finding joy and intellectual stimulation even in her 92nd year.

Mary Somerville's *Recollections* were first published as a book edited by her daughter Martha in 1873, one year after her death, with several editions following its initial release. There are seven recorded editions in the nineteenth century: 1873, 1874, 1875, 1876, 1879, 1885, and 1895, published in London and Boston, but only two in the twentieth century, published in 1975 and 1986 in New York. While receiving several favourable reviews at the time of its publication,[43] Somerville's memoir was mostly forgotten in the twentieth century, until it was reprinted in 1975 by the American Mathematical Society. Although Somerville's position in the history of science was mostly marginalised in the twentieth century, it did become the object of a small number of studies in Victorian culture and the history of science.[44]

What is also interesting is that there are two undated versions of the manuscript of Somerville's autobiography, housed at the Bodleian Libraries in Oxford: a first draft handwritten by Mary Somerville and a second draft partly in her own hand and partly in the hands of her daughters Martha and Mary.[45] In editing her mother's memoir, Martha followed the advice of several friends, but she kept within the conventions of Victorian autobiographies and enwrapped the whole memoir within the discourse of 'the angel in the house', topped up with the ability to solve mathematical problems and tackle scientific questions.

Over the years, scholars have engaged with parts of the first and second drafts that were deselected for the 1873 publication and its subsequent editions. Moreover, there is a new interest in Somerville's life and work in the new millennium, with a critical body of literature situated mostly in the fields of the histories of science and mathematics, but also in the interdisciplinary area of science and literature entanglements.[46] In 2001, Dorothy McMillan edited a new publication of the *Recollections*, which includes extracts from the first and second drafts along with a comprehensive introduction on her editorial acts and decisions (McMillan 2001). Brigitte Stenhouse's (2021) research in recovering Somerville's position in the history of mathematics is particularly important in this body of literature. Somerville's *memory machine* is thus alive and kicking, and in this section, I will consider her *mnemonic technologies* in the act of writing and editing her memoir. As Virginia Woolf has influentially put it, 'words are full of echoes, of memories, of associations'.[47] As we, therefore, shift and shuffle either their form, content, and/or context, we create different meanings and store different memories for the archives of the future.

As already noted in the Introduction, in the first draft of her autobiography, Mary Somerville explicitly stated her purpose for writing a memoir: to advocate for women's education and demonstrate the strength of the female mind, declaring, 'my only motive in writing it, is to show my country women that self-education is possible under the most unfavourable and even discouraging circumstances' (1872, 1). However, this statement was omitted from the published memoir, replaced by an introductory note from her daughter and editor, Martha, which framed her life as the one devoted to family and scientific pursuits, emphasising domesticity over intellectual advocacy: 'The life of a woman entirely devoted to her family duties and to scientific pursuits affords little scope for a biography' (Somerville 2001, 38).

This editorial decision effectively erased Somerville's clear support for women's education from the opening of her memoir, a particularly striking omission given her active support for higher education initiatives for women, such as Girton College, founded in 1869. In later chapters, Martha did acknowledge her mother's enthusiasm for extending classical and scientific education to women, noting Somerville's donation of her scientific library to Girton College as a testament to her dedication (Somerville 2001, 291).

Two questions arise from this editorial choice: Was Martha deliberately downplaying her mother's support for women's education to avoid the stigma of being self-educated and perceived as an amateur? This seems unlikely, given that the early nineteenth century embraced amateurism in science (see Patterson 1983, ix). Alternatively, was Martha reinforcing traditional gender norms by presenting her mother as a domestic figure who balanced family responsibilities with scientific work? Alice Jenkins suggests that Martha's edits adhered to the 'old-world theory', prioritising domestic achievements over intellectual pursuits (1999, 172). It is also possible that the final version of the memoir was shaped by multiple editorial voices and decisions, reflecting a complex interplay of intentions and compromises. By emphasising Somerville's domesticity, the published memoir aligned with prevailing discourses of gender, science, and respectability, even as it obscured the pioneering spirit of Somerville's self-education and advocacy for women's intellectual advancement.

Here, Rigney's fourth modality of modelling and remodelling (2005, 21) sheds light on the editorial decisions taken regarding the publication of the memoir. In examining Somerville's *Recollections* as a case study in Victorian print culture, Jenkins has focused on 'questions of mediation, relation and originality, as they were manifested in Mary Somerville's situation as a writer who was not male, not English, and yet an iconic figure in Victorian British culture' (1999, 163). Jenkins has examined Somerville's *Recollections* within the broader conventions of Victorian autobiography, comparing it to the autobiographies of Frances Power Cobbe, Harriet Martineau, and Margaret Oliphant. Both Cobbe and Martineau were Somerville's friends, though their interactions were primarily through correspondence. Notably, Cobbe played a key role in the final editing of *Recollections*, as revealed in her letter to Martha Somerville on 11 July 1873. In this letter, she praised Martha's introduction as 'touching' and 'genuine', predicting it would resonate with readers. While she admitted to making minor editorial changes, such as rephrasing colloquialisms and adjusting sentence structure, she emphasised that she avoided altering the substance of the text. Cobbe expressed confidence in the work, believing the book would enhance public affection for Somerville while maintaining admiration for her intellectual accomplishments (McMillan 2001, 12).

Jenkins' analysis situates Somerville's *Recollections* within a network of intellectual and editorial collaboration, highlighting how the work reflects both Victorian norms and the influence of Cobbe's editorial interventions.

In preparing her memoir, Somerville sought input not only from her daughter Martha, who later edited the final version, but also from trusted friends. Among them was Sir John Herschel, her long-time mentor and advisor, who reviewed the first draft. Herschel praised the simplicity and beauty of Somerville's personal narrative, particularly her account of her early passion for science and her self-taught journey under challenging circumstances. However, he recommended removing lengthy scientific passages that might alienate a broader audience and advised against immediate publication. In a letter dated 14 March 1869, he described the memoir as 'pleasing and simple' but urged caution regarding its timing (McMillan 2001, 10–11). Somerville heeded his advice, both in content and by delaying publication until after her death.

This delay was partly pragmatic. As Stenhouse notes, Somerville intended the *Recollections* to provide financial support for her daughters, given the family's reliance on income from her publications (2021, iii). Despite postponing publication, Somerville played an active role in shaping the memoir, making key editorial decisions about its structure and content. These included selecting letters from prominent scientists, writers, and artists from across Europe and the USA. These additions transformed the memoir into an archival repository, commemorating her connections and cementing her status as a respected scientist of her time. Indeed, Somerville is often hailed as 'the world's first scientist', a title linked to the supposed coining of the term 'scientist' by William Whewell in his 1834 review of her book *On the Connexion of the Physical Sciences*. While the *Oxford English Dictionary* credits this as the first use of the term, Stenhouse has clarified that Whewell did not actually use 'scientist' in his review or in any other context relating to Somerville's work (2021, 1n1). In this light, the omission of certain statements from the published version of her *Recollections* is understandable given her legacy as a renowned Victorian scientist. Notably, a passage expressing her view on originality and gender was excluded. In it, Somerville reflected on her lack of personal discoveries, attributing this to a perceived absence of 'genius' in women, contrasting her perseverance and intelligence with the 'spark from heaven' that she believed was denied to her sex: 'I was conscious that I had never made a discovery myself, that I had no originality', she wrote (1872, 168). This self-deprecating sentiment aligns with Ruth Messbarger's notion of a 'double-voiced discourse' (2005, 18), where Somerville both defends women's education and scientific participation and apologises for perceived deficiencies. Although this extract was not included in the published version, it remains a component of her *memory machine* as well as a discursive trace of the complexities entangled around women's 'automathographies', then and now.

Similarly, other passages were left out, including one where Somerville humorously discusses her struggles with language, admitting difficulties in writing English, and spelling (1872, 94–5). Despite this, she acknowledged the praise she received for her English and expressed amusement over the linguistic differences she shared with her friend Joanna Baillie: 'On comparing notes with my dear old friend, Joanna Baillie the poetess, who agreed with me in points much more important than that of spelling, we were not a little amused to find that in writing a letter we made use of words we could spell, not those we should have used in conversation' (94–5). Being an accomplished artist herself, Somerville also chose to highlight her ignorance of 'the highest branch of art', acknowledging her limited understanding of fine art despite extensive travel through Europe's great museums. In an extract omitted from the published version, she explained that while she could recognise beauty, her appreciation of art was intuitive rather than learned (96).

In addition to her intellectual limitations, Somerville's political views, particularly her stance on anti-slavery, were also excluded. Despite her well-known abolitionist sentiments, a passage in which she critiques British colonial violence and the American Civil War was left out (132–3)—perhaps because it presented her mature political opinions in a way that was less personal and more politically charged than her earlier, more symbolic actions like refusing sugar in her tea (2001, 153), as part of the abolitionist sugar boycott.[48] These omissions reflect editorial decisions that both protect and shape Somerville's public image, emphasising her intellectual achievements while downplaying personal vulnerabilities and controversial political stances.

As already noted in the Introduction, the Somervilles decided to move to Italy around 1838 due to financial difficulties. Their departure was rather rushed, and their belongings, including their library, letters, and manuscripts, were scattered among relatives in London, as the following omitted extract from the first draft of the *Recollections* shows: 'Our departure was so sudden, that everything was left in confusion. I sent my cabinet of shells to my niece, and I left in the care of my son and daughter-in-law our books, the cabinet of minerals, a valuable collection of original letters, and several manuscripts' (Somerville 1872, 193). Since they lived nomadically in Italy from 1838 onwards, the Somerville collection at the Bodleian Libraries only includes a limited number of letters and family documents as well as the handwritten drafts of the two versions of her autobiography. Somerville's grave is in the English cemetery in Naples, which has now become a public park. Her sculpture on the grave was created by the Calabrian artist Francesco Jerace and portrays her sitting on a chair, looking away at an ideal horizon.[49] Apart from her family house in Burntisland and Edinburgh, which have been duly marked, there are no other physical *lieux de mémoire*.[50]

Somerville's embodied memories, however, have been inscribed in a series of portraits, including her self-portrait at Somerville College in Oxford, where she depicts herself sitting and holding a book and a quill in her hand. She faces the viewer directly, and her expression is pensive, conveying a sense of her intellectual status. The neutral background behind her allows the focus to remain on her. This self-portrait, the only one we have created by a woman mathematician and scientist in the nineteenth century, is different from her other portraits as a young woman by John Jackson, by Thomas Phillips in 1834, and by James Rannie Swinton in 1848, where she appears more elaborately dressed and with more fashionable hairstyles.[51]

Indeed, Somerville took pride in the effortless way of delving into her mathematical and scientific work, while also engaging with embroidery and fashion: 'I am very much delighted to have a cap made by you and the more so as it shows [that] we mathematicians can do other things besides studying xes and ys' (LB 174, ff. 19–20)[52], Somerville wrote to Ada Lovelace in 1834. When commissioning her first portrait as a young married woman to the renowned artist Margaret Carpenter, Lovelace also asked Somerville's expert advice, as she was very much concerned about her portrayal: 'The head of my picture is now completed, but I would rather not give my opinion of it, that your judgement may be unbiased. Mrs Carpenter would be very glad of your opinion and so should we' (MSC, c367, f.53), Lovelace wrote to Somerville in November 1835, while asking her to pay a visit to Carpenter's studio 'to have a look' at her portrait. Women mathematicians were thus busily engaged in the *machine* of their memory work, as I will further discuss in the next section, where I consider Lovelace's *mnemonic technologies* as well as the memory wars that raged after her passing.

Ada Lovelace: memory wars

'I begin to understand *Death*; which is going on quietly & gradually every minute, & will never be a thing of one particular moment' (LB44, f.146)[53], Lovelace wrote to her mother in an undated letter sometime in August 1852, three months before she died from uterine cancer at the age of 36, on 27 November 1852. Shortly before her death, Lovelace expressed her wish to be buried beside her father, the famous romantic poet Lord Byron, in the Byron family vault at St. Mary Magdalene Church in Hucknall, Nottinghamshire. In a brief note dated 30 September 1852, she specified the details of her burial and the inscription she desired. She requested a modest yet elegant marble monument with a specific epitaph from St. James's Epistle (Chapter 5, Verse 15), emphasising the phrase, 'and if he have committed sins, they shall be forgiven him' (LB 175, f.146). She also wished to include a 14-line sonnet she had written about 'The Rainbow', followed by her signature and the date. Lovelace envisioned similar monuments in other locations, including East Horsley, Oakham, and Kirkby, with the one at Hucknall surrounded by flowers, funded for their upkeep.

However, Lovelace's sonnet was not included on her tombstone as she had requested. Her funeral became a source of contention, with Mary Russell Mitford describing it as 'too ostentatious' (cited in Langley Moore 1977, 333), reflecting Lord Lovelace's tastes rather than Lady Byron's. Despite not attending her daughter's funeral, Lady Byron later honoured Lovelace's wishes by erecting a memorial at Kirkby Mallory, where Lovelace's 14-line poem and the prayer from James's Epistle were finally inscribed:

> The Rainbow
> Bow down in hope, in thanks, all ye who mourn
> Where'en that peerless arch of radiant hues
> Surpassing earthly tints,—the storm subdues!

Of nature's strife and tears 'tis heaven-born
To soothe the sad, the sinning and the forlorn;-
A lovely loving token; to infuse;
The hope, the faith, that Pow'r divine endures
With latent good the woes by which we're torn.
'Tis like a sweet repentance of the skies,
To beckon all by sense of sin opprest,
Revealing harmony from tears and sighs;
A pledge,—that deep implanted in the breast
A hidden light may burn that never dies,
But burst thro' storms in purest hues exprest.

(LB 175, 226–229)[54]

Lovelace's sonnet engages memory as a dynamic force of becoming, where the rainbow operates as an *assemblage* of nature, emotion, and transcendence. Emerging from the chaos of 'nature's strife and tears', the rainbow reterritorialises the storm's destructive rhythms, opening a space for new connections and intensities. Memory, in this sense, is not a static repository but a plane of immanence, where grief and harmony coexist and continually transform. The 'hidden light' Lovelace invokes is a virtual force, latent yet always ready to actualise as the rainbow's 'purest hues', bursting through the storms of life. This poem resists linear interpretations of memory, instead presenting it as a field of affective resonance, folding the past and future into the present moment of aesthetic and emotional intensity. The rhythmic structure of the sonnet itself performs this process, oscillating between strife and reconciliation, creating a movement that dismantles fixed oppositions and generates new configurations of meaning. *The Rainbow* becomes a site of becoming, where memory, nature, and spirit intersect in a perpetual flow of creation and renewal, expressing a vibrancy that cannot be contained within static forms or singular narratives.

As I will further explore in Chapters 5 and 6, Lovelace seamlessly blended poetic imagination with mathematical insight, using both to investigate patterns, relationships, and the interplay of logic and intuition. This section, however, focuses on how Lovelace's memory has been designed, constructed, mediated, and circulated over time, examined through the lens of Rigney's (2005, 16) concept of *mnemonic technologies*.

In the time leading up to her death, Lovelace meticulously arranged not only her burial and memorials but also how she wished to be remembered by her friends. She wrote several notes specifying that she wanted each of them to have a keepsake—a book or other personal item she had used, or even money. She acknowledged that she might not be able to determine exactly what each person should receive but wanted to ensure they had something to remember her by (LB 175, f.157). One of the most sentimental bequests was made to a mysterious figure, referred to as 'T', in an undated note. She bequeathed him her gold pencil case, marked with her name and a coronet, 'with the request that he will use it habitually, in remembrance of the many delightful & improving hours we have jointly passed in various literary pursuits' (f.161). Lovelace also requested that T select twelve works from her personal library—books, not volumes—that he found most meaningful so that his choice would hold personal value to him. Additionally, she left T her old, worn, dark-coloured leather writing box, which had been a constant presence on her writing table for many years. Although it contained little of material value—just scraps of paper, pins, and other bits of 'rubbish'—she wanted it to be given to 'T' exactly as it was, with its contents untouched, as the sentimental value of the box lay in its unaltered state (f.161).

Apart from the many speculations that have revolved around the mysterious 'T', as well as the fact that the bequest was folded in a paper addressed to another mysterious 'B'[55], what is interesting in this note from the perspective of Rigney's (2005, 16) *mnemonic technologies* is the material and spatial way that Lovelace wanted to be remembered, which included not only objects—her pencil case, her leather writing box, her seals, papers, pins, minerals, etc.—but also the unruliness of their arrangement, to the point that 'putting it tidy' (LB 175, f.161) would annihilate the memory work.

The fact that she sat for a portrait depicting her at the piano, where she used to play duets with her daughter in the very last months of her life, when she was in great pain, is also amazing in terms of the materiality and embodied nature of her *memory machine*.[56] Apart from this last portrait, as well as a drawing on her deathbed sketched by her mother, there are a series of portraits and daguerreotypes carrying visual traces of Ada Lovelace from her childhood to young age, maturity, and finally her death.[57] It was not unusual for members of the aristocracy to have a series of portraits during their lives, and Lovelace followed this tradition. What is interesting in these portraits, however, is that they depict her beauty, aristocratic status, and even her musical talent, but there is no hint of her standing as a mathematician or scientist, as was the case with Du Châtelet's portraits.

In early September 1852, shortly after Lady Byron took control of Lovelace's household to care for her dying daughter, an unusual request emerged. On 1 September 1852, Lovelace wrote a final wish: 'It is my earnest and dying request that all my friends who have letters from me will deliver them to my mother, Lady Noel Byron, after my death'.[58] Several of her friends honoured this request and returned her letters, but Charles Babbage did not comply. He argued that Lovelace had named him as her executor in a letter dated 12 August 1852, shortly before her mother took charge and restricted their contact (LB 138, f.3). In that letter, Lovelace gave Babbage clear instructions in the event of her sudden death. She requested that he obtain £600 from her mother for a specific purpose she had privately discussed with him, instructed him to access her account at Messrs Drummond's and collect all her old drafts, and directed him to carefully review and then dispose of all papers and property in his possession.

Lovelace's letter ended with a trusting appeal to Babbage to fulfil these wishes. However, after Lovelace's death, Lady Byron contested the legality of the letter and refused to honour her daughter's wishes. She destroyed many of Lovelace's personal letters from Babbage, though she kept the scientific correspondence. Lacking funds to execute Lovelace's will, Babbage returned all bills and receipts he had been entrusted with but withheld the letters. In response to Lady Byron's request, Babbage sent a deposition to her lawyers, asserting his right to handle the collection of papers and letters, including extensive correspondence with Lovelace. He stated that Lady Lovelace's family's treatment of him released him from any concerns about delicacy in dealing with the materials, and he believed her testamentary letter gave him full authority over them (LB 138, f.110).

We don't know how many of Lovelace's letters and papers Babbage kept or destroyed, since only very few have been preserved, and most of them were deposited in the British Library by his son after his death (BL, MS 37190-2).[59] There are twenty-nine letters from Lovelace to Babbage, dating between 1836 and 1851, which are mostly records of their working relationship, their mathematical exchanges, and scientific interests, but also their friendship. Finally, there are a few letters that some commentators have observed include cryptic messages about Lovelace's gambling and betting schemes.[60] These letters might explain why Lovelace wanted Babbage to take care of her financial issues after her death and had entrusted him with bills and receipts that he eventually returned to her mother, given her refusal to follow her daughter's instructions in her testamentary letter of 12 August 1852.

What we therefore have in the archives of the Bodleian Libraries is a collection of papers that initially belonged to Lovelace's mother, Annabella, Lady Byron, and Baroness Wentworth, and was therefore scrutinised, organised, and controlled by her until her passing in 1860. At that point, they passed to her heir, Ralph, Lovelace's younger son, who later became the 2nd Earl of Lovelace. It is unclear how many of Lovelace's letters were preserved or destroyed by Ralph, who worked through them and used them in his own memoirs. After Ralph's daughter Ada Mary died without heirs, the Wentworth title passed to his sister, Anne Blunt, and eventually to her grandson, John, Earl of Lytton. Lytton deposited the papers in the Bodleian Libraries in the 1970s, where they remain today.[61]

Knowing the provenance of archival documents is very important in mapping the discursive constraints and limitations of any archival collection, a crucial component of 'the archival sensibility' (Moore et al. 2016) through which I have approached women mathematicians' archives. On this epistemological plane, we can understand that what was kept and what was destroyed in Lovelace's papers and letters was not accidental but the effect of *memory war machines* that went into full operation after her death. These papers were used and perhaps abused in a series of memoirs that Lovelace's heirs produced in writing and revising the family's turbulent history, as we have seen above.

Lovelace's *memory machine* has also produced a dispute about her mathematical abilities, with opinions ranging from 'genius' to 'charlatan' (Hollings, Martin, and Adrian Rice 2017, 222). Early supporters, like Babbage and Augustus de Morgan, praised her intellectual prowess, with Babbage describing her as a 'magical' force in mathematics and de Morgan acknowledging her extraordinary abilities for a beginner (James 1996, 164). However, her work was rediscovered in the 1940s when the concept of programming gained prominence, particularly through Alan Turing's reference to 'Lady Lovelace's Objection' on Artificial

Intelligence (1950, 450), meaning that AI can't originate anything. In the 1950s, Bertram Bowden revived Lovelace's memory in the computing world, calling her a competent mathematician whose ideas were ahead of her time. He included her translation of Menabrea's paper in his 1953 book and found her granddaughter to share more of Lovelace's papers (Bowden 1953).

In the 1970s, after Lovelace's papers were deposited at the Bodleian Libraries, biographies began to challenge her mathematical abilities. Dorothy Stein (1985) argued that her understanding of mathematics was limited, and some critics, like Bruce Collier (1990), portrayed her as overrated. Seen through the lens of Lovelace's *memory machine*, what could be a better exemplar of Rigney's five *mnemonic technologies*: 'the selectivity of recall, the convergence of memories, the recursivity in remembrance, the recycling of models of remembrance and memory transfers' (2005, 16)? It is while thinking with Rigney's metaphor of the memory vortex (2005, 18) that I will now excavate *mnemonic technologies* (16) in Sofia Kovalevskaya's *memory machine*.

Sofia Kovalevskaya: events in-between archives and meta-archives of memory work

> Each of Sofia's numerous friends retained a different image of her in their memory, as she presented herself differently and even uniquely to each one. However, there was no falsehood in this; it was only her rich nature that endowed the person she found interesting at that moment with what seemed to her appropriate for that person. Engaging with a particular individual was exclusive to her, excluding simultaneous engagement with others. I never saw her in a state, if I may express it thus, of equilibrium, and she couldn't somehow develop the ability to constantly and multifariously interact with various people at the same time (Mendelson 1911, 32–3).

This extract from Marie Jankowska-Mendelson's[62] reminiscences—one of many that her friends wrote about her after her untimely death—aptly summarises Kovalevskaya's multifaceted character and personality that kept fascinating her friends and collaborators during her life and beyond. How then does Kovalevskaya's image emerge from the archive? Life stories create a design, a discursive unity that carries meaning, Hannah Arendt has argued, although this meaning is never fixed, but rather felt and lived: 'it is true that storytelling reveals meaning without committing the error of defining it, that it brings about consent and reconciliation with things as they really are' (1968, 105). As was the case with other women mathematicians, however, Kovalevskaya's life narrative has been drawn, redrawn, translated, and transcribed so many times that it has eventually become a collage of moments, a labyrinth of winding passages, narrow alleys, and abrupt deadlocks, a cartography of words and names.

It was Kovalevskaya herself who put in motion her *memory machine.* In 1889 having already secured a tenured position at Stockholm University in Sweden as the first woman professor in modern Europe,she published the first version of her autobiography. Titled, *Life in Russia: The Sisters Rajevski* (Kovalevskaya 1895 [1889]), it was written in the third person and was disguised as a novel. Just a year later, in 1890, Kovalevskaya published the second version of her autobiography, *A Russian Childhood* (Kovalevskaya 1978 [1890]), which was written in the first person and appeared in two issues of the journal *Vestnik Europy* [Messenger of Europe].

We do not know with certainty what made Kovalevskaya write her life in two versions, first as autofiction and then as a classical autobiography. Perhaps her reservations about revealing herself in the public arena receded, as her fame as an accomplished mathematician rose. Or perhaps the immediate success of the first publication in Sweden made Kovalevskaya bolder and more fearless as an author. Her second autobiography was also a literary event of great significance, particularly since only six months after its publication, Kovalevskaya died from pneumonia on 10 February 1891.

Apart from the two versions of her autobiography, a short autobiographical sketch appeared in 1891, the year of her death, in the first issue of *Ruskaya starina* (Kovalevskaya 1891). The date of this text is 29 May 1890, but it was prepared for publication by her brother Fyodor after her passing (see Kovalevskaya 1978, xi). This autobiographical essay was included in the Russian publication of *Memories of Childhood and Autobiographical Essays* in 1945 as well as in her *Memories and Letters* in 1951 (Kovalevskaya 1945, 1951). What is particularly interesting in this short autobiographical sketch is the way Kovalevskaya highlights her mathematical work, making it the centrepiece of how she wanted to be remembered. Indeed, in the very first line of this autobiographical essay, she declares her love for mathematics and attempts to trace its genealogy: 'As far as I can recall it now, *my love of mathematics,* first showed itself in the following manner' (Kovalevskaya 1978, 213).

In her memorial essay on Kovalevskaya, Ellen Key[63] has noted that shortly before her death, Kovalevskaya had in mind to add a chapter to her memoirs, with the intended title 'How I Became a Mathematician' (1891, 71). This chapter was never written, but it seems that the posthumously published 'Autobiographical Sketch' was perhaps a preliminary draft of it. What is further intriguing in this short text is that Kovalevskaya slightly revises some aspects of her autobiography, which had only been published a year before, particularly in relation to her first mathematical tutor. 'It is to Yosif Ignatievich Malevich that I am indebted for my first systematic study of mathematics' (Kovalevskaya 1978, 216), she writes in the autobiographical sketch, while in Chapter 3 of her autobiography, she had noted that although the tutor 'was a quiet, knowledgeable man who taught superbly [...] his actual influence on my education was slight' (79).

According to Stillman's commentary, Kovalevskaya must have felt the need for this slight revision for two reasons. First, because the editor of *Russkaya Starina*, where the essay was first published, had himself been Malevich's pupil; second and more importantly, because Malevich had published in *Russkaya Starina* in 1890 his own reminiscences of his nine years as a tutor in Kovalevskaya's family, with a detailed presentation of his course of study, as a response perhaps to Kovalevskaya's demeaning remark about his influence on her education (Kovalevskaya 1978, 229, n5). There are again some mnemonic disputes and little wars raging in the subtext of autobiographical narratives, triggering Rigney's (2005) modalities of recursivity and remodelling in Kovalevskaya's *memory machine* from the very beginning.

Kovalevskaya's first biography *Sonja Kovalevsky: What I Experienced Together with Her and What She Told Me about Herself*, appeared shortly after her untimely death, in 1892. It was written by her friend and co-author Anna Carlotta Leffler, Duchess of Cajanello (Leffler 1892). According to Leffler, Kovalevskaya had a premonition that she would die young and had repeatedly asked her to promise that she would write her biography . As Leffler argued in the Introduction, Kovalevskaya considered her friend to be her *'second I'* (1895, 156, emphasis in the text) and thus her authorial decision was 'to write, so to speak, under her suggestion' (156) and to present her biographical subject 'as far as possible, in the light in which she showed herself to me' (156). If we consider Mendelson's commentary on Kovalevskaya's tendency to reveal herself differently and uniquely to each of her friends, our reading of Leffler's memoir is limited within the discursive constraints of 'Leffler's Kovalevskaya' and therefore the various critiques it received as 'subjective' are not a defect, but rather the inevitable effect of mnemonic pen portraits and sketches.

While Leffler's narrative unfolds in the Cavarerian scene of the I/you relation (Cavarero 2000), it is nevertheless populated by other voices, which erupt from letters and memoirs that have been enwrapped in the biographer's discourse. These voices include, among others, lines from George Eliot's diary recounting Kovalevskaya's visit to her house in London in 1869 (Leffler 1895, 172), her friends' Iulia's reminiscences of their university days in Heidelberg and Berlin, and discussions that Leffler had with Kovalevskaya's doctoral supervisor, Karl Weierstrass.

Leffler's depiction of her friend somehow succeeded in bringing out the woman in the mathematician, but 'she did not show the Mathematician in the woman', Key, who knew both women well, wrote in her review of Leffler's biography (in Björk 2002, 41). Although Key highlighted the importance of Leffler's biography in terms of preserving intimate memories from Kovalevskaya's life that would have been otherwise lost, her critique was difficult since Leffler herself died unexpectedly in Naples in 1892, soon after she had finished Kovalevskaya's biography, which was published posthumously. Key wrote a biographical sketch for Kovalevskaya soon after her death, which was initially published in the Swedish journal *Dagny* (Key 1891) and was according to Kovalevskaya's daughter Fufa, 'one of the best biographies of my mother' (in Kochina 1985, 321).

These initial biographies were just the beginning of a series of reminiscences, memoirs, and biographical sketches, which started at the time of her death, the turn of the nineteenth century, and have reached our days. Initially written in Russian, Swedish, German, Polish, and French, these biographies and memoirs have been translated into different languages, often with both semantic and stylistic variations, not only between languages but also within the same language, as is always the case with translations.[64] Some of these memoirs have been repeatedly cited and/

or paraphrased in the biographical literature around Kovalevskaya, while others have scarcely been mentioned, like her partner Maxim Kovalevsky's (1951) reminiscences of her. I have found this omission particularly intriguing since Maxim's pen-portrait of Kovalevskaya's life, wide intellectual interests, and state of mind in Sweden is unsurprisingly quite distinct in relation to what her friends, family, and colleagues in Russia, Sweden, and France have written about her.

Maxim's memoir portrays Kovalevskaya as a profoundly interdisciplinary thinker, well ahead of her time. She was not only a mathematician with literary talent but also deeply interested in natural sciences, history, and social sciences (390). Maxim also highlighted her political engagement, both ideologically and through action. At her funeral, Maxim's eulogy reflected her complex relationship with her homeland. He praised her as a source of pride for Russia, mourning the fact that she had to work abroad in Sweden due to circumstances in her native country. He remarked on her unwavering connection to 'young Russia', a symbol of the progressive, just, and free future she had envisioned, contrasting it with the oppressive realities of 'old Russia'. Maxim himself shared her fate as an émigré, having left Russia after losing his academic position in 1887, despite his prominence as a sociologist (396).

This tension between progress and tradition is further reflected in Elizabeta Litvinova's account, which highlights the role of the progressive 'Young Sweden' movement. Advocates of reform, this group not only sought to modernise science and education by challenging the stagnation of traditional universities like Lund and Uppsala but also supported women's advancement (2014 [1894], 95–6). Kovalevskaya admired their commitment to action, writing that they were not content with mere words but consistently translated their convictions into tangible change (96).

I very much doubt that Maxim's idea of the 'Young Russia' would coincide with Soviet Russia, but it was in this regime that Kovalevskaya's *memory machine* was put in motion from the top when the then Soviet Academy of Sciences published her autobiographical writings, including a selection from her letters and diaries, as well as her published and unpublished literary writings in three major collections, with interesting editorial histories (Kovalevskaya 1945, 1951, 1974). Their own discursive limitations notwithstanding, these edited volumes have never been translated into English or any other European language, with the exception of her *Memories of Childhood* and the autobiographical sketch, as already discussed above. Moreover, Kovalevskaya's correspondence with Weierstrass, or rather his letters to her, have been published in German and have only been translated in Russian (see Bölling 1993; Kochina 1973).

In the midst of all these gaps, it is only inevitable that a lot has been lost either in translation or in the gaps of the numerous selections and deselections. Since Kovalevskaya's biographies include extracts and fragments of her 'documents of life' (Plummer 2001), her textual self eventually emerges as a palimpsest in the midst of multifaceted layers of auto/biographical documents and their various selections, transcriptions, and translations. In this backdrop, Michèle Audin's *Remembering Sofya Kovalevskaya*, first published in French in 2008, is an exemplary textual archive of Kovalevskaya's *memory machine*, with the refrain 'I remember Sofya' becoming the connecting thread of dispersed and entangled memories, repeatedly assembled and reassembled (Audin 2011).

In the context of Kovalevskaya's multi-lineal textual archive, my visit to Stockholm in June 2024 was an event that shed light on the world of her grey documents and opened up a path of material and spatial mnemonic traces. My first encounter with traces and vibes of her 'real' world was in the premises of the Mittag-Leffler Institute. It was a sunny morning when I arrived there, and as I was walking up the path from the gate to the main building, which used to be Mittag-Leffler's house, Kovalevskaya's bust was waiting for me, literally at the door. The days I spent at the Institute's Library, reading, copying, and photographing her correspondence with Mittag-Leffler and Weierstrass were filled with mnemonic vibes and experiences of the 'living archive' (Tamboukou 2014). The décor and intellectual ambiance of the house have been preserved even today, and the fact that I was working in the library where Kovalevskaya must have spent days working and talking with Mittag-Leffler created a particularly sensitive atmosphere for my own research and mnemonic practices. The fact that there were very few people around and that I was certainly the only researcher at the library added to the thrill of such space/time encounters. It was the first time I could see and touch her letters—or rather copies of them—and have an immediate experience of her handwriting, knowing at the same time that it was her handwriting in the French and German Latin alphabet and not the Cyrillic script of her mother tongue that I could have access to.[65] And yet, the epistolary addresses of her letters also sent me down memory paths and walks in Stockholm. Humlegärden must have been her favourite park, since all her epistolary addresses were around it. While strolling there, taking photographs, I imagined her crossing its various paths, walking to her apartments at 10 Kommendörsgatan, 14 Humlegårdgatan, and 18 Villagatan, which was particularly impressive.[66]

The visit to her grave at the Norra begravningsplatsen cemetery, literally 'The Northern Burial Place', was deeply emotional. This is a huge cemetery, and it took me some time and some Google map links to find her grave, which I knew how it looked from photographs. I was searching on my own, and there were scarcely any people there, but when I found it, it was one of those moments when I felt profoundly moved and elated. To my surprise, there were fresh red roses on and around the grave, a strong sign not only that her memory is alive and kicking but also that it is being crafted as I write.

Thus, while considering Kovalevskaya's 'real' and textual archive and 'meta-archive', I have addressed questions arising from working with fragmented, inaccessible, translated, and dispersed auto/biographical documents, mostly digital and published, but also analogue from the Mittag-Leffler archive. What I have found in my extended engagement with Kovalevskaya's archive and meta-archive is the futility of any search, past or present, for coherence in her stories and in her life (see Tamboukou 2022). The fragmented and incomplete state of her archive has actually confirmed my conviction that the search for coherence is absolutely pointless and conceptually misguided when trying to understand processes of becoming a woman mathematician. In disentangling such mythographies in Kovalevskaya's *memory machine*, I have been assembling my own meta-archive, which keeps enlarging the more I descend into the archive, an ongoing process, since memory work is never done, as I will further discuss in the concluding section.

Events and **space/time/matter entanglements in archives and memory work**

Throughout this chapter, I have approached memory work in the archives as a process, a becoming. In doing so, I have considered Foucault's (1969, 1986) archaeological and genealogical excavations and the ways they have informed influential scholarship in the field of memory studies (Rigney 2005; Basu 2009). I have further made connections with how Deleuze's and Guattari's (1988) notion of the *assemblage* has created a plane for materialist and machinic approaches to be considered, thus offering the notion of the *memory machine* as an onto-epistemological and methodological tool. My previous scholarship in theorising the archive as a living organism in the context of Whitehead's process philosophy has thus become a component of various *memory machines* within the archive and beyond (Tamboukou 2014, 2016b).

Juxtaposing his philosophy of the organism with philosophies of substance, Whitehead has pithily observed that the latter 'presuppose a subject which then encounters a datum and then reacts to the datum' (1985, 155). My experience with women mathematicians' archives, meta-archives as well as with spatial and material mnemonic practices could not fit this configuration, since 'my datum' was lost, blurred, and fragmented. The only way of reacting to it would be either simulation of already existing discourses and narratives or desperation and abandonment of the effort altogether. However, in my perplexity with this impossible 'datum', I was drawn to the insights of process philosophies which 'presuppose a datum, which is met with feelings, and progressively attains the unity of a subject' (155). It was by way of feeling women mathematicians' auto/biographical documents through the sounds of words, the rhythm of language, and the vibratory energy of re-imagining women in mathematics that my perception and understanding eventually took form. It is only within my entanglement in the maze of their archival documents, memorabilia, and *lieux de mémoire* that I have seen and recognised them.[67] In lieu of the separation between myself as the subject/researcher and women mathematicians as 'the object' of the research, we have become 'superjects' (155) entangled in the cogs of memory machines that I have mapped throughout this chapter.

As I will further discuss in Chapter 4, *events* rather than material entities or 'substances', as Whitehead calls them (155), are the fundamental elements in the constitution of reality. *Events* are not mere happenings or occurrences but eruptions that mark discontinuities and ruptures in our habitual modes of being, seeing, and understanding. *Events* happen and disappear, but they also leave traces behind them in places, objects, bodies, novellas, and stories. When we see, read, or touch such traces of *events*, we feel that we have somehow grasped 'the real', no matter how fleeting or ephemeral such experiences have been. And yet the idea of 'touching the real' is an illusion, Arlette Farge has pithily noted: 'No matter how much the real seems to be there, visible and tangible, it reveals nothing more than its physical presence, and it is naïve to believe that this is its essence' (Farge 2013, 11). The importance of the archival object, be it a story, an administrative document, a photograph, a portrait, a grave, or even a piece of string that slips out of an envelope, 'lies in the interpretation of its presence, in the search for its complex meaning, in framing its "reality" within systems of symbols—systems for which history attempts to be the grammar', Farge argues (12).

Here is precisely where creative imagination works: it disentangles the researcher from the material and affective forces of *memory machines* and facilitates 'the researcher's cut' (Tamboukou 2014, 10), put simply intellectual interventions and insights erupt in the middle of *machinic* entanglements and open up vistas beyond the stubbornness of common-sense notions and perceptions within the archive and beyond. In this light, creative imagination enables the researcher to see things differently and ultimately creates conditions of possibility for interpretations, as well as 'exaggerations', to use Adriana Cavarero's epistemological vocabulary. It is through exaggerations that novel and original ideas emerge when we think with others and push their thoughts to the extreme, Cavarero argues (2016, 14), be they philosophers, theorists, narrators, or women mathematicians in my case. Since 'the past is not always what we expect' (Holmes 2026), here as elsewhere in my work, I have drawn on new materialist approaches in feminist science studies to theorise the archive as a laboratory of memory [and forgetting], but also as an experimental time-space continuum where memory and imagination are brought together in the study and understanding of *events* in the past (see Tamboukou 2019).

Its complexity and multifaceted nature notwithstanding, archival research needs to be meticulous, consistent, and systematic, as we have argued in our book *The Archive Project* (Moore et al. 2016), particularly highlighting the importance of shedding light on provenance. In short, 'how a document or a collection was assembled, when and by whom, and how it ended up in the particular archival location' (164). Such questions are embedded in the analysis that will derive from the archive and shape understandings and interpretations. Throughout this chapter, the search for provenance has unveiled some extraordinary histories and has mobilised the *machinic* approach to the work of memory around women mathematicians. But while thinking with Foucault, I have considered 'why it is necessary to understand how catalogues work and what are the effects of such understandings in analysis, interpretation, and the multiple rewritings of research results' (165). The mingling of analogue and digital archival documents and collections has brought new questions to the fore.

As I have shown throughout this chapter, the discursive components of women mathematicians' *memory machines* have included analogue, digital, digitised, and published documents. In this context, what is the difference between reading a handwritten letter in the archive, and later its image version in the researcher's files, a digitised version of a handwritten letter in an online library or archive, a published version of handwritten letters, or a transcribed and published letter in both historical and contemporary publications?

Here, as elsewhere in my work, I have considered these differences through the lens of diffraction. As an optical phenomenon in classical physics, diffraction follows the bending of waves as they move through passages or encounter small obstacles: think of how a rainbow of different colours is formed when white light enters a prism. What happens when two or more waves arrive at the same point is also interesting since, unlike particles, waves can overlap at the same point in space, creating *superposition* effects. Consider the all-too-familiar moment when we throw pebbles into a lake: the waves that appear as an effect of this disturbance interfere with each other in different ways, depending on their properties and corresponding forces: they can create a more intense wave together, producing constructive interference, or they can cancel each other out in the mode of destructive interference. The superposition of most waves, however, produces a combination of constructive and destructive interferences, which vary from place to place and time to time. For feminist philosopher Donna Haraway, diffraction becomes a metaphor for the effort to see difference and its effects at the bottom of the mirror or on the surface of a pond: 'diffraction patterns record the history of interaction, interference, reinforcement, difference. Diffraction is about heterogeneous history, not about originals', she has written (1997, 16).

Following a vibrant body of feminist literature that has deployed diffractive methodologies in research,[68] I have taken diffraction as an approach for reading insights through one another, attending to and responding to the nuances of theoretical, philosophical, and epistemological differences. In the case of this chapter, these differences pertain to the modalities and forms of archival documents. Put simply, diffractive ways of reading and analysing archival documents are not only about how I understand the content and discourse of any text as a static entity or meaning. It is also about how I tune into the various rhythms of meaning and understanding. In the light of wave phenomena, the meanings that are enacted through the reading and analysis of different types of archival documents keep creating a variety of superposition effects, producing constructive and destructive interferences. Finally, diffraction changes not only the way I see differences in the documents by tracing their archival tracks but also how I mark and change my position vis-a-vis the documents I am studying in my attempt to understand the process of becoming a woman mathematician in the eighteenth and nineteenth centuries.

In the mode of diffraction then, touching actual letters in physical archives and forms has triggered intense feelings of connection with corporeal and material traces, in short, the tactile properties of the epistolary act—the paper, the envelope, the ink, and the handwriting. Touching is a pathway to feeling and understanding, although tactile ways of knowing have been downplayed and devalued, since 'visual detachment has replaced tactile engagement, entrenching the divide between body and mind', as textile artist Solveigh Goett has remarked (2010, 170). The tactile and visual properties of the archival documents are thus entangled in the researcher's perception and understanding, since while an archival document is always visible, it also has a specific smell and even a specific sound as its pages and leaves are turned, while the researcher flicks through them, arranges and rearranges them, and immersing themselves in a particularly tactile process of making sense. But 'touch is slow, sight is fast', Goett has further observed ((170), and indeed the slowness of the tactile process runs in parallel with the slow rhythms of reading and copying in the archive that Farge has influentially theorised:

In the digital age this act of copying can seem quite foolish. [...] And it would indeed be foolish, stubborn, maybe even pridefully obsessive...if this exact recopying of words did not feel somehow necessary, an exclusive and privileged way of entering into the world of the document, as both accomplice and outsider [...] This flow of words can sweep you off into unexpected directions, taking you to a place poised somewhere between the familiar and the exotic (2013, 16).

Elsewhere in my work, I have written about trance situations of making meaning and understanding, triggered by the slow process of touching, reading, and copying documents in the archive (see Tamboukou 2016b, 154), but I have also admitted that, since the researcher's budget and time are always limited, the 'real document' will inevitably be scanned or photographed, and it is to its visual image that the researcher will return. This is where the visual and the tactile create superposition effects in the researcher's diffractive reading, as it is in the middle of this slow process of touching, reading, copying, and, in effect, rewriting texts that the researcher's understanding emerges, and new ideas and insights arise. In this context, the texts themselves are being transformed, not just as analogue or digitised documents, but rather as components of complex entanglements between the analogue and the digital within the space/time/matter of archival research.

But while memories of engaging with real documents at the archive are diffused and diffracted with reading and writing practices at the researcher's desk, what happens if the researcher only has access to online or published documents, often edited and inevitably selected? Since 'the digital revolution' is well underway, we need to accept that in the future more and more researchers will only have access to digitised or published documents, whether online or on paper. Even when having the opportunity to visit an archive or a library, the analogue manuscripts might not be available for reasons of preservation and protection, which is what happened with Kovalevskaya's letters at the Mittag-Leffler Institute, where I was only given access to copies and not 'the real' documents, apart from three photographic albums. I have to admit that when engaging with manuscripts, notebooks, papers, and 'real' letters of women mathematicians, either in the Ambrosiana Biblioteca in Milan, the Moreniana Biblioteca in Florence, the British Library in London, or the Bodleian Libraries in Oxford, I could not believe that I was still allowed to unbox, touch, and flick through such important documents without restrictions. I was almost feeling guilty that I was not asked to wear gloves, and I was frightfully apprehensive about the way I handled these documents. Above all, I was feeling really lucky that I was still able to do it. For how much longer though, I kept wondering.

There is no question that the researcher's feeling of touching the imaginary real cannot be substituted by the visual or the digital. But since digitisation is increasingly becoming 'the new real', we just need to accept it and get on with it. Thus, while considering the epistemological consequences of working with digitised documents—that we hope will become better curated, organised, and presented—the first thing to consider here is that 'wholeness' or the idea of 'touching the real' (Farge 2013) is never achievable anyway, even when working with analogue and/or unpublished archival documents: 'you find nothing in the Archive but stories caught halfway through: the middle of things: discontinuities', Carolyn Steedman has influentially written (2001, 45). In this light, the letter that was kept and found in the archive should always be read with the letter that was lost or destroyed in mind, and in the same way that we interpret voices, we should perhaps start interpreting silences and absences or somehow include them in our analysis and understanding.

As I have shown throughout this chapter, in the case of women mathematicians' letters, it is not only the differences between analogue and digitised letters that we should consider, but also that many of their letters have only been found in draft forms, that many letters have been lost, damaged, or destroyed, and that even in several extant letters, we have very few traces of personal feelings and thoughts. And yet, multi-modal, fragmented, and discontinuous as they are and will always be, analogue, digitised, edited, translated, and published letters have left traces of 'epistolary waves', which as they ebb and flow, create a pattern that has meaning in understanding the process of becoming a woman mathematician in post-Enlightenment Europe. Here, the researcher's imagination is important and elsewhere in my work, I have used the motifs of epistolary waves and rhythms in analysing the letters of four revolutionary women in the nineteenth and twentieth centuries (see Tamboukou 2023).

What is also important when working with fragmented letters and documents in the archive is that the researcher should not prolong and further obscure their fragmented state but should rather attempt to mark and trace their 'archival trails' (Nygaard 2024) in the process of reassembling the archive, thus addressing questions of provenance and contextual loss, as I have done throughout this chapter. In introducing their book *Archiving Activism in the Digital Age*, Anne Rigney and Daniele Salerno have observed that 'the role of archives has been persistently flagged in memory studies as a key factor in cultural memory-making but it has rarely been studied in its own right' (2024, 8), with the exception of Aleida Assmann's (1999) conceptualisation of the relationship between archive and memory, which we have already discussed at the beginning of this chapter. It is in this lacuna in the literature that this chapter is situated, by charting *memory machines* in women mathematicians' archives.

Notes

1 See van den Elzen 2021 for a comprehensive review of this literature.
2 See Andrew Brown, discussing Du Châtelet's manuscripts, 17 December 2020, https://www.youtube.com/watch?v=V8zLoA9uxBM [10.15-13.36]

3 See *Kölving and Brown*, Lettre 77, vol.1, 230 (Voltaire to Du Châtelet); Lettre 303, vol.1, 597 (V-DC); Lettre 387, vol.1, 99; (DC-V) Lettre 512a, vol.1, .274 (V-DC).

4 See, https://nlr.ru/voltaire/RA417/history-library-Voltaire [Accessed 4 April 2024]

5 See Natalia Speranskaja from the Voltaire Collection of the NLR discussing Voltaire's collection acquisition, 17 December 2020, https://www.youtube.com/watch?v=V8zLoA9uxBM [4.10-8.25]

6 Voltaire was denied a Christian burial in Paris because of his criticism of the church, which he refused to retract before his death. His remains were transferred to the Pantheon after the French Revolution (see Pearson 2005, 386–416).

7 See Paillard 2008, particularly Chapter 6.

8 See, https://nlr.ru/voltaire/RA417/history-library-Voltaire

9 See https://historyofwomenphilosophers.org/stp/ [Accessed 3 March 2024]

10 See Brown, https://www.youtube.com/watch?v=V8zLoA9uxBM [13.04-13.24]

11 See https://historyofwomenphilosophers.org/project/the-st-petersburg-manuscripts/ [Accessed 18 April 2024]

12 See http://rodama1789.blogspot.com/2015/02/found-and-lost-manuscripts-of-madame-du.html

13 See Brown, https://www.youtube.com/watch?v=V8zLoA9uxBM [13.25-16.40]

14 See Brown, https://www.youtube.com/watch?v=V8zLoA9uxBM [13.25-16.40]

15 For a full list of Du Châtelet's archival locations, see Kölving and Brown 2018, vol. 2, 503–10.

16 See https://gallica.bnf.fr/services/engine/search/sru?operation=searchRetrieve&version=1.2&startRecord=15&maximumRecords=15&page=2&query=%28%28dc.creator%20all%20%22Gabrielle-%C3%89milie%20Le%20Tonnelier%20de%20Breteuil%20Du%20Ch%C3%A2telet%22%20or%20dc.contributor%20all%20%22Gabrielle-%C3%89milie%20Le%20Tonnelier%20de%20Breteuil%20Du%20Ch%C3%A2telet%22%29%29 [Accessed 4 April 2024]

17 See https://gallica.bnf.fr/ark:/12148/btv1b6000754n/f1.item.r=maupertuis [Accessed 18 April 2024]

18 See https://sites.google.com/site/mariatamboukou/the-book-archive/mapping-the-seamstress-1 [Accessed 18 April 2024]

19 See https://www.chateaudecirey.com/en [Accessed 17 April 2024]

20 https://chateauluneville.meurthe-et-moselle.fr/un-peu-dhistoire/les-personnes-illustres-li%C3%A9es-%C3%A0-la-cour/emilie-du-ch%C3%A2telet [Accessed 17 April 2024]

21 The Peace of Vienna in 1738 recognised Augustus III as king of Poland but allowed Stanislas to keep his royal titles while granting him the provinces of Lorraine and Bar for life. See Zinsser 2006, for more details on Du Châtelet's relationship with the Lorraine court, particularly Chapter 6.

22 https://sites.google.com/view/numbersandnarratives/a-feminist-genealogy-of-automathographies/%C3%A9milie-du-ch%C3%A2telet/archives-of-material-culture

23 For a discussion of these portraits, see Fara 2002; Sheriff 2007. See these portraits at: http://rodama1789.blogspot.com/2015/02/madame-du-chatelet-portraits-of-femme.html [Accessed 23 November 2022]

24 See https://www.ambrosiana.it/en/ for more details about the history of this cultural institution [Accessed 10 January 2024]

25 See Agnesi's catalogue at the Ambrosiana Biblioteca: https://ambrosiana.comperio.it/manoscritti/search?q=Maria+Gaetana+Agnesi&sort=items [Accessed 13 January 2023]

26 Veneranda Biblioteca Ambrosiana, Agnesi, Maria Gaetana <1718–1799>, 0.180- 0.204. SUP, thereafter [VBA, followed by the codice number Unless otherwise indicated all translations from Latin and Italian are mine. , see https://ambrosiana.comperio.it/opac/detail/view/ambro:catalog:83002 [Accessed 18 March 2024]

27 see https://ambrosiana.comperio.it/opac/detail/view/ambro:catalog:83003

28 Extract from the Viennese journal, *the Neue Freie Presse*, included in the annexes of the new edition of Germain's philosophical work (1896, 374).
29 *Neue Freie Presse,* in Germain 1896, 374.
30 For details on Libri's political activities during this period, see Maccioni-Ruju and Mostert 1995, particularly Chapter 5, 'Revolutionary in Paris, Insurrectionist in Florence'.
31 Also known as the 'bourgeois monarchy', this is the period of the reign of Louis-Philippe (1830–1848), who was brought to the throne after the 1830 July Revolution that led to the abdication of Charles X and the fall of the Bourbon monarchy. For historical studies about the *July Monarchy* 1830–48, see amongst others, Pinkney 1973.
32 For more details about Libri's illegal practices see Maccioni-Rujo and Mostert 1995, particularly Chapter 9; Del Centina and Fiocca 2018; Musielak 2020, particularly Chapter 11.
33 There is a rich body of historical studies around the February 1848 Revolution, but the classic source remains, de Tocqueville 2009 [1969].
34 See, Correspondance de Sophie GERMAIN avec les mathématiciens et savants Cauchy, Delambre, Fourier, Gauss, Le Gendre, d'Ansse de Villoison, etc., BnF, Cote: Français 9118, https://archivesetmanuscrits.bnf.fr/ark:/12148/cc542255, also available in Gallica, https://gallica.bnf.fr/ark:/12148/btv1b9007572p.r=sophie%20germaine?rk=42918;4 [Accessed 10 April 2024]
35 See, Papiers de Sophie GERMAIN. — Recueil de dissertations et problèmes mathématiques et physiques. Français 9114–116, 1801–900, http://archivesetmanuscrits.bnf.fr/ark:/12148/cc54223p/ca59902191713965, also available in Gallica, https://gallica.bnf.fr/ark:/12148/btv1b90724875?rk=21459;2 [Accessed 10 April 2024]
36 See Musielak 2020, particularly Chapter 13, 'Unanswered questions'.
37 See Del Centina and Fiocca 2018, 672–74.
38 See Del Centina and Fiocca 2018, 595–602, for a detailed history of the discovery of Germain's correspondence as well as its different editorial projects.
39 See amongst others, Müller 2021; O'Sullivan 2022; Potts 2021.
40 Bucciarelli and Dworsky 1980; Musielak 2020.
41 See images of these visits at: https://sites.google.com/view/numbersandnarratives/newsletter/2025/january-2025
42 See the bust and other artistic representations of Sophie Germain at the website of the project: https://sites.google.com/view/numbersandnarratives/a-feminist-genealogy-of-automathographies/sophie-germain/archives-of-material-culture [Accessed 24 January 2025]
43 See McMillan, *Queen of Science*, 9.
44 See among others: Baker 1948; Jenkins 1999; Swindells 1999; Patterson 1983; Sanderson 1974.
45 Oxford, Bodleian Libraries, Mary Somerville Collection/MSC: Dep. c.356, fol. MSAU-2; fol. MSAU-3, [OBL/MSC/Dep.c.356/f,MSAU-2&3]
46 See among others, Neeley 2001; Brock 2006; Jenkins 2007; Arianrhod 2012; Secord 2014; Lamprecht 2015;Chapman 2015; Boswell 2017; Wiegand Brothers 2015; Strickland 2016; Stenhouse 2021.
47 'On Craftmanship', Virginia Woolf's only surviving recording of her voice, https://soundcloud.com/brainpicker/words-the-only-surviving [Accessed 23 June 2024]
48 There is a rich body of literature around sugar and the antislavery movement in Britain, which explores the complex interconnections between the British Empire's economic foundations and the exploitation of enslaved labour in the sugar colonies. See Midgley, 1995.
49 See https://www.famedisud.it/mary-somerville-riposa-a-napoli-una-delle-prime-astronome-della-storia/ [Accessed 20 April 2024]
50 See her house in Burntisland, https://www.burntisland.net/mabon.htm [Accessed 22 April 2024]

51 See her portraits at the website of the project Numbers and Narratives: https://sites.google.com/view/numbersandnarratives/a-feminist-genealogy-of-automathographies/mary-fairfax-somerville/archives-of-material-culture [Accessed 26 April 2024]
52 Oxford Bodleian Libraries, Archive of the Noel Byron and Lovelace Families, Dep. Lovelace Byron. In the text and thereafter, LB followed by the folder and folio number.
53 All emphases in Lovelace's notes herein are in the text.
54 There are four handwritten versions of the sonnet in Lovelace's manuscripts at the Bodleian Libraries. The sonnet has been published with slight differentiations in Stein 1985, 237; Toole 1992, 307. The line 'Revealing Harmony from sighs and tears' in folios 227–9 has replaced the line 'Proclaiming Loveliness from sighs and tears' in folio 226.
55 See Stein 1985, 235–36, but also Langley Moore 1977, 385.
56 See https://blogs.bodleian.ox.ac.uk/adalovelace/2015/10/14/only-known-photographs-of-ada-lovelace-in-bodleian-display/ [Accessed 7 May 2024]
57 See Lovelace's series of portraits at the website of the project "Numbers and Narratives', https://sites.google.com/view/numbersandnarratives/a-feminist-genealogy-of-automathographies/ada-lovelace/archives-of-material-culture [Accessed 7 May 2024]
58 Cited in Stephen Wolfram, 'Untangling the Tale of Ada Lovelace', https://writings.stephenwolfram.com/2015/12/untangling-the-tale-of-ada-lovelace/ [Accessed 1 May 2024]
59 See British Library, Augusta Ada King, daughter of Lord Byron; wife of William, Lord King, 1st Earl of Lovelace (1838): Letters to C. Babbage: 1836–51, Add MS37190-37192. [BL/MS37190-2], see also https://www.youtube.com/watch?v=Q-wl-4WAREY [Accessed 1 May2024]
60 See Stein 1985, 259–63.
61 See Langley Moore 1977, 356–60; Stein 1985, 271–2; Toole 1992, 314.
62 Marie Jankowska-Mendelson was a Polish revolutionary whom Kovalevskaya met in Paris in 1882, and they became close friends. See Audin (2011, 152) for details of their friendship.
63 Ellen Key (1849–1926) was a Swedish feminist and writer whose ideas on children's education, sex, family, and marriage were largely influential in Sweden and beyond. For her biography, see Nyström-Hamilton 1913.
64 See among others, Key 1891 (Swedish); Korvin-Krukovskii 1951; Kovalevskii 1951; Kovalevskaya [Fufa] 1951; Lermontova 1951; Litvinova 1894 (Russian); Mendelson 1879 (German), 1911 (Polish); Mittag-Leffler 1892 (French).
65 Kovalevskaya's correspondence with Mittag-Leffler is mostly in French and with Weierstrass in German.
66 See, https://sites.google.com/view/numbersandnarratives/newsletter/2024/june-2024 for images of these places [Accessed 26 January 2025]
67 See Tamboukou 2016b, for an elaboration of the idea of 'feeling the archive'.
68 See among others, Bozalek and Zembylas 2017; Goodman 2019.

References

Ahmed Sara. 2017. *Living a Feminist Life.* Durham and London: Duke University Press.
Anzoletti, Maria Luisa. 1900. *Maria Gaetana Agnesi.* Milan: Cogliati.
Arendt, Hannah. 1968 [1955]. *Men in Dark Times.* New York: Harvest Books.
Arianrhod, Robyn. 2012. *Seduced by Logic: Émilie Du Châtelet, Mary Somerville and the Newtonian Revolution.* New York: Oxford University Press.
Assmann, Aleida. 1999. *Erinnerungsräume: Formen und Wandlungen des kulturellen Gedächtnisses.* Munich: Beck.
Assmann, Aleida. 2012. *Cultural Memory and Western Civilization: Arts of Memory.* Cambridge: Cambridge University Press.

Audin, Michèle. 2011. *Remembering Sofya Kovaleskaya*. Translated by M. Audin. London: Springer.

Baker, John N. L. 1948. 'Mary Somerville and Geography in England'. *The Geographical Journal* 3 (4/6): 207–22.

Barad, Karen. 2007. *Meeting the Universe Halfway: Quantum Physics and the Entanglement of Matter and Meaning*. Durham, NC: Duke University Press.

Basu, Laura. 2009. 'Towards a Memory Dispositif: Truth, Myth and the Ned Kelly *Lieu de Mémoire*'. In *Mediation, Remediation, and the Dynamics of Cultural Memory*, edited by Astrid Erll and Ann Rigney, 139–56. Berlin: De Gruyter.

Besterman, Theodore, ed. 1958. *Les Lettres de la Marquise Du Châtelet*. 2 Vols. Geneva: Institut et Musée Voltaire.

Bölling, Reinhard, ed. 1993. *Briefwechsel Zwischen Karl Weierstrass Und Sofja Kowalewskaja*. Berlin: Academie Verlag.

Boswell, Michelle. 2017. 'Poetry and Parallax in Mary Somerville's *On the Connexion of the Physical Sciences*'. *Victorian Literature and Culture* 45 (4): 727–44.

Bozalek, Vivienne Bozalek, and Michalinos Zembylas. 2017. 'Diffraction or Reflection? Sketching the Contours of two Methodologies in Educational Research'. *International Journal of Qualitative Studies in Education* 30 (2): 111–27.

Boucard, Jenny. 2020. 'Arithmetic and Memorial Practices by and Around Sophie Germain in the 19th Century'. In *Against All Odds. Women's Ways to Mathematical Research Since 1800*, edited by Eva Kaufholz-Soldat and N. Oswald, 185–230. Cham: Springer.

Bowden, Bertram, Vivian, ed. 1953. *Faster Than Thought: A Symposium on Digital Computing Machines*. London: Sir Isaac Pitman & Sons.

Braidotti, Rossi. 1991. *Patterns of Dissonance*. Cambridge: Polity Press.

Braidotti, Rossi. 2001. *Patterns of Dissonance*. Cambridge: Polity Press.

Brock, Claire. 2006. 'The Public Worth of Mary Somerville'. *British Journal of the History of Science* 39 (2): 255–72.

Bucciarelli, Luis L., and Nancy Dworsky. 1980. *Sophie Germain: An Essay in the History of the Theory of Elasticity*. Boston, MA: Dordrecht.

Björk, Jan-Erik. 2002. 'Sonja Kovalevsky: Her Life and Professorship at Stockholm'. *Operator Theory: Advances and Applications* 132: 11–53.

Cavarero, Adriana. 2000 [1997]. *Relating Narratives: Storytelling and Selfhood*. Translated by Paul A. Kottman. London: Routledge.

Cavarero, Adriana. 2016. *Inclinations: A Critique of Rectitude*. Translated by Amanda Minervini and Adam Sitze. Stanford: Stanford University Press.

Chapman, Allan. 2015. *Mary Somerville and the World of Science*. Cham: Springer.

Chidgey, Red. 2018. *Feminist Afterlives: Assemblage Memory in Activist Times*. Palgrave Macmillan.

Colebrook, Clare. 2002. *Gilles Deleuze*. London: Routledge.

Collier, Bruce. 1990. *The Little Engines That Could've: The Calculating Machines of Charles Babbage*. Garland Publishing.

Cupillari, Antonella. 2007. *A Biography of Maria Gaetana Agnesi, an Eighteenth-Century Woman Mathematician: With Translations of Some of her Work from Italian into English*. New York: Edwin Mellen Press.

De Tocqueville, Alexis. 2009 [1969]. *Recollections: The French Revolution of 1848*, edited by J. P. Mayer and A. P. Kerr, and translated by G. Lawrence. New Brunswick: Transaction Publishers.

Dean, Mitchell. 1994. *Critical and Effective Histories: Foucault's Methods and Historical Sociology*. London: Routledge.

Del Centina, Andrea. 2005. 'Letters of Sophie Germain Preserved in Florence'. *Historia Mathematica* 32 (1): 60–75.

Del Centina, Andrea, and Alessandra Fiocca. 2004. *L'archivio di Guglielmo Libri dalla sua dispersione ai fondi della Biblioteca Moreniana* [The Archive of Guglielmo Libri from its Dispersal to the Collections at the Biblioteca Moreniana]. Firenze: Olschki.
Del Centina, Andrea, and Alessandra Fiocca. 2018. 'On the Correspondence of Sophie Germain'. In *Mathematical Correspondences and Critical Editions,* edited by Maria Teresa Borgato, Erwin Neuenschwander, and Irène Passeron, 147–66. Cham: Birkhäuser.
Delambre, A. Choron, Jh. Fourier, Sophie Germain, De La Grange, Lalande, Guillaume Libri, and C. Henry. 1879. 'Les manuscrits de Sophie Germain et leur récent éditeur-documents nouveaux'. *Revue Philosophique de la France et de l'Étranger* 8: 619–41.
Deleuze, Gilles. 1990. *Negotiations, 1972-1990.* Translated by Martin Joughin. New York: Columbia University Press.
Deleuze, Gilles. 1992. 'What Is a Dispositif?' In *Michel Foucault, Philosopher: Essays Translated from the French and German*, edited by Timothy J. Armstrong, 159–68. London: Harvester Wheatsheaf.
Deleuze, Gilles, and Felix Guattari. 1984. *Anti-Oedipus*. Translated by Robert Hurley, Mark Seem, and R. Lane Helrn. London: The Athlone Press.
Deleuze, Gilles, and Felix Guattari. 1988. *A Thousand Plateaus: Capitalism and Schizophrenia.* Translated by Brian Massumi. London: The Athlone Press.
Derrida, Jaques. 1995. *Archive fever: A Freudian Impression.* Translated by Eric Prenowitz. Chicago: University of Chicago Press.
Fara, Patricia. 2002. 'Images of Émilie du Châtelet'. *Endeavour* 26 (2): 39–40.
Farge, Arlette. 2013 [1989]. *The Allure of the Archives* [Le goût de l'archive]. Translated by T. Scott-Railton. New Haven: Yale University Press.
Findlen, Paula. 2014. 'Listening to the Archives: Searching for the Eighteenth-Century Women of Science'. In *Writing about Lives in Science: (Auto)Biography, Gender and Genre*, edited by Paola Nini and Zelda Alice Franceschi, 87–115. Göttingen: V&R Unipress.
Foucault, Michel. 1969. *L'Archéologie du savoir.* Paris: Gallimard.
Foucault, Michel. 1977. *Discipline and Punish: The Birth of the Prison.* Translated by Alan Sheridan. London: Allen Lane, Penguin.
Foucault, Michel. 1978. *The History of Sexuality, An Introduction,* vol. 1. Translated by Robert Hurley. New York: Pantheon.
Foucault, Michel. 1980. 'The Confession of the Flesh, a Conversation'. In *Power/Knowledge: Selected Interviews and other Writings 1972–1977,* edited by Colin Gordon, 194–228. London: Harvester Wheatsheaf.
Foucault, Michel. 1986. 'Nietzsche, Genealogy, History'. In *The Foucault Reader*, edited by Paul Rabinow and translated by Donald F. Bouchard and Sherry Simon, 76–100. Harmondsworth: Peregrine.
Foucault, Michel. 1988 'Technologies of the Self'. In *Technologies of the Self*, edited by Luther H. Martin, Huck Gutman, and Patrick H. Hutton, 16–49. London: Tavistock.
Frisi, Antonio Francesco. 1799. *Elogio storico di D^A Maria Gaetana Agnesi, Milanese Dell'Academia Dell' Instituto Delle Scienze, e Lettrice Onoraria di Matematiche nella Università di Bologna.* Milan: Presso Giuseppe Galeazzi. *Reprint* with an introduction and additional notes by Arnaldo and Giuseppina Masotti, Milan: Scuola Tipografica del Pio Istituto pei Figli della Provvidenza, 1965. Translated by Antonella Cupillari in *A Biography of Maria Gaetana Agnesi, an Eighteenth-Century Woman Mathematician: With Translations of Some of her Work from Italian into English*, 33-106. New York: Edwin Mellen Press, 2007.
Germain, Sophie. 1896 [1879]. *Œuvres philosophiques de Sophie Germain, suivies de pensées et de lettres inédites. Et précédées d'une notice sur sa vie et ses œuvres par H^te Stupuy (Nouvelle Édition)* [Philosophical Works of Sophie Germain, Followed by *Thoughts* and Unedited Letters. Preceded by a Notice on Her Life and Works by H^te Stupuy (New Edition)]. Paris: Firmin-Didot.

Goett, Solveigh. 2010. 'Linking Threads of Experience and lines of Thought: Everyday Textiles in the Narration of the Self'. PhD thesis, University of East London.

Goodman, Joyce. 2019. 'Willystine Goodsell (1870–1962) and John Dewey (1859–1952): History, Philosophy and Women's Education'. *History of Education* 48 (6): 837–54.

James, Frank A J L., ed. 1996. *The Correspondence of Michael Faraday*, vol. 3. Stevenage: Institution of Engineering and Technology.

Jenkins, Alice. 1999. Writing the Self and Writing Science: Mary Somerville as Autobiographer'. In *Rethinking Victorian Culture*, edited by Juliet John and Alice Jenkins, 162–78. London: Macmillan.

Jenkins, Alice. 2007. Alice, *Space and the 'March of Mind': Literature and the Physical Sciences in Britain, 1815–1850*. Oxford: Oxford University Press.

Jorati, Julia. 2019. 'Du Châtelet on Freedom, Self-Motion, and Moral Necessity'. Journal of the History of Philosophy 57: 255–80.

Halbwachs, Maurice. 1994 [1925]. *Les Cadres sociaux de la mémoire*. Paris: Albin Michel.

Haraway, Donna. 1988. 'Situated Knowledges: The Science Ques tion in Feminism and the Privilege of Partial Perspective'. *Feminist Studies* 14 (3): 575–99.

Haraway, Donna. 1997. *Modest_Witness@Second_Millennium. FemaleMan_Meets_ OncoMouse: Feminism and Technoscience*. New York: Routledge.

Harlor. 1953. 'Sophie Germain: Mathématicienne et Philosophe'. *Revue des Deux Mondes (1829–1971)* (1 September): 134–45.

Halbwachs, Maurice. 1997 [1950]. *La Mémoire collective*. Paris: Albin Michel.

Hollings, Christopher, Ursula Martin, and Adrian Rice. 2017. 'The Early Mathematical Education of Ada Lovelace'. *BSHM Bulletin: Journal of the British Society for the History of Mathematics* 32 (3): 221–34.

Holmes, Mary. 2026. 'How We Feel: The Mass Observation Archive, Elias and Doing Sociology'. In *Doing Feminist Research with Texts: Thinking with Liz Stanley*, edited by Órla Murray and Maria Tamboukou. London: Routledge.

Keller, Fox, and F. Evelyn. 1987. 'Women Scientists and Feminist Critics of Science'. *Daedalus* Fall 116 (4), Learning about Women: Gender, Politics, and Power: 77–91.

Key, Ellen. 1891. 'Sonja Kovalewsky'. *Tidskrift Dagny* 3: 1–17. h. Accessed 15 May 2024. ttps://gupea.ub.gu.se/bitstream/handle/2077/53531/gupea_2077_53531_1.pdf?sequence =1&isAllowed=y.

Kochina, Pelageya, ed. 1973. *Briefe von Karl Weierstrass an Sofie Kowalewskaja in 1871–1891*. [Correspondence from Karl Weierstrass to Sofie Kowalewskaya]. Moskau: Nauka (In German and Russian).

Kochina, Pelageya. 1985. *Love and Mathematics: Sofia Kovalevskaya*. Translated by Michael Burov. Moscow: Mir Publishers.

Kölving, Ulla, et Andrew Brown, eds. 2018. *La Correspondance d' Émilie Du Châtelet*. 2 vols. Ferney-Voltaire: Centre international d'étude du xviiie siècle (vol. 1, pp. 1–633; vol. 2, pp. 1–593).

Korvin-Krukovskii, Fyodor, V. 1951 [1891]. 'Memories of my sister [*ВОСПОМИНАНИЯ О* СЕСТРЕ'. In Kovaleskaia, Sofe, *Memories and Letters, 369–74*. Moscow: AN SSSR.

Kovalevskaya, Sofia. 1891. 'An Autobiographical Sketch'. In *Russkaya starina*, No. 1, 1891, 450–63, translated by Beatrice Stillman in Kovalevskaya, Sofia, 1978. *A Russian Childhood*, 213–29. New York: Springer-Verlag.

Kovalevskaya, Sofia. 1895. 'Reminiscences of Childhood [Life in Russia: The sisters Rajevski]'. In *Sonia Kovalevsky. Biography and Autobiography*, translated by Louise Von Cossel,167–317 London: Walter Scott.

Kovalevskaya, Sofia. 1945. *Childhood Memories and Autobiographical sketches* [Воспоминания детства и автобиографические очерки], edited and annotated by S. Y. Shtraikh. Moscow: USSR Academy of Sciences. Accessed 24 May 2024. https://www.prlib.ru/item/404617.

Kovalevskaya, Sofia. 1951. *Memories and Letters* [Коваевской, Софье, В. Воспоминанця и письма]. Moscow: USSR Academy of Sciences.
Kovaleskaya, Sofia. 1974. *Воспоминания* повести [Recollections, Novellas]. Edited by Pelageya Kochina. Moscow: Nauka.
Kovalevskaya, Sofia. 1978 [1895]. *A Russian Childhood.* Translated by Beatrice Stillman. New York: Springer-Verlag.
Kovalevskaya, Sofia Vladimirovna (Fufa). 1951. 'Memories of My Mother' [Воспоминан ця о материъ]. In *Memories and Letters*, edited by Sofia Kovaleskaya, 360–8. Moscow: USSR Academy of Sciences.
Kovalevskii, Maxim, M. 1951. S. Kovalevskaya: A Friend's Memories' [*ВОСПОМИНАНИЯ ДРУГ А*]. In *Memories and Letters*, edited by Sofia Kovaleskaya, 388–407. Moscow: USSR Academy of Sciences.
Lamprecht, Elizabeth Ann. 2015. 'The Life of Mary Fairfax Somerville, Mathematician and Scientist: A Study in Contrasts'. *Michigan Academician* 42 (1): 1–25. https://doi.org/10.7245/0026-2005-42.1.1
Langley Moore, Doris. 1977. *Ada Countess of Lovelace, Byron's Legitimate Daughter.* New York: Harper & Row.
Leffler, Anna Carlotta. 1892. *Sonja Kovalevsky. Hvad jag upplefvat tillsammans med henne och hvad hon berättat mig om sig själf* [What I Experienced Together with Her and What She Told Me about Herself]. Stockholm: Albert Bonniers. Accessed 15 May 2024. https://arkivkopia.se/sak/littbank-LefflerAC_SonjaKovalevsky.
Leffler, Anna Carlotta. 1895. 'A Biography'. In *Sónya Kovalévsky, Her Recollections of Childhood: With a Biography*, translated by A. M. Clive Bayley, 154–303. New York: The Century Co.
Lermontova, Iulia, V. 1951. 'Memories of Sofia Kovalevskaya' [Воспоминанця О Софье Ковадевской']. In *Memories and Letters*, edited by Sofia Kovaleskaya, 375–87. Moscow: USSR Academy of Sciences.
Libri, Guglielmo. 1832. 'Sciences Mathématiques-Mlle Germain'. *Journal Des Débats*, 18 May 1832: 1–2.
Litvinova, Elizaveta F. 2014 [1894]. *Софья Ковалевская. Женщина - математик* [S.V. Kovalevskaya (a Woman Mathematician), Her Life and Scientific Work]. Moscow: Direct Media.
Maccioni-Rujo, P. Alessandra, and Marco Mostert. 1995. *The Life and Times of Guglielmo Libri (1802–1869) Scientist, Patriot, Scholar, Journalist and Thief: A Nineteenth-Century Story.* Hilversum: Verloren Publishers.
Mahon, Michael. 1992. *Foucault's Nietzscean Genealogy, Truth, Power and the Subject.* Albany: State University of New York Press.
Masotti, Arnaldo. 1940. '"Maria Gaetana Agnesi", Rendiconti del Seminario Matematico e Fisico di Milano'. *Reports of the Mathematical and Physics Seminar in Milan* 14: 89–127.
Mazzotti, Massimo. 2007. *The World of Maria Gaetana Agnesi, Mathematician of God.* Baltimore: Johns Hopkins University Press.
McMillan, Dorothy. 2001. 'Introduction'. In *Queen of Science: Personal Recollections of Mary Somerville*, edited by Dorothy McMillan, 7–39. Edinburgh: Canongate.
Mendelson, Marie. 1897. 'Briefe von Sophie Kowalewska' [Letters from Sofia Kovalevskaya]. *Neue Deutsche Rundschau* 6: 589–614 (in German).
Mendelson, Marie. 1911. *Wspomnienia o Zofii Kowalewskiej* [Memories of Sofia Kovalevskaya]. Kraków : Księgarnia D.E. Friedleina (in Polish).
Messbarger, Rebecca. 2005. 'The Italian Enlightenment Reform of the Querelles des Femmes'. In *The Contest for Knowledge*, edited and translated by Rebecca Messbarger and Paula Findlen, 1–22. Chicago and London: The University of Chicago Press.
Midgley, C. 1995. *Women Against Slavery: The British Campaigns, 1780–1870.* London: Routledge.

Mittag-Leffler, Gösta. 1892. 'Sophie Kovalevsky. Notice Biographique'. *Acta Mathematica* 16: 385–92.

Moore, Niamh, Andrea Salter, Liz Stanley, and Maria Tamboukou. 2016. *The Archive Project: Archival Research in the Social Sciences*. London: Routledge.

Mozans, H. J., (John Augustine Zahm, C.S.C.), J. I. Jenkins, and Sabrina Schnell. 2022 [1913]. *Woman in Science* (Version 1). University of Notre Dame. https://doi.org/10.7274/24727551.v1

Müller, Katja. 2021. *Digital Archives and Collections, Creating Online Access to Cultural Heritage*. New York: Berghahn books.

Musielak, Dora. 2020. *Sophie Germain: Revolutionary Mathematician*. 2nd ed. Cham: Springer.

Neeley, Kathryn A. 2001. *Mary Somerville: Science, Illumination, and the Female Mind*. New York: Cambridge University Press.

Nora, Pierre. 1989. 'Between Memory and History: *Les Lieux de Mémoire*'. *Representations* 26, Special Issue: Memory and Counter-Memory: 7–24.

Nygaard Pernille S. 2024. Home Economics as Space for Women's Politics 1875–1961: Following the Archival Trails of Hulda Pedersen and the Establishment of Home Economics at Aarhus University, Denmark'. Working paper.

Nyström-Hamilton, Louise. 1913. *Ellen Key, Her Life and Her Work*. Translated by A. E. B. Fries. New York: G. P. Putnam's Sons.

O'Sullivan, James. 2022. *The Bloomsbury Handbook of the Digital Humanities*. London: Bloomsbury.

Paillard, Christophe. 2008. *Jean-Louis Wagnière, secrétaire de Voltaire*. SVEC (Studies on Voltaire and the Eighteenth Century) 12. Oxford: Voltaire Foundation.

Patterson, Elizabeth Chambers. 1983. *Mary Somerville and the Cultivation of Science 1815–1840*, The Hague: Martinus Nijhoff.

Patton, Paul. 2000. *Deleuze and the Political*. London: Routledge.

Pearson, Roger. 2005. *Voltaire Almighty: A Life in Pursuit of Freedom*. London: Bloomsbury.

Pinkney, David H. 1973. *The French Revolution of 1830*. Princeton, NJ: Princeton University Press.

Plummer, Ken. 2001. *Documents of Life 2*. London: Sage.

Potts, John, ed. 2021. *Use and Reuse of the Digital Archive*. Cham: Springer.

Rigney, Anne. 2005. 'Plenitude, Scarcity and the Circulation of Cultural Memory'. *Journal of European Studies* 35 (1): 11–28.

Rigney Anne, and Salerno Daniele. 2024. 'Archival Activism in the Digital Age: Introduction'. In *Archiving Activism in the Digital Age*, edited by Salerno Daniele and Ann Rigney, 5–24. Amsterdam: The Institute of Network Cultures.

Sanderson, Marie. 1974. 'Mary Somerville: Her work in Physical Geography'. *Geographical Review* 64 (3): 410–20.

Schiebinger, Londa. 1987. 'The History and Philosophy of Women in Science: A Review Essay'. *Signs* 12 (2), Reconstructing the Academy: 305–32.

Secord, John. 2014. *Visions of Science: Books and Readers at the Dawn of the Victorian Age*. Oxford: Oxford University Press.

Sheriff, Mary. 2007. 'The Naked Truth? The Allegorical Frontispiece and Woman's Ambition in Eighteenth-Century France'. In *Early Modern Visual Allegory: Embodying Meaning*, edited by Cristelle Baskins and Lisa Rosenthal, 243–64. Burlington, VT: Ashgate.

Somerville, Mary. 1872. Oxford Bodleian Libraries, Dep. c. 355, MSAU-2, fols. 72–363.

Somerville, Mary. 2001 [1873]. *Queen of Science: Personal Recollections of Mary Somerville*. Edited by Dorothy McMillan. Edinburgh: Canongate.

Soprani, Anne. 1997. *Lettres d'amour au marquis de Saint-Lambert*. Paris: Éditions Paris-Méditerranée.

Stanley, Liz. 2004. 'The Epistolarium: On Theorizing Letters and Correspondences'. *Auto/Biography* 12 (3): 201–35.
Steedman, Carolyn. 2001. *Dust.* Manchester: Manchester University Press.
Stein, Dorothy. 1985. *Ada, a Life and a Legacy.* Cambridge, MA: The MIT Press.
Stenhouse, Brigitte. 2021. 'Mary Somerville: Being and Becoming a Mathematician'. PhD thesis The Open University. https://doi.org/10.21954/ou.ro.0001346f
Strickland, Elisabetta. 2016. *The Ascent of Mary Somerville in 19th Century Society*. Cham: Springer.
Stupuy, Jean-Léon-Hippolyte. 1879. 'Notice sur la Vie et les œuvres de Sophie Germain'. In *Œuvres philosophiques de Sophie Germain, suivies de pensées et de lettres inédites*, 1–92. Paris: Paul Ritti.
Stupuy, Jean-Léon-Hippolyte. 1896. 'Preface de la nouvelle edition'. In *Œuvres philosophiques de Sophie Germain, suivies de pensées et de lettres inédites*, v–vii. Paris: Firmin-Didot.
Swindells, Julia. 1999. 'Other People's Truths? Scientific Subjects in the Personal Recollections of Mary Somerville'. In *Women's Lives into Print*, edited by Pauline Polkey, 96–108. Basingstoke: Macmillan.
Tamboukou, Maria. 2003. 'Interrogating "the Emotional Turn": Making Connections with Foucault and Deleuze'. *European Journal of Psychotherapy, Counselling and Health* 6 (3): 209–23.
Tamboukou, Maria. 2008. 'Machinic Assemblages: Women, Art Education and Space'. *Discourse: Studies in the Cultural Politics of Education* 29 (3): 359–75.
Tamboukou, Maria. 2010. *In the Fold Between Power and Desire; Women Artists' Narratives.* Newcastle-upon-Tyne: Cambridge Scholars Publishing.
Tamboukou, Maria. 2014. 'Archival Research: Unravelling Space/time/matter Entanglements and Fragments'. *Qualitative Research* 14 (5): 617–33.
Tamboukou, Maria. 2015. *Sewing, Fighting and Writing: Radical Practices in Work, Politics and Culture*. London: Rowman & Littlefield.
Tamboukou, Maria. 2016a. 'Archival Rhythms: Narrativity in the Archive'. In *The Archive Project: Archival Research in the Social Sciences*, edited by Niamh Moore, Andrea Salter, Liz Stanley, and Maria Tamboukou, 71–95. London: Routledge.
Tamboukou, Maria. 2016b. 'Feeling Narrative in the Archive: The Question of Serendipity'. *Qualitative Research* 16 (2): 151–66.
Tamboukou, Maria. 2016c. *Gendering the Memory of Work: Women Workers' Narratives.* London: Routledge.
Tamboukou, Maria. 2017. 10. 'Reassembling Documents of Life in the Archive'. *The European Journal of Life Writing* 6: 1–19. doi.org/10.5463/ejlw.6.215.
Tamboukou, Maria. 2019. 'New Materialisms in the Archive: in the mode of an œuvre à faire'. In *Mai: Feminism and Visual Culture*. https://maifeminism.com/new-materialisms-in-the-archive-in-the-mode-of-an-oeuvre-a-faire/.
Tamboukou, Maria. 2020. 'Archives, Genealogies and Narratives in Women Workers' Education'. *Women's History Review* 29 (3): 396–412.
Tamboukou, Maria. 2022. 'Traces in the Archive: Re-imagining Sofia Kovalevskaya'. *Life Writing* 19 (3): 341–56.
Tamboukou, Maria. 2023. *Epistolary Narratives of Love, Gender, and Agonistic Politics.* London: Routledge.
Tamboukou, Maria. 2024. 'Epistolary Becomings: Archival Traces and Diffractive Readings'. *Tautosakos Darbai [Folclore Studies]* 68: 15–27. https://doi.org/10.51554/TD.24.68.02
Tello, Verónica. 2022. 'Counter Memory and and–and: Aesthetics and Temporalities for Living Together'. *Memory Studies* 15 (2): 390–401.
Tilche, Giovanna. 1984. *Maria Gaetana Agnesi, Mathematica e Compassione*. Roma: Castelvecchi.

Toole, Betty Alexandra. 1992. *Ada, the Enchantress of Numbers.* Moreton-in-Marsh: Strawberry Press.
Turing, Alan. 1950. 'Computing Machinery and Intelligence'. *Mind* 59: 433–60.
Van den Elzen, Stella Sophie Maxime. 2021. 'Antislavery in the Transnational Movement for Women's Rights 1832–1914: A Study of Memory Work'. PhD thesis. Utrecht University, The Netherlands.
Vanacker, Beatrijs, and Lieke van Deinsen, eds. 2022. *Portraits and Poses: Female Intellectual Authority, Agency and Authorship in Early Modern Europe.* Leuven: Leuven University Press.
Wade, Ira, O. 1947. *Voltaire and Madame du Châtelet and Studies on Voltaire with Some Unpublished Papers of Mme Du Châtelet.* Princeton: Princeton University Press.
Whitehead, Alfred, North. 1985 [1929]. *Process and Reality* [Corrected Edition]. Edited by David Ray Griffin and Donald W. Sheburne. New York: The Free Press.
Wiegand Brothers, Dometa. 2015. *The Romantic Imagination and Astronomy.* Basingstoke: Palgrave Macmillan.
Zerubavel, Eviatar. 2003. *Time Maps: Collective Memory and the Shape of the Past.* Chicago: Chicago University Press.
Zinsser, Judith P. 2006. *Daring Genius of the Enlightenment, Émilie du Châtelet.* New York: Penguin.

2 Epistolary Entanglements

Between the Personal and the Scientific

Letters, in general, and epistolarity, in particular, have been at the heart of my research for several years, and I have written extensively about the intricacies of reading letters in tracing hidden genealogies in women's lives (see Tamboukou 2020). What I have observed, however, is that the rich body of literature around letters has not considered the importance of scientific correspondences, in general, and mathematical correspondences, in particular, in the genealogical excavation of the female self. I have been struck by this omission, and this chapter revolves around the importance of letters in writing a feminist genealogy of *automathographies*.

Here, it is important to note that I am not suggesting that women mathematicians' letters have not been discussed at all. Historians of mathematics have often drawn on letters to examine women's contribution to the history of mathematical sciences, and there have been some publications in this area, although most of them have not been translated into English yet. As already noted in the previous chapters, Kovalevskaya's correspondence with her doctoral supervisor Karl Weierstrass is in German (Bölling 1993) and has only been translated in Russian (Kochina 1973), Maria Gaetana Agnesi's correspondence with the Riccati family (Mazzone et al. 2010) is in Italian, and Émilie Du Châtelet's correspondence is entirely in French (Kölving and Brown 2018). What we have instead in the anglophone literature are some fragments of their letters translated for the purpose of being used as evidence or illustrations of various biographical studies and/or other publications around their work. But such work has been limited in this area, and women mathematicians' letters have not been examined or considered on the analytical plane of Liz Stanley's (2004) influential concept of the 'epistolarium', as I will further discuss, particularly focusing on entanglements between 'the personal and the scientific'.

Scientific correspondences were indeed central in processes of knowledge production and dissemination in the eighteenth and nineteenth centuries.[1] In the context of epistolary worlds and the digital turn in archival research, as already discussed in Chapter 1, letters have become particularly important in throwing light on women's engagement with science in the eighteenth and nineteenth centuries, and there is an increasing interest in scientific correspondences through a gendered lens.[2] As we have already seen in the Introduction, women often studied mathematics in domestic settings, given their exclusion from universities and other formal scientific societies and institutions up until the turn of the nineteenth century. Moreover, cultural historians have been interested in women's epistemological and

DOI: 10.4324/9781003533993-3

intellectual involvement in the making of scientific knowledge, which included the development of mathematical sciences but was also expanded into the wider cultural formations of modernity. As Schurch has persuasively argued, 'the methods of scientific pursuit, literary creativity and cultural productivity were enmeshed and could, at moments, elide with each other' (2019, 18–19).

In thus looking at the letters that women mathematicians wrote, I trace ways in which women's epistolary writing contributed to knowledge and research in mathematical sciences, but also reveal the minutiae of their constitution as subjects in science. In doing so, I am interested in how the literary structures, persistent patterns, as well as formal characteristics of the letter, what Janet Altman (1982) has theorised as 'epistolarity', are entangled in cultural *assemblages* in mathematics. As already discussed in Chapter 1, the notion of the *assemblage* is taken from Deleuze and Guattari's philosophical vocabulary (1988) as a configuration denoting the complexity of discursive and non-discursive components and formations in the constitution of knowledge, culture as well as gendered subjectivities in mathematics.

Analysing the letters of women mathematicians through the lens of epistolarity offers a unique way to understand their lives and narratives in relation to other autobiographical documents. This approach draws heavily on Stanley's (2004) influential work on letters, which emphasises the rich insights correspondences can provide. Stanley identifies three key dimensions of epistolary narratives that reveal their dynamic nature. First, letters are inherently dialogical. They open channels of communication and reciprocity, not just between the correspondents but also between the writer and any future reader. This dialogical aspect creates a layered interaction that extends beyond the immediate exchange. Second, letters are perspectival, meaning their structure and content evolve based on the recipient and the passage of time. They are deeply context-sensitive, reflecting the shifting circumstances of both writer and recipient. Lastly, letters possess emergent properties, resisting researcher-driven interpretations. Instead, they reveal their own preoccupations, conventions, and ethical frameworks, offering insights on their own terms (202–4). At the heart of Stanley's analysis is the concept of the *epistolarium*. She defines this in three interconnected ways: as a historical record available for future study; as the complete collection of surviving correspondence from a particular writer; and as the 'ur-letters' that emerge through transcription, editing, and publication. Each of these forms brings distinct epistemological complexities and implications, highlighting the multifaceted nature of letters as both historical artefacts and evolving texts (218).

In mapping the archival documents of women mathematicians, including their letters, I have already considered these complexities. Building on Stanley's 'epistolarium', I have explored the methodological and epistemological effects of working with different types, forms, and collections of letters. This approach underscores the dynamic and self-revealing nature of epistolary documents, emphasising their ability to transcend simple historical records and instead act as living, interactive narratives within broader autobiographical and historical contexts. Here, as elsewhere in my work, I deploy the notion of *epistolary sensibility* (Tamboukou 2020) as an epistemological and methodological move that goes against the dominant trend of using letters as mere 'sources' or 'data' in socio-historical research and analysis. While recognising the evidentiary value of letters, the analysis is deeply engaged with pertinent ontological, epistemological, and ethical questions revolving around what it is exactly that we do when using letters and correspondences to derive meaning about subjects, their lived experiences, their relation to the world and others, as well as their entanglement in processes of knowledge production and circulation. How is this *epistolary sensibility* to be configured? Drawing on my work with women's letters in different social, cultural, and historical contexts, I have made a cartography of *epistolary sensibility* that includes, amongst other practices, striving for understandings that are driven by the letters and collections under investigation, considering the content, form, and context of letters and analysing them in their interrelation.

On this plane of analysis, letters are analysed as important auto/biographical documents and not as mere illustrations, or as captions of images and other visual artefacts. In light of the *epistolary sensibility*, letters are not simply treated as 'a document of life' (Plummer 2001) amongst others. The analysis pays close attention to their epistolary traits, which include, among others, their relational and dialogic character, their ephemerality, the power dynamics of the epistolary relationship, the importance of salutations and greetings, the different time frames of writing a letter, posting it (or not), reading it and responding to it (or not), the materiality of the epistolary act, and the narrative extravagances of letter writing.

Since letters are always already dialogic, there is a particular analytical interest in the I/you/we epistolary relation and in how this relation shapes the emergence, development, and dissemination of mathematical knowledge, but also becomes constitutive of the female self in science and mathematics. Epistolary exchanges usually imply distance and therefore linguistic, cultural, and in the case of women mathematicians' correspondence, gender differences. As I will further discuss in Chapter 3, the problematics of language and translation become central in the search for meaning and understanding, particularly if we consider that the majority of eighteenth- and nineteenth-century women mathematicians' philosophical and literary writings, as well as their letters, autobiographies, journals, and diaries have not been translated into English.

Women mathematicians used correspondence as a powerful medium to question, challenge, and position themselves within the scientific networks and communities of their time. Through their letters, they imagined and established themselves as mathematicians, navigating the fluid boundaries between private and scientific discourse. For many women excluded from formal education and academic institutions, letters became essential tools for learning, teaching, and contributing to the production and dissemination of mathematical knowledge.

Sophie Germain, for example, used correspondence to conceal her gender, engaging with professors at the École Polytechnique and later Karl Friedrich Gauss. This allowed her to participate in the 'Republic of mathematics' (Oosterhoff 2016) , creating a space for scientific communication that circumvented the gender biases of her time (see Schurch 2019). Even for women with formal tutors, such as Émilie Du Châtelet, Maria Gaetana Agnesi, and Ada Lovelace, mathematical education often relied on correspondence, as their tutors were prominent mathematicians who frequently travelled. Similarly, Sofia Kovalevskaya's correspondence with Karl Weierstrass extended beyond her formal studies, transforming into a long-term mentorship that supported her navigation of European academic circles. What began as a pedagogical relationship often evolved into collaborative scientific exchanges, encompassing peer review, critique, and even disagreements.

This interplay between education and scientific dialogue reveals how correspondence functioned as both a refuge from societal constraints and a platform for intellectual collaboration. The following sections will delve into the specifics of these six women mathematicians' correspondence and examine their interconnected dynamics.

Epistolary laboratories in Du Châtelet's correspondence

Throughout her life, Du Châtelet corresponded with numerous eminent mathematicians across Europe, 'all as prestigious as each other: all scholars, academicians and university professors' as Roland Bonnel has commented (2000, 82). Besides her first mathematics tutor, Pierre Louis Moreau de Maupertuis, her correspondents included his friend and colleague Alexis-Claude Clairaut, who also became her tutor and eventually managed the publication of 'her Newton' after her death. She also corresponded with Swiss mathematicians Leonhard Euler, Johann Bernoulli, and his son Johann II, the German Christian Wolff, as well as the English scientist and physician James Jurin, and the French mathematician and physicist François Jacquier.[3] Finally her public and polemic correspondence with Jean Jacques Dortous de Mairan, Secretary of the Royal Academy of Sciences in Paris, is well known in the literature around her scientific work, as I will further discuss in Chapter 4.

As Judith Zinsser has noted, Du Châtelet's letters played a crucial role in maintaining her connections with the scientific community and engaging in their debates: 'she consulted with her correspondents, made clear her disagreements—later footnoting their works in her writings—and sent them her own publications' (2006, 118). In this light, her scientific correspondence serves as an exemplary showcase of the complex interplays among networks of scientists, mathematicians, and philosophers across Europe. This correspondence also reflects her passion for study and reveals the cultural contingencies influencing the production and dissemination of scientific knowledge and research, illustrated by the concept of imaginary belonging in the Republic of Letters (Goodman 1994).

In this context, Du Châtelet's extensive correspondence with Maupertuis, which went through 1730s to 1748, serves as an exemplary Stanleian *epistolarium* (2004), showcasing the evolution of their pedagogical and later scientific relationship, their friendship, but also their disagreements, both scientific and personal. Maupertuis first accepted to become Du Châtelet's tutor around 1734. In her first letter to him, we understand that he had given her two manuscripts, which she had read diligently: 'I believed, sir, that in order to be worthy of responding to the letter you wrote, I needed to have read your work. I was very pleased with your two manuscripts, and I spent yesterday evening benefiting from your lessons', she wrote to him in January 1734 (BnF, fr.12269, f. 3).[4]

In this early phase of their correspondence, Du Châtelet wrote letters to raise questions, discuss problems, or report on her readings and mathematical work with exercises that were included in their lessons: 'I sent someone to look for you at the Academy and at your home, sir, to let you know that I would be spending the evening at my place today. I spent it with binomials and trinomials. I can no longer study if you do not give me a task, and I have a great desire for it', she wrote in the first months of 1734 (f. 11). Being aware of her intellectual weaknesses, Du Châtelet hoped that her desire for learning would compensate for her lack of rigorous mathematical education.

Their lessons took place in person, often at her Parisian home, but also at the Jardin des Plantes or at Maupertuis' residence in Mont Valérien. She would therefore write letters to arrange their meetings or send short notes to inform him that she would be waiting in her carriage outside the Royal Academy of Sciences after its sessions, or outside the Café Gradot, a well-known Parisian venue where intellectuals met to discuss, but where women were not allowed to enter. Du Châtelet's letters also show that their lessons would be an opportunity for social exchanges, but also an outlet for personal feelings to be expressed: I hope I will have the honour of seeing you on Wednesday, when you leave the Academy [I will wait for you at my place, where I expect you will want to spend the evening]', she wrote to him in January, 1734 (ff. 3-4). When later in the same year, she suffered the bereavement of the death of her second son, she turned to Maupertuis for consolation: 'My son died last night sir; I confess, I am extremely distressed; I will not go out, as you can understand. If you want to come console me, you will find me alone. Although my door is closed for visitors, I feel that there is no time when I do not find extreme pleasure in seeing you.' (f. 15) Writing from her mother's residence at Créteil, not very far away from Paris, on 15 September 1735, her letter below carries traces of personal and affective intra-actions:

> If you still love me a little, you will come see me; you know my mother well enough for that; furthermore, if you wish, she won't know you are here. In any case, I must see you. I will spend a few hours in Paris tomorrow, Friday, and leave at ten in the evening to return here. If you are willing to wait for me at Café Gradot, I will meet you there between five and six, and you will come back here with me, or at least we will travel together (f. 46).

Du Châtelet's correspondence with Maupertuis has been read and interpreted from a variety of angles and perspectives, including the assumption that they were sexually involved.[5] Zinsser, however, has persuasively argued that the affectionate and emotional tones of Du Châtelet's epistolary discourse were 'the style of the era' and that her 'melodramatic' language can be traced throughout the body of her correspondence (2006, 72). I will come back to Du Châtelet's 'extravagant narratives' in Chapter 5, where I discuss her amorous correspondence with St. Lambert, but what I want to discuss further is the mingling of the personal and the scientific in her correspondence with Maupertuis.

As noted in the Introduction, Du Châtelet and Voltaire transformed the Château de Cirey into a dynamic scientific retreat, immersing themselves in reading, writing, and hosting scholars from across Europe. Their lives at Cirey were a rich blend of philosophy, culture, and mathematics, as vividly expressed in a letter Du Châtelet wrote to Maupertuis on 1 December 1736, during his Lapland expedition to determine the Earth's shape (BnF, fr.12269. ff. 61–2). In the letter, she shared how they had 'used your absence to render the people who inhabit Cirey worthy of you' (f. 61), adding, 'we have become real philosophers' (f. 61). She proudly mentioned that Voltaire had written an introduction to Newtonian philosophy, dedicated to her, complete with a poetic frontispiece, which she enclosed: 'I believe that you will find the verses worthy of the philosopher of whom they speak, and of the poet who made them' (f. 61).

Du Châtelet also credited Maupertuis for inspiring Voltaire's transformation, writing, 'You have for a very long time wanted to make a philosopher of the first of our poets, and you have succeeded' (f. 62). Reflecting on her own intellectual journey, she expressed gratitude for Maupertuis's earlier mentorship, declaring, 'I enjoy a great advantage over the greatest philosophers: that of having had you as my master' (f. 62). The letter offers a glimpse into the intellectual vitality of Cirey and the deep admiration Du Châtelet held for Maupertuis, whose influence resonated in their philosophical and scientific endeavours.

Despite the happy days at Cirey, Du Châtelet was still missing the opportunity to travel, like her tutor, given the stringent gender-based restrictions on travelling abroad, regardless of social class and status. As Zinsser has observed, 'a marquise could not travel alone' (2006, 116), and yet, travelling was essential for the Republic of Letters, allowing scholars to meet 'the learned with whose work you were already familiar' (Jordan 1735, cited in Zinsser 2006, 116), which was especially important for mathematicians, who often work and think collaboratively. 'I am in despair over my ignorance and all the things that prevent me from overcoming it. If I were a man, I would be on Mont Valérien with you, abandoning all the useless things of the world' (BnF, fr.12269. f. 116), Du Châtelet wrote to Maupertuis from the Château de Cirey on 24 October 1738, while trying to persuade him to go there.

In the course of his travels, Maupertuis visited the Château de Cirey in the winter of 1738/1739–9. It was during this period that scientific frictions in their relationship erupted, particularly concerning the 'vis-viva' debate, as well as the question of metaphysics.[6] After this visit, and given the emergence of dissonances in their scientific relationship, Maupertuis promised to arrange for Johann Bernoulli the younger to become Du Châtelet's tutor. However, this arrangement never materialised. Instead, Samuel König, a friend of Maupertuis and a Leibnizian adherent, arrived in Cirey at the end of April 1739. König and Du Châtelet did not get along well. On 27 September 1739, she wrote to Maupertuis: 'Have you been kind enough to write to Mr Bernoulli? This has become more necessary for me than ever because I despair of keeping König' (BnF, fr.12269, f. 141). It appears from his correspondence that Maupertuis actually dissuaded Bernoulli from travelling to Cirey (see Hagengruber 1999, 193). In a letter written to Bernoulli on 1 February 1740, Du Châtelet seems to be aware of Maupertuis' intervention, which aroused her anger and created additional tensions in their personal relationship:

> I knew, sir, that the same people who had suggested you come to my place have now dissuaded you, so I would have expected the letter I just received from you, if I had not believed I could count enough on the strength of your character to think that their advice would arrive too late and that you would not want to break such a firm commitment as that of your last letters (in Kölving and Brown 2018, vol.1, 568–9).

Matters worsened when Du Châtelet's *Institutions de Physique* was first published in 1740, and König accused her of plagiarism. These disputes led to a brief but complete rupture in her relationship with Maupertuis during this period. Since Maupertuis's letters to Du Châtelet are either missing or not yet accessible, his epistolary responses to her comments, suggestions, questions, and complaints can only be inferred indirectly from her side of the correspondence.

Du Châtelet's position in the epistolary relationship with Maupertuis had variations and changes over time. As Roland Bonnel has aptly observed (2000, 83) in the beginning of their correspondence, she positioned herself as 'a schoolgirl' [écolière], despite her aristocratic status: 'I have studied a lot and I hope you will be a bit less displeased with me than last time' (BnF, fr.12269, f. 5), she wrote to him on a Monday, between January and February 1734. In a later letter during the same period, she complained about being neglected: 'You don't want to encourage a schoolgirl, because I still don't know if you found my work good' (f. 12), she wrote.

The external reader can trace a playful tone in these epistolary exchanges, which went on throughout their correspondence. Despite Maupertuis's eminent status as a mathematician, significant social differences existed between them, and such intersections of class and gender had an effect upon the form and style of Du Châtelet's epistolary narratives. Mary Terrall has observed, however, that Du Châtelet's relationship with Maupertuis cannot be easily framed within 'the patronage model', namely a personal relationship between a noblewoman and a male expert, 'entailing asymmetrical status and some sort of recipient obligations between the two parties' (1995, 286). Maupertuis was already on the rise in the French scientific scene, and their relationship was mostly coloured by their unconventional interest in Newtonian mathematics.

Later on, when Du Châtelet's mathematical knowledge advanced and her critical thought deepened, Bonnel has observed (2000, 89) that the tutor became a scientific advisor: 'I have taken up, Sir, the sweet habit of writing to you for all the posts, and I cannot detach myself from it, even today that I have no pretext, to send a letter to you. I confess that I am a little annoyed to consult you from afar' (BnF, fr.12269, f. 131). This is what she wrote on 26 January 1739, while sending Maupertuis a long list of questions about his book *La Figure de la Terre,* which had been published in 1738. Du Châtelet had become more confident about her knowledge and subsequently more critical vis-à-vis the work of others, Maupertuis included.

In the final phase of their correspondence, when Du Châtelet was preparing the publication of the *Institutions de Physique*, they eventually became colleagues, who were discussing, debating, and had ultimately agreed to disagree: 'I did not expect you to become Leibnizian, nor that the monads would conquer you. I do not know, however, if the metaphysical ideas, which are at the beginning of the book, do not at least deserve to be known' (f. 168), she wrote from Brussels on 8 August 1741.

In examining the entirety of Du Châtelet's scientific correspondence, Bonnel has identified three epistolary sub-genres that frequently intermingle or transition into one another, sometimes even within the confines of a single letter. The 'dispute letters' are those in which she responds to questions about her work and presents, explicates, and supports her scientific ideas. Writing to Frederick of Prussia on 16 February 1739, she defended her work on the nature and propagation of fire (Du Châtelet 1738) and explained her views by giving examples, providing evidence and making references to the existing literature. 'There is no way to support the argument that forests are sustained by the wind since Your Royal Highness persists in believing it impossible, and even Mr. Voltaire is against it' (in Kölving and Brown 2018, vol. 1, 488), she wrote at the beginning of her argument, while being modest about the evidentiary value of her scientific work: 'I find that those who enquire about this matter with Your Royal Highness will be better informed than through my work' (488). And yet, she went on to give specific examples, since science is about data, evidence, and research: 'I am more confident about the Rhine freezing in Switzerland because I have asserted only what Scheuchzer reports: that in the bishopric of Basel, there is a river that freezes in the summer and flows in the winter' (488).

Within the body of her scientific correspondence, there are also letters in which she discusses new topics and original ideas emerging in the scientific field, a kind of a scientific newsletter or the 'letter-gazette' in Bonnel's typology. Her letter to Maupertuis written on 9 May 1738 includes, amongst others a criticism of his theory of attraction, in a section that exemplifies elements of the 'letter-gazette'. At the beginning of this critical section, she starts with a disclaimer, namely the adequacy of her algebra knowledge: 'I ventured to read your Memoir, given in 1734, on the different laws of attraction, and I frankly admit that I do not know enough algebra to have been able to follow you everywhere' (BnF fr. 12269, f. 84). Her deficiencies notwithstanding, she is bold enough to raise her concerns and explain why:

> But allow me to tell you that I find what is written in French a little obscure. For first, you do not say (in French) why an attraction in direct reason of the simple distance, which also has so many advantages, would not have that of the agreement of the same law in the parts and in the whole; you do not say either why the inverse reason of the square of the distances has this advantage, and you will see that you are very wrong, because there are many ignorant people like me, and everyone will hear it in his way (f. 84).

What lies at the heart of her scientific correspondence, however, in Bonnel's (2000) tripartite schema, is 'the laboratory letter', which takes the form of an examination, where previous literature on the topic is critically reviewed, constructive criticisms are made, hypotheses are configured, thought experiments are tested, and new ideas and findings are put forward (81–2). In Bonnel's classification, the 'laboratory letters' bend the conventions of the epistolary form, which is not just about communication anymore. Instead, these letters capture the evolution of thought, as the writer engages in a dialogue not only with the recipient but also with herself—formulating hypotheses, reasoning through ideas, drawing conclusions, and testing intuitions. This process mirrors the work of a scientist conducting experiments in a physics lab, Bonnel pithily notes (81).

Bonell has observed that Du Châtelet's 'laboratory letters', where she first tested ideas, hypotheses, and intuitions, would be expanded upon in the body of her *Institutions de Physique*. He has further shown that her laboratory work was methodical, chronological, and went through her scientific work from her first published essay on the nature of fire (Du Châtelet 1738), her magnum opus, *Institutions de Physique* (1740) and finally her swan song, the translation of and commentary on Newton's *Principia* (1756). Taking this idea further, Arianne Nicole Margolin (2008) has dissected the notion of Du Châtelet's 'epistolary laboratory' by meticulously showing how the laboratory—the site par excellence of the scientist's work—was transposed within the context of the epistolary form. In considering the form and context of Du Châtelet's letters, Margolin has shown that they unfold within the dominant philosophical discourses of the time—didactics and polemics. Margolin (38) asserts that didactics is significant because it accomplishes an educational objective, but given the intense debates and rivalries between the Cartesians, the Newtonians, and the Leibnizians in the eighteenth-century French scientific community, Du Châtelet's letters were also polemical.[7] As Marie-Claire Grassi has argued, three features can be traced in the polemical epistolary form: an open dialogism, an ironic tone, but also and perhaps most importantly, a simple style and refined argumentation, since the aim of the polemical letters is, after all to convince (1998, 109). Combined together, the dialogic and the polemical epistolary elements present an intellectual crisis to their audiences, argues Margolin, while simultaneously articulating a response, 'showing that the author-scientist has mastered the subject and can express themselves in a pleasant and elegant way' (2008, 38).

My reading of Du Châtelet's letters to Maupertuis, as well as to other significant mathematicians and scientists of her time, has traced many more entangled components—beyond the dialogic and the polemical— including self-reflections, playfulness, but most importantly ambivalence, particularly in the dangerous territory between physics and metaphysics. 'You will have undoubtedly found my question quite ridiculous when I asked you how it followed that the same quantity of motion could subsist in the universe, supposing that the force of bodies in motion is the product of their mass by the square of their speed' (BnF, fr.12269, f. 79), Du Châtelet noted in a letter to Maupertuis, written from her château in Cirey on 30 April 1738, while she was preparing the publication of her *Institutions de physique*. While playfully admitting that 'you are the master in Israel and I am ignorant and seek to instruct myself trembling before you' (f. 79) she also adds that ignorant as she was, she had managed to find the answer to her 'ridiculous' question: 'Since I wrote to you, I read what M. Leibniz gave in the Acta Eruditorum on forces vives, and I saw that he distinguished between the quantity of motion and the distinct quantity of force; and then I found what I needed' (f. 79). Despite her discovery, however, she still needed Maupertuis' advice on the metaphysical problem of freedom. 'I believe myself free and I do not know if this quantity of force, which is always the same in the universe, does not destroy liberty' (f. 79), since 'if we have not the power to begin motion, we are not free' (f. 79), she observed while asking a mathematician to enlighten her, given the pure truth that only mathematical analyses could ever convey.

Apart from the discussion of scientific and philosophical issues, this letter contains a series of textual traces that reveal the unfolding of Du Châtelet's way of thinking, developed through curiosity, reflection, study, conversations, and disputes. This is why Bonnel argues that her letters look like a laboratory, 'which acts as a space of research, a matrix of reflection and not only a place of communication' (2000, 81). Through her letters, Du Châtelet would throw herself into a perpetual dialogue, not only with her epistolary interlocutors, but also with herself, as we have seen in her metaphysical questions above. Her critique develops as a reflection, but also with 'formulations of hypotheses, conditional proofs, deductions from conclusions, writing down of intuitions' (81). In this sense, her letters virtually recreate the space of a laboratory, while her recipients become collaborators in an experiment within a physics cabinet, as Bonnel has aptly commented (81).

Overall, the scientific and philosophical epistolary exchanges between Du Châtelet and Maupertuis not only reflect significant debates in eighteenth-century science and philosophy but also serve as evidence of Du Châtelet's originality and independence as a thinker, as Ruth Hagengruber (1999, 2022) has notably demonstrated. Their epistolary discussions about the 'vis viva' controversy[8] and the role of metaphysics in science illustrate the complexities of integrating empirical observations with rational and philosophical inquiry. While Maupertuis's methodology was more empirical and pragmatic, Du Châtelet believed that understanding the fundamental nature of scientific principles required a systematic and coherent metaphysical framework. For her, physics could not be fully comprehended without a metaphysical underpinning that provided rational justification for observed phenomena: 'I am convinced that physics cannot do without metaphysics, on which it is founded' (in Kölving and Brown (2018, vol. 1, 576–77), she wrote to Frederick of Prussia on 25 April 1740. Thus, in her letters, Du Châtelet expressed frustration with Maupertuis's dismissive attitude towards the metaphysical aspects of science. She felt that a comprehensive scientific theory needed to address not only empirical data but also the underlying philosophical and metaphysical principles. This difference in perspective underscores a fundamental debate in the philosophy of science that reaches our days: the role of metaphysics in science and mathematics, an area that was also at the heart of Agnesi's mathematical and philosophical work and scientific correspondence, as I will discuss next.

God, truth, and mathematics: Agnesi's correspondence as a review process

We don't have many letters from Agnesi to her tutors, but her notebooks at the Ambrosiana Biblioteca in Milan provide a clear view of how her education was structured. This organisation involved the exchange of educational texts, essays, and written exercises, which could be easily transmitted via correspondence. From the few extant letters of this early phase, her personal letter below to her philosophy tutor Michele Casati emits signs of her frustration with the filial duty to display her knowledge:

> Everything went according to His Highness' wishes. The evening after our arrival, the prince came to our house; he heard from me to his great enjoyment, a dispute that I made with Father Branconi about the cause of the motion of the planets, and another with Father Stampa, about the nature of colours. Then he heard Teresa singing and playing the harpsichord; and then both of us together also played for him a melody on that sweet musical instrument, which borrows its name from love, so that he could sleep well at night having filled his eyes with pleasurable visions (VBA. 0201. SUP).[9]

Agnesi wrote this letter on 9 December 1739, describing an evening where she and her younger sister Maria Teresa had entertained the prince of Braunschweig-Wolfenbüttel, who had visited their city and had asked to attend a literary and musical soirée with the two Agnesi sisters performing for him. Although both sisters were on winter vacation at their countryside villa in Machiago, they were hastily summoned back since such gatherings were important in her father's efforts to gain entry into the circles of the Milanese nobility.

Agnesi, however, got quite tired of the soirées her father was organising in their Palazzo in Milan to display his daughter's extraordinary abilities in philosophical and scientific debates, thus transform her knowledge into a spectacle 'for educated travellers passing through Italy from northern Europe' (Cavazza 2014, 74). Although these *conversazioni*, part of the broader 'culture of scientific spectacle',[10] made Agnesi famous in eighteenth-century Europe, they also disrupted her desire to work on mathematics in solitude and tranquility, interfering with her preferred method of esoteric contemplation and intellectual practice, according to Mazzotti (2014, 131).

When Agnesi eventually succumbed to the pressure of these commitments, such events became infrequent and tightly regulated. It is in this final phase of her serious engagement with mathematics that her correspondence with her tutor Ramiro Rampinelli and with the Riccati family sheds light on the importance of mathematical correspondence in a critical period of her work, when she was preparing the manuscript of her much-celebrated *Analytical Institutions for the Use of the Italian Youth*, published in 1748 (Agnesi 1748). This correspondence is a testament not only to the difficulties she faced but also to the hard work she put into preparing her manuscript, as well as the recognition of her originality by well-known mathematicians of her time (see Roero 2015; Soppelsa 1985).

Agnesi was introduced to her tutor Rampinelli around 1740, during a period when she decided to commit seriously to the study of mathematics. He had just arrived in Milan to teach geometry at the Monastery of San Vittore al Corpo, having previously studied and collaborated with renowned mathematicians in Bologna and Padua. As Mazotti has noted, mathematics was considered essential for the cultural and spiritual development of Catholic youth within the ecclesiastical circles where Agnesi was educated (2007, 105). She thus asked Rampinelli to instruct her in Cartesian algebra and the new mathematical techniques of calculus, as she had already conceived the idea of writing a textbook, as a concise introduction to mathematical analysis, as well as a clear and methodical exposition of the recent discoveries in infinitesimal calculus, which, in her view, had been 'disconnected, disordered and scattered across the works of many authors' (Agnesi 1748, 2). In writing the Preface of this book, Agnesi duly acknowledged Rampinelli's critical influence on her mathematical education.

> Notwithstanding the strong inclination I had to this science and the great application I made use of to acquire it, I might still have been lost in a maze of inextricable difficulties had I not been assisted by the secure guidance and sage direction of the very learned Father Don Ramiro Rampinelli (xxi).

It was Rampinelli who first introduced Agnesi's mathematical work to Jacopo Riccati—an important figure in the Italian sciences of the early eighteenth century—and his two sons, Giordano and Vincenzo. He had met the Riccati family while residing in the Veneto area between 1727 and 1731, and he had himself benefited from their in-depth knowledge of Leibnizian and Newtonian analyses. 'I assure you that I have always been and still am amazed by seeing such talent and such depth of knowledge in a woman as would be remarkable in a man, and in particular by seeing this accompanied by quite remarkable Christian virtue' (in Mazzone et. al. 2010, 325), [11] he wrote to Giordano Riccati about Agnesi, while asking the Riccati family to review her treatise on mathematical analysis, a task that he felt was above his abilities:

> Her Ladyship has noted the great amount of twaddle heard from my lips about Analysis, she has greatly improved, tidied and added to it with her own work, and with a reading of books; in substance she has compiled a body that can be called an accomplished Introduction to Analysis. Her father is very keen to publish this little work, but, as I both lack the necessary ability and also have had a small part to play in its composition, I feel unable to judge its merits; therefore I ask you and Count Jacopo to kindly accept the task of endeavouring to read the writings I shall send you one part at a time when I have apprised that you are willing to perform this favour, which I beg you do (325).

Encouraged by Rampinelli's warm introduction, Agnesi wrote her first letter to Jacopo Riccati on 20 July 1745: 'I beg your most excellent Lordship to honour me with your sage criticisms and at the same time oblige me by placing your corrections in the margins' (329), she wrote, given that her tutor felt that he was too close to the work to pass judgement. Riccati's response was swift and enthusiastic in his first letter to Agnesi, dated 18 August 1745: 'I was amazed when casting my eye over an accomplished treatise on Cartesian analysis that a young lady could reach these heights in such delicate and abstruse matters' (333), he wrote, but despite his initial enthusiasm, he promised 'to fill the shoes of a critic completely against my inclination, and I will fill them in such a way that I may seem indiscreet' (333).

While Jacopo Riccati's first impression was that 'there is very little to be removed or added' (333), the manuscripts would be scrutinised by his sons, Giordano and Vincenzo, who were also Lecturers in Mathematics at the University of Bologna. Giordano was as enthusiastic as his father in his first impression of the work. In a letter to Rampinelli, dated 19 August 1745, he reassured him that 'we read the most learned writings with great avidity and found many reasons to admire the Lady's great intelligence, the accuracy of her method and the clarity of her explanations' (334), further promising that 'we will soon give her our respectful opinion' (334). While waiting for the examination of her work, Agnesi was bold enough to write to Jacopo Riccati at the end of September 1745, asking for some clarification, given that her tutor was away:

> Since Father Rampinelli has left for the countryside and is continually traveling, I am unable to communicate some difficulties I am encountering in drafting the differential calculus that I am currently working on. Therefore, I take the liberty of asking Your Most Illustrious Lordship to provide some guidance on the matter so that I may continue my work without hindrance (337).

What is interesting in this correspondence is the way mathematical knowledge, information, and advice circulate between Rampinelli, the Riccatis, and Agnesi in a relationship that is highly structured by gender, knowledge, as well as kinship and social hierarchies and differences: 'I beg you anew to express your sentiments in all freedom, and if you have trouble explaining yourself directly to the Lady you may do so to me because, as the aforementioned Lady's father means to print the work, it is a very peculiar matter' (340), Rampinelli wrote to Giordano Riccati on 7 October 1745. In a letter written to Rampinelli on 13 April 1746, Giordano reiterated his commendation for the clarity and rigour of Agnesi's work: 'I can assure you in all sincerity that my esteemed father's sentiment and my own is that the manuscripts are more than worthy of seeing the light of publication, as throughout the work are found truth, method and admirable clarity' (343). He further reassured Rampinelli that 'my excellent father examined the work on Integral Calculus, and he will very shortly finish drafting some notes he deems necessary' (343). Given that Rampinelli had raised a number of questions regarding Agnesi's work, he finally added that he had asked his brother Vincenzo to consider them: 'As I was unable to meditate on the contents of your letter [...] I sent it to my brother, Father

Vincenzo, who, upon thinking it over, found that the issue was much more complicated than he first thought' (343).

We can see in the above epistolary exchanges how Agnesi's work was scrutinised and reviewed by some of the most eminent mathematicians in the field of analysis in the Italian states. In this light, her own letters to Rampinelli and the Riccatis are traces not only of her mathematical knowledge and hard work but also of how she positioned herself in the mathematical circles of her time through correspondence. In writing to Jacopo Riccati on 22 June 1746, she sent him the last part of her work, asking for his expert opinion, as she was already preparing her publication:

> Before you make your learned annotations, I must beg you to review them and give me your opinion regarding the overall work. I do not wish to proceed with printing the first part (which I intend to begin upon receiving your initial writings accompanied by your learned annotations and those of Count Giordano, your most worthy son) without first obtaining your esteemed general opinion, even concerning the remainder (345, my translation).

Indeed, her letters reveal that she was bold enough not only to converse with highly esteemed mathematicians but also to situate her request for help within the temporal conditions of her own personal circumstances. In the letter below, written to Jacopo Riccati on 23 August 1746, she attempts to coordinate his response to the risks of the postal services, her vacation plans, as well as the publication preparations:

> I have a great fear that these writings may arrive at a time when I am absent from Milan and could therefore be lost which, as you can imagine, would be a great affliction for me. I am most sorry to have [...] to bother your Lordship in this way, but as your Lordship has been so good as to assist me I hope that he will pardon my impertinence given my reasons above, especially as for me it is enough to have those that you have already perused, which I shall endeavour to read during my vacation in order to profit from your most learned teachings. Meanwhile I shall make ready to print the work, which task I intend to begin at the end of the vacation, while your Lordship may peruse the remaining pages at leisure and then send them after granting me the honour of making your wise observations upon them, together with those of your excellent son, Giordano (349).

As Maria Laura Soppelsa has noted in editing and presenting the Riccati-Agnesi correspondence, 'the unique function of the "letter" is highlighted as an irreplaceable key for understanding the genesis of various works and projects with significant scientific and cultural implications' (1985, 119). Moreover, Clara Silvia Roero has argued that delving into the details of this mathematical correspondence throws light on the exchange of rich scientific ideas in the period immediately preceding the final drafting of the *Analytical Institutions*, and 'can help us understand the role that Agnesi's work played in the mathematics and general culture in Italy at the time' (2015, 296).

Roero has further observed that a key insight from this correspondence is that, while Agnesi valued the wisdom of her epistolary advisors, her primary focus was on the pedagogical significance of her mathematical works (207). Her goal was to teach young learners mathematical analysis without introducing unnecessary complexities that might hinder their comprehension. This is what she wrote to Riccati in a letter dated 1 October 1746:

> As for differential polynomials, I really haven't done any, believing that the method was too complex and at the same time very limited; concerning the squaring of various curve lines your Lordship mentioned, in truth I had not considered many of these, but those that depend on a knowledge of physical matters I deliberately left alone because, as your Lordship noted, I did not wish to embroil myself in questions of physics; therefore I ignored all the problems related to that subject in order not to extend the work beyond pure Analysis and its applications in geometry (in Mazzone et al. 2010, 352).

Here, however, we also need to note that Agnesi's mathematical correspondence reveals that she was taking up different positions vis-à-vis her addressees. While graciously accepting their advice and actively responding to their comments within the frame of her writing goals, she was at the same time positioning herself as a mathematician and scientist who was interested in the most recent developments within her field, a member of the Republic of Letters, a phrase she actually uses in her letter to Jacopo Riccati dated 19 August 1748:

> I feel great shame at having troubled your Lordship when so busy with domestic matters and literary occupations, so much so that I can find no words corresponding to my inner feelings of gratitude with which to properly express my thanks. Oh how obliged I shall be to you if you will do me the honour of informing me of your discoveries! Oh how I long to see complete your work on the principles and methods of Physics! How much profit and pleasure I will derive from it! And how much the literary Republic [Repubblica letteraria] will gain! If only I were able to return your kindness with some support of my own, worthy of your consideration (452).

When the work was finally published in 1748, Agnesi did become a celebrated member of the Republic of Letters, as we have seen in the Introduction, but her correspondence with the Riccatis went well beyond the publication of her work. As Roero has commented, 'the mathematicians of the Riccati family dedicated a number of studies to her and cited her remarks' (2015, 308–9). Agnesi's mathematical correspondence with Rampinelli and the Riccati family has thus become an *epistolary assemblage* wherein mathematical education and research are effortlessly entangled. Since Agnesi had benefited from the careful reading and examination of her work by eminent mathematicians of her times and geographies, it was her turn to be invited as a reviewer. Mazzotti has noted that 'as late as 1762, the collected works of young mathematicians from a newly founded scientific academy in Turin, including one Giuseppe Luigi Lagrange, were sent to the palazzo for criticism and encouragement' (2007, 145). Although there is no evidence that she ever responded, Lagrange would eventually become one of the greatest mathematicians in Europe, and as a professor of the newly founded *École Polytechnique* in Paris, one of Germain's first scientific correspondents, as I will discuss next.

Sophie Germain: number theory in letters

As we have already seen in the Introduction, Germain's mathematical education started in her home library, but was by no means restricted within it. When the École Polytechnique opened its doors in 1794—for men only, of course—she managed to get access to the professors' lecture notes and, what is more, she also found a way of submitting her written responses and observations at the end of their courses, as was required at the time. Her method was old and well-rehearsed in the gendered politics of European cultural history: adopting a male pen name. It was thus as Antoine-August LeBlanc, the name of a student at the École Polytechnique, that she first wrote to the famous Lagrange, professor of analysis at the École, who got interested in the student's comments, praised them highly, and after discovering the true identity of the student, visited her to express his astonishment and admiration. As Libri has related in his obituary, 'the emergence of a young female geometer caused quite a stir; and Mademoiselle Germain soon found herself visited by savants of superior merit, whose conversations provided nourishment for her mind' (1832, 1).

When Adrien-Marie Legendre, a prominent mathematician in the Parisian academic community, published his *Essai* on number theory in 1798, Germain committed herself to studying this field. There is no record of how their relationship began; most likely, Germain wrote him a letter with questions about his work, as Musielak has noted (2020, 157). Musielak has further surmised (40) that it was likely through Legendre that Germain was introduced to Gauss' *Disquisitiones Arithmeticae*, published in 1801, and delved deeply into its study. This interest led her to initiate a correspondence with Gauss, initially using the pseudonym LeBlanc, as discussed above.

There are fourteen extant letters in their correspondence, ten from Germain and four from Gauss, all housed at the Lower Saxony State and University Library in Göttingen.[12] The first one is dated 21 November 1804, and the last one, 28 March 1829, with a period of ten years without epistolary communication between 1809 and 1819. What characterises this correspondence from the very first letter is Germain's attitude towards Gauss as a fellow mathematician and not as a schoolgirl seeking advice: 'Your *Disquisitiones Arithmeticae* has been the object of my admiration and of my studies for a long time' (Boncompagni 1880, 1), she wrote in the first letter. While praising his 'beautiful theorem' in the last chapter of his book, she was also keen to point out that this theorem could be generalised, enclosing two proofs of this generalisation in the letter (1).

Apart from sending her own comments and proofs, Germain was further bold enough to disclose that 'I did this work with all the more pleasure' (1), linking this unusual feeling of pleasure in doing mathematics with an informal learning process: 'it gave me the opportunity to familiarize myself with this method which, I have no doubt, will be in your hands the instrument of new discoveries' (1), she wrote. Despite the assertive tone of the letter in responding to Gauss's problems and solutions with her own ideas and proofs, Germain was also aware of their difference, given Gauss's status in the mathematical field in juxtaposition to her own self-identified position as 'an amateur': 'I take the liberty of submitting these essays to your judgment', she wrote towards the end of the letter, 'hoping that you will not disdain to enlighten with your opinion an amateur enthusiast of the science which you cultivate with such brilliant successes' (3). In finally signing the letter as 'LeBlanc', Germain also indicated to her recipient that should he find it worth the candle to reply, he should direct his letter to M. Silvestre de Sacy, a member of the National Institute, rue Hautefeuille in Paris. De Sacy was a linguist and orientalist, but there are no details about his friendship with Germain in her historical biographies. It was Musielak, who first noted this unusual arrangement in Germain's epistolary relationship with Gauss, namely a third person complicit in this correspondence (2020, 201). De Sacy was the recipient of the letter as an object to be transferred, as well as a secret to be preserved, a go-between in Germain's epistolary relationship with Gauss. His intermediary role differed from Rampinelli's interventions in the Agnesi-Riccatis correspondence, as discussed earlier. However, we can observe that the correspondence of women mathematicians was expanding the traditional I/you epistolary relationship into an enlarged I/you/we/they Republic of Mathematics.

Germain's correspondence with Gauss—initially through LeBlanc's pen name—has been well treated in the literature in terms of the mathematical problems they tackled.[13] What has not been discussed, however, is the effortless way that 'the personal and the scientific' are entangled in their correspondence. While in their first letters, they simply exchanged ideas about mathematical formulas and theorems, the Napoleonic Wars opened up the scene for personal relationships to unfold, and eventually for the revelation of Germain's gender identity. When the French army invaded Prussia, Germain got worried about Gauss's life: according to her biographers, her fear was linked to her childhood fascination with Archimedes, who was killed by a Roman soldier during the siege of Syracuse while being absorbed in his mathematical problems.[14] In the same vein, Germain probably thought that Gauss might not have even realised the danger of the Prussian-French war, in which even his patron, the Duke of Brunswick, had died. Through family contacts, she thus asked for Gauss's military protection. But when the famous mathematician was reached in Brunswick, he seemed totally perplexed by the news that an unknown Mademoiselle Germain was concerned about his safety. The confusion was cleared through Germain's revelation of her true name and gender identity in a letter she wrote on 20 February 1807, in which she explained that 'fearing the ridicule ascribed to the title of learned woman, I borrowed the name of M. LeBlanc to write to you and to communicate to you notes which, doubtless, did not deserve the indulgence with which you were kind enough to respond to them' (Boncompagni 1880, 13–14).[15]

Although the circumstances of the revelation of her gender identity to Gauss have often been cited and discussed in the literature, what has gone unnoticed is Germain's explanation that she had not hidden her gender identity because she was a woman, but because she wanted to detach herself from the figure of 'the learned woman', which she conceived as ridiculous and unworthy of her position in the field of mathematical sciences. The ridicule attached to the figure of 'the learned woman' [femme savante] comes, of course, from Molière's 1672 comedy, *Les femmes savantes*.[16] A letter of apology from the astronomer Lalande, who had also published a scientific book for ladies, the *Astronomie des dames* in 1785, testifies to Germain's indignation at not being taken seriously as a scientist and mathematician: 'You told me that you had read Laplace's *Système du Monde* and that you did not wish to read my short work on astronomy. I said that I thought that you could not understand the one without the other. I assume that it was this suggestion that caused your anger. For this, I apologize', Lalande wrote to young Germain on 4 November 1797 (Delambre et al. 1879, 635).[17]

Thus, in lieu of the literary image of the 'learned woman', whose obsession with learning is superficial and pretentious, Germain wanted to constitute herself as an independent mathematician in conversation and collaboration with her contemporaries. Germain's fear of being associated with the image of the 'learned woman' sits uncomfortably, however, with her identification as a scientist amateur, which we have already read in her first letter to Gauss, but which also reappears in the 'revelation' letter. Beyond merely explaining her use of a male pseudonym, Germain expressed her deep concern for Gauss's safety, tying it to her admiration for his significant contributions to mathematics and her gratitude for his encouragement of her own mathematical pursuits. This sentiment is captured in her letter below, dated 20 February 1807:

> The gratitude I owe you for the encouragement you have given me, by showing me that you count me among the amateurs of the sublime arithmetic whose mysteries you have unrevealed, was for me a particular reason for seeking news from you in a moment when the troubles of war could inspire some fears, and I have learned with real satisfaction that you have remained at home as undisturbed as circumstances could permit (Boncompagni 1880, 13).

It is important to note, however, that the subject position of the amateur—an independent scholar working outside the confines of institutions—held different connotations compared to both the degraded notion of 'the amateur' as we understand it today and 'the ridicule attached to the learned woman', at least in Germain's perception. As Gianna Pomata has pointed out, non-institutionalised scholarship was common in the ancient regime in France, 'the rule rather than the exception' (2013, 197). The figure of the 'amateur', as distinct from the academic scientific persona, emerged in the nineteenth century in the process of professionalisation in the sciences that included the organisation of national and international conferences, the foundation of scientific journals, as well as the creation of disciplinary benchmarks. Germain could thus identify herself as an independent scholar—an amateur of her times—but not as a 'learned woman'.

Having revealed her gender identity to Gauss, Germain was also keen to emphasise her eagerness to continue with her mathematical studies, since his work was motivating: 'I have an appetite for analysis that I developed after reading your work, and which inspired me and gave me the confidence to send you my feeble attempts' (Boncompagni 1880, 14), she wrote in the same letter cited above. She also expressed her hope that their correspondence would continue despite the revelation, asked for his response as an acknowledgement, and enclosed an addendum of four pages with mathematical proofs (15):

> I hope that the information, that I have today confided to you, will not deprive me of the honour that you have granted me under a borrowed name, and that you will not disdain to devote a few moments to send me your news. Believe Monsieur, the interest I attach to it and receive the assurance of sincere admiration with which I have the honour to be,
>
> Your very humble servant,
>
> Sophie Germain

Germain need not have worried, as Gauss's response was warm and enthusiastic: 'The scholarly notes with which all your letters are so richly filled have given me immense pleasure. I have studied them attentively and greatly admire the ease with which you master all branches of arithmetic, as well as the wisdom with which you generalize and perfect' (Boncompagni 1879, 2).

Gauss went on to express his gratitude for her 'precious friendship' and his admiration for her 'noble courage, extraordinary talents, and superior genius' (2), acknowledging the challenges and prejudices she faced as a woman. Following this warm and enthusiastic introduction, Gauss continued his letter with three densely written pages of mathematical formulas. Historians of mathematics have since regarded this letter as highly significant due to the complex mathematical problems he communicated to Germain.[18]

But as already noted above, what has not been discussed in the literature around the Germain-Gauss correspondence is the effortless mingling of 'the personal and the scientific' in the letters they exchanged. This trait is recurrent in Germain's correspondence but has also been traced in the letters of all six women mathematicians, as I argue in this chapter. Such entanglements between 'the personal and the scientific' show that the epistolary conventions in scientific letters in general and mathematical correspondences in particular—such as formal structures, objective tones, clarity and precision, and rigorous logic amongst others—are bent when gender relations and scientific communications override. It also shows that the communicative, dialogic, and relational nature of letters is stronger and more visceral than the mere exchange and circulation of knowledge in the context of the usual peer review process that reaches our days. Germain's letter to Gauss dated 22 May 1809, beautifully encompasses this harmonious coexistence of 'the personal and the scientific', echoing Germain's overall philosophical approach to science, which was further elaborated in her treatise, as I will further discuss in Chapter 6:

> I do not pretend to fathom the profundity of your research. I sense that my intellect is far removed from yours, although our tasks are similar, since I, as you, have a great predilection for arithmetic problems. I find this part of science susceptible to a particular kind of elegance, which is not attained in the mathematical-physical sciences. It appears that in everything the interest of ideas is in inverse proportion to the usefulness they have in practice. This is not surprising when we consider that the human intellect, when working for its own satisfaction, should encounter the greatest intellectual beauties rather than when guided by an external motive (NS-UG/MS Gauss/Bref-Germain).

Germain's correspondence with Gauss ceased in 1809, after which she shifted her scientific focus to uncovering the mathematical laws underlying the physics of acoustics. This effort culminated in her unprecedented achievement of winning a prestigious mathematics prize from the French Academy of Sciences in 1816, as noted in the Introduction. However, on the same day she received this award, the Academy announced Fermat's Last Theorem as the topic for the 1818 mathematics contest, rekindling Germain's interest in number theory. It was in this backdrop that she resumed her correspondence with Gauss after a ten-year hiatus, writing to him on 12 May 1819, to introduce her work on proving Fermat's Last Theorem with the words, 'Here is what I have found' (NS-UG/MS Gauss/Bref-Germain). In the same letter, she confided to Gauss that number theory had always been her preferred area of study.

> Although I have worked for some time on the theory of vibrating surfaces (to which I would have much to add if I had the means to conduct the experiments I have envisioned concerning cylindrical surfaces), I have never stopped thinking about number theory. To give you an idea of my preoccupation with this type of research, I confess that even without any hope of success, I still prefer it to other work which might interest me while I think about it, and which is sure to yield results (NS-UG/MS Gauss/Bref-Germain).

We do not have Gauss's response, and it is unknown whether he ever replied to this letter or to Germain's final correspondence with him ten years later, on 28 March 1829. In that letter, she reiterated her disappointment over the missed opportunity to exchange mathematical ideas with him: 'Regardless of what I could learn from you, I still regret not being able to submit to your judgment many ideas that I have not published and that it would take too long to write down' (NS-UG/MS Gauss/Bref-Germain).

Unlike Gauss's intermittent and infrequent responses, Legendre became Germain's true mentor, tutor, friend, and eventually collaborator throughout her mathematical career and academic writing (see Musielak 2020). Their correspondence reveals that he was genuinely interested in her work and deeply respected her intellect and mathematical mind. His memoir, *Recherches sur quelques objets d'analyse indéterminée et particulièrement sur le théorème de Fermat*, presented to the Academy in September 1825, contains a footnote on page 17 stating that the proposition is due to Mlle Germain.[19] Legendre's reference to what is now known as 'Germain's theorem' (see Laubenbacher and Pengelley 2010) was a significant acknowledgement of her contributions to number theory. Despite his unwavering support for Germain's work, Legendre was also a rigorous critic, offering both harsh and constructive feedback during her long struggle to win the prestigious mathematical prize, as I will further discuss in Chapter 4. Although it is not known when exactly Germain discussed Fermat's Last Theorem with Legendre, their correspondence shows that she wrote to him on number theory in late 1819. Legendre's epistolary response on 31 December 1819 was rather discouraging:

> I must inform you that since I first mentioned this method of research to you, my initial belief in its potential for success has greatly diminished, and I now think it will be as unproductive as many others. Therefore, it would be wise for you not to spend any more time on it, to avoid wasting time that could be much more usefully employed on other research (Germain 1896, 311).

Legendre's reservations did not stop him from referencing her work in his own memoir and book, as we have seen above. Germain knew the difficulties she was grappling with, and despite her renewed focus on number theory, she also began preparing her publications from her work on elastic theory. Her first publication appeared in 1821 and was self-funded, but still gave her honour and established her as an author. During this period, she developed a friendship with Joseph Fourier, and it is in their correspondence that we further follow tracks and traces of the long and difficult process of being established as a published mathematician and scientist. Their correspondence began in 1816, and although their relationship was more personal than professional, Fourier's position as a permanent secretary of the Academy of Sciences enabled Germain to become more integrated into the Parisian scientific community. In a letter sent on 30 May 1823, Fourier invited Germain to become a permanent attendee of the public meetings of the Institute, a privilege which was preserved only for the wives of the members:

> I have the honour of informing you that every time you wish to attend the public meetings of the Institute you will be admitted to one of the reserved seats in the centre of the hall. The Academy of Sciences wishes to demonstrate, by this distinction, all the interest that your mathematical works inspire, especially the scientific research that it has crowned through the award to you of one of its annual, grand prizes (Germain 1896, 323).[20]

As already mentioned in the previous chapters, Germain first met the Italian mathematician Guglielmo Libri in Paris in 1825, at one of the parties that the astronomer François Arago used to hold at the Observatory every Thursday evening. They became interested in each other's work, and Germain invited Libri to her home for lunch. This is how their friendship flourished, mostly through correspondence. 'I cannot express to you how much I regret having to leave Paris without having the honour of seeing you' (Henry 1879, 636),[21] Libri wrote from Florence on 17 November 1825, explaining that he had to leave Paris in haste due to his mother being unwell. In the same letter, he mentioned that he had sent her a paper that she was interested in: 'Mr. Arago, who was kind enough to spend a few days here, was kind enough to take it upon himself to give you, on my behalf, a copy of Riccati which you seemed to wish to possess: I hope it will help you remember a person who has the deepest feelings of esteem for you', Libri wrote (636). Their letters have become important in the appreciation of Germain's contribution to mathematical sciences, but they are also revealing of her personal feelings and thoughts. It was to Libri that she could talk openly about her frustration at being marginalised in the Parisian scientific circles, as in the following letter sent on 26 September 1826:

I am not surprised at your eagerness to renew conversations that cannot be found elsewhere than in Paris, all the doors are open to you. As I am not allowed to go to any sessions, I find myself almost as foreign to the movement of sciences as if I lived in another country. And yet, I prefer to be here even more than elsewhere because sometimes it happens to me to find by chance an opportunity to instruct myself (MBF/FPL/143).[22]

This intimate letter reveals indeed how women mathematicians felt knowing that it was only through letters that they could have a sense of belonging to the academic community, although they continuously felt the effects of exclusion. And yet they would stick to these rare epistolary opportunities to maintain some kind of belonging to the academic community and of course to advance their knowledge, as well as get feedback about their work, as the following letter to Augustin-Louis Cauchy, dated 18 July 1826, indicates: 'I have followed your judgment, by publishing. I would be happy if you would take the trouble of communicating your observations to me. There are in this small work three things that seem to me to be of some importance' (1896, 328).[23]

Germain wrote this letter to Cauchy after submitting to the Academy a memoir she published in 1826 to further elucidate her ideas on vibrating surfaces. When the Academy asked Cauchy to review this memoir, Germain thought that this would be an opportunity for her to get feedback from the scientific community, which would be otherwise difficult to obtain. As Musielak has commented, the context of this letter reveals Germain's uncomfortable position in the Parisian scientific world (2020, 105). First of all, she wrote a memoir and published it without the Academy's approval, but then submitted it to them for a review, following Cauchy's advice, as the letter indicates. Bucciarelli and Dworsky have further observed that Cauchy must have given her this strange advice in order to avoid the difficult and unpleasant task of having to reject it, a process that would be only natural with any male mathematician: 'they could not treat her as a full professional colleague, as they would any man, by simply rejecting the work. But how was she to know this? Once again, respect for her sex led to such disrespect for her person that she was allowed to appear ridiculous' (1980, 106–7).

Despite her efforts not to be identified with the ridiculed figure of 'the learned woman', Germain must have often felt that she was treated like one, but in her correspondence with Libri, she felt freer to express both her mathematical ideas, as well as her personal feelings and fears. Even when writing letters during the painful time of her illness, she would find a way to convey some new ideas to her friend and correspondent. Thus, in the first letter where she informed Libri of her illness, she mentioned that she had been suffering greatly and, as a result, was unable to study: 'I have lost the ability to bring together the most common ideas, I have felt that my intelligence has been destroyed' (MBF/FPL/f432).[24] However, in the same letter, she returned to a mathematical problem she had been contemplating before her illness: 'a short time before the start of all these problems I had undertaken to write on a subject [which] has been rolling around in my head for a long time, it is about average curvatures whose notion derives from the comparison between the curvature of the sphere and that of the surfaces' (f. 432).

Despite the fact that she was still suffering, she wrote to Libri about her struggle to clarify her ideas and invited him to visit her so that they might work together. But in a letter dated only a year later, on 18 April 1831, Germain confessed to her friend that her illness had advanced and had become unbearably painful:

> My health is in a frightful state. A prompt death would be a relief to me, because I suffer from unbelievable pain, which leaves me not a moment's rest. I wanted to read at least the third volume of De LaCroix, but I cannot. I remain shut up. I see neither M. Legendre nor my other friends, except for St. Amant, who is always concerned about you and my sister. I am told that my condition is not desperate, but I am warned of long suffering (Henry 1879, 636).[25]

This letter was written only two months before her death, and yet she did find the courage to inform her friend about the publication of their work: 'I have received from M. Crelle the issue that contains one of your memoirs and mine' (636), she wrote. In the same letter, she also conveyed to her friend news about the Parisian scientific community: 'There is definitely a fate hanging over all mathematicians—your unhappy preoccupation, that of Cauchy, the death of M. Fourier. Finally, that student Galois, who in spite of his impertinence displays a good disposition, has managed to be expelled from the Ecole Normale (636).

Germain's letters to her mentors, advisors, and friends in the European mathematical community are the only autobiographical documents we have from her, as she did not leave behind an autobiography, diary, or journal. These letters are therefore extremely valuable as 'documents of life' (Plummer 2001), elucidating the agonistic process of becoming a woman mathematician, which I will explore further in Chapter 4. Entanglements of personal and scientific components are crucial in analysing her correspondence and have indeed illuminated the obscured life of a nineteenth-century woman mathematician. As noted in the Introduction, despite their connections with the European mathematical community in general and the French academicians in particular, Germain and Somerville never met. Here lies an interesting twist: although Somerville wrote an autobiography that has significantly informed the biographical literature about her, it is within her scientific correspondence that her love for mathematics is most evident, as I will discuss next.

Excavating the archive of Mary Somerville's mathematical correspondence

When Mary Somerville's translation of Laplace's *Mécanique Céleste* was published in 1831, it received widespread acclaim not only in Britain but also across Western Europe. Somerville was celebrated as a highly skilled mathematician, as her work included a substantial preliminary section on French mathematical analysis. As she wrote in her autobiography: 'at this period mathematical science was at a low ebb in Britain; reverence for Newton had prevented men from adopting the "Calculus", which had enabled foreign mathematicians to carry astronomical and mechanical science to the highest perfection'[26] (Somerville 2001, 114). Somerville's inclusion of her 70-page 'Preliminary Dissertation' at the start of *Mechanism of the Heavens* elevated the work well beyond a mere translation, a point I will examine further in Chapter 3.

But how was it possible for an autodidact woman in science and mathematics to reach such a high level of competence to the point of Laplace saying that 'Mrs Somerville is the only woman who understands his work' (Somerville 2001, 182). Although the different versions of Somerville's autobiography have significantly influenced the literature on her life and work, they provide limited insight into her early mathematical endeavours, as Brigitte Stenhouse has astutely noted (2020, 2021). To address this gap, Stenhouse has focused on Somerville's mathematical correspondence with John and William Wallace, who provided guidance during her initial foray into mathematics. In Chapter 4, I will further discuss how they supported her in her first publications, which were solutions to mathematical problems raised in a mathematical journal edited by William Wallace, and how she subsequently won a prize in the overall *agonistic* process of engaging with mathematics.

While Somerville's correspondence with the Wallaces was formative in the beginning of her mathematical education, her correspondence with John Herschel is especially illuminating, particularly for the period she was preparing the translation of Laplace's *Mécanique Céleste*. As Stenhouse has commented, this correspondence 'shines a light on the questions Somerville herself grappled with in translating Laplace, and the solutions she found' (2021, 88). It was William

Wallace who first introduced Somerville to the Herschel family, and her lifelong friendship with John Herschel turned out to be catalytic in the long process of becoming a mathematician.

The Herschel Papers contain nineteen letters exchanged between Somerville and Herschel between March 1829 and June 1831, concerning the translation of Laplace's work. This correspondence, now digitised and available online, is preserved in the archives of the Royal Society in London.[27] Stenhouse has transcribed these letters and not only analysed them in her significant study of Somerville's mathematical contributions but also included them as an appendix in her 2021 PhD dissertation (2021, 81–202).

In her first letter to Herschel, dated 31 March 1829, Somerville is apologetic for disturbing his time, 'which would be so much more agreeably spent with your fair and charming Bride' (HS/16/327, f. 1). The letter was written shortly after Herschel's wedding earlier in March, and since the Somervilles had not yet met his wife, the botanical artist Margaret Brodie Stewart, the whole letter is both a request for academic help, as well as an invitation for social intermingling: 'It will be my pride to consider your wife as my friend and I trust to your giving us early notice of your return to London that you may spend a quiet day with us at Chelsea to make us acquainted' (f. 2). Indeed, when the two women met, they formed a warm friendship, their correspondence lasted for over forty years, and in 1844, Somerville became the godmother of the Herschels' daughter Matilda Rose (Stenhouse 2021, 102).

It is in this first letter that we understand that her work with Laplace's text was still in the early stages. Although Somerville had committed herself to this translation two years earlier, she seems ambivalent about its progress: 'I esteem it a very great mark of friendship that you should donate any of your time to my paper' (HS/16/327, f. 1), she writes, as she is still unsure about its development into a manuscript. Despite her hesitations, she acknowledges the value of the fair comments and criticisms she had already received from her recipient. By October, she refers to her work as her 'MMS', indicating progress and growing confidence on her part: 'If you will have the goodness to send me the last 8 or 10 sheets of my MMS together with the one I sent you the other day. I think I can both make an improvement and save you some trouble', she wrote in a short note on 20 October 1829 (HS/16/328, f. 1).

Herschel's tone in this correspondence highlights Somerville's status as an equal, rather than a 'pupil' or 'apprentice'. In an undated letter, during the same year, he shares updates on his own research and writing, noting that the work turned out to be more challenging than anticipated: 'I have at lengths sent an end to a work which (though not much in itself) yet gave me a great deal of trouble, much more so than when I entered on it I could have imagined possible' (OBL/SC/Dep. c. 370/F.MSH-3, f. 310)[28]. This undated letter makes clear that they are in regular communication about their work, with Herschel continuing to send Somerville his latest publications: 'It is an essay on *sound,* as a companion to the work on Light of which you have a copy, as I mean you shall of its fellow—as soon as it appears' (f. 310), he adds. He explains that his own research has delayed his review of her writing, but assures her, 'I shall now be able to bestow my undivided attention on your work, which hitherto I have found impossible to do though I have often begun it and as often been forced to desist' (f. 310). Confident that Somerville will understand the delay, he notes, 'To you, who knows what it is to *write* such a work, I need offer little excuse for the delay' (f. 310). The letter also reveals the candid nature of their scientific relationship, as Herschel admits, 'I *cannot* pretend to read the entire work—there will be parts, very large ones, on which I cannot give an opinion' (f. 310).

As the Somervilles were friends with not only John Herschel and his wife, Margaret, but also with his parents, astronomer William Herschel and Mary Baldwin, the letter concludes with some family news. John updates Somerville on his mother's health, noting that her 'constitution is wonderfully vigorous, and though it has endured a fearful struggle, I still trust that much of her strength remains unsubdued' (f. 310). The personal and the scientific are effortlessly intertwined in their correspondence, whether it is a letter from Somerville or Herschel, throughout their epistolary communication.

Despite his own research commitments and writing, Herschel kept his promise, becoming closely involved in reviewing and even proofreading Somerville's manuscript. 'I send you for Mr Richards the first 40 pages of your MS' (HS/16/329, f. 1), he wrote to her on 23 February 1830. In addition to making 'some pencil notes which you can easily rub out' (f. 1) in the manuscript, he also offered broader comments on the work, including advice on tailoring it to her intended audience:

> Allow me too to observe that you might take for granted that your reader understand a good deal of algebra, & geometry, &c. In Consequence, were I to advise, I should dispense with all those papers where very elementary abstract properties, and very common methods are stated, such as the passages I have marked in pages 21, 17, 19, &c. (f. 3).

The letter concludes with some editing notes from his own experience as a scientific writer: 'I think you would find a regular system of numbering the paragraphs (each paragraph to be numbered as it begins on a fresh line in a page) extremely convenient for reference. I find such an aid invaluable' (f. 4). He acknowledges that he may come across as 'a rough critic' but emphasises that his intention is to shield her from 'some sneering reviewers' (f.4) in the future. The letter concludes on a personal note, with his wife sending 'her kind regards' (f. 4).

Somerville's response to the February letter was swift and appreciative: 'I beg you will accept of my very sincere thanks for your criticism which is as just as it is friendly' (HS/16/330, f. 1), she wrote on 1 March 1830. Just a week later, on March 6, she wrote again, resending him the first few pages of her manuscript, as she wanted him to have a final look at her revisions:

> I really should feel uneasy without your opinion as to the changes I have made. I think it is now more consistent, and regular, perhaps I have omitted some things that ought to have been inserted, and probably have not

> sufficiently explained the more obscure passages—others I may have misapprehended. At all events your opinion would be invaluable and I shall write it again with pleasure should you think it necessary (HS/16/331, ff. 1–2).

Somerville's response shows that she has critically engaged with the suggested revisions, adapting them as it seems to her own overall idea about how her work should look, but still needs her critical reader's approval. It is in the same letter that she writes that she now considers her work completed, but the publication process would depend on Herschel's final opinion about the work: 'I have finished the moon and satellites which completes what I intended, but whether the whole or any part ever goes to the press will depend on your advice which I know will be given with the truth of a friend and for which I never can be sufficiently thankful' (f. 2).

In response, Herschel sent her a translation of the *Mécanique Céleste* 'with a running comment by (apparently) an able hand' (HS/16/332, f. 1) on 9 March 1830. Having just received this work from the author, he promptly forwarded it to Somerville, hoping it would prove useful for her to consult at such an early stage. Although he praises her as 'indefatigable' (f. 1) and believes 'the retouching you have given it has been most effective' (f. 1), he still expressed reservations about the mathematical work, which he detailed further in his letter (ff. 1–4). Somerville's response is appreciative, but she clearly states that her approach is different and that she is eager to see her work go to press sooner rather than later:

> I have gone through the commentary as far as the time has permitted and excellent as the notes are, I confess I am not dismayed as I rather wish to state principles clearly, and to arrive at the results by as easy methods as possible, than to enter into all the mathematical detail. I daresay you think me very bold, but I do feel inclined to proceed and to get it into the press as soon as possible (HS/16/333, f. 1, undated).

In a letter dated 31 March 1830, Herschel mentioned that he had reviewed the latter part of Somerville's manuscript and was pleased to report that, 'with the exception of one or two slight observations (which I have noted), I don't think it can be materially improved' (HS/16/334, f. 1). At the end of this brief letter, he also shared the news of the birth of his first daughter. Somerville's prompt response on 5 April 1830 focused on congratulating the new parents, while also expressing her appreciation for his attention to her work during such a significant and life-changing event: 'I trust they are both gaining strength that your new honour of being a Father may have no allay. Your thinking of my affairs at such a moment is a mark of friendship truly valuable and gratifying' (HS/16/335, ff. 1–2).

There was no further communication between them for the rest of April, reflecting Somerville's respect for Herschel's responsibilities as a new father. However, on 15 May, she wrote to him again, this time sending the entire manuscript for a final review before it went to press, noting, 'I dare not print without knowing that it does not contain any great blunders' (HS/16/336, f. 1). She reassured him that she didn't expect him to read the entire manuscript and had marked the passages where she was uncertain to save him time, adding that 'the introduction is altogether new' (f. 2). Even amidst the excitement of preparing her first publication, Somerville remained eager to see Herschel's new work: 'I cannot let you forget your promise of giving me your two new works, I value those I have got so highly that I look forward with real pleasure to every thing that comes from your pen' (f. 2). The letter, of course, concludes with warm greetings for the baby and the new mother.

It took a month for Herschel to respond to her letter, and by way of apology, he suggested returning her manuscript in person during a visit to her family in London. What stands out in this letter is Herschel's humble acknowledgement of his own limitations in certain areas of science and mathematics related to her work: 'The truth is I had hopes to have found leisure to have read more deeply on some of the subjects treated [...], of which I never possessed more than a superficial knowledge so as to become better qualified to criticize what you have said respecting them' (HS/16/337, f. 1).

Despite these limitations, he remained thorough in his final review: 'Where I *have* seen my way clearly I have not spared remark and even where I have *thought* I perceived room for objection I have considered it my duty to point it out for your own judgement' (f. 2, emphasis in the text). These exchanges reveal a clear pedagogy of epistolary scientific practice: candid opinions shared openly, the admission of knowledge gaps, and thoughtful discussions on debatable topics between the correspondents.

Due to Herschel's ongoing critique, the route to the press was delayed for one more year with a flurry of letters exchanged between the two parties until the very last moment. Even when the book was practically in press, Somerville sent Herschel her important preliminary dissertation for review. On 28 May 1831, she wrote: 'I am tempted by your constant kindness to me to submit to you a few preliminary observations which are to be prefixed to my book now ready for publication' (HS/16/344, f. 1), asking him to 'mark on the blank pages any thing that strikes you without which I should be unwilling to let it appear before the public' (f. 1).

Even in his last letter sent on 11 June 1831, Herschel was meticulous in his review, and their correspondence regarding her first publication happily concludes with a family announcement: 'You will be happy to hear that Mrs H on Sunday presented me with a fine girl & is doing perfectly well' (HS/16/345, f. 2). The cycle of this particular correspondence was thus marked by a book, one wedding, and two births, effortlessly encompassing 'the personal and the scientific'.

Stenhouse (2021, 101–14) has thoroughly examined the mathematical work exchanged in the nineteen extant letters between March 1829 and June 1831. Her study is significant in highlighting Somerville's contribution to the history of the reform of British mathematics. What I find particularly interesting in this intense exchange of letters during this critical period is how Somerville emerges as a mathematician through a relational process, which can be traced in her scientific correspondence with an eminent Cambridge scholar. She is acutely aware of the gravity and magnitude of the work she has undertaken, and while she acknowledges the challenges and deeply appreciates her critical reader, she remains steadfast in her vision for the work and continually repositions herself in relation to her correspondent.

What is also important to highlight is that while Somerville primarily addresses Herschel as a friend in their correspondence about her work, his expertise in analytical mathematics and physical astronomy made his final opinion crucial to the publication process, which turned out to be less smooth than initially expected. When Henry Brougham first asked her to translate Laplace's work for the book series of the *Society for the Diffusion of Useful Knowledge*, Somerville accepted with some reservations as she feared that the *Mécanisme Céleste* could never be 'popularized' (see Patterson 1983, 50). Indeed, when the manuscript was finished, Brougham clearly saw that both the length of the treatise and its superior quality were not suitable for the intended publication series of bringing 'sound literature and self-improvement within the reach of all' (50).

This decision was not a disaster, however, but rather a new beginning for the book and Somerville herself. As I will further discuss in Chapter 3, Somerville's excellent treatise was a living example of what a reformed science culture in Britain could achieve. It was therefore thanks to Herschel's enthusiastic endorsement that the publisher John Murray agreed to publish the work in August 1830 under favourable conditions (Patterson 1983, 75). As Patterson has further commented, both Herschel and Babbage were evidently determined to ensure that Laplace's translation reached English readers (73).

Although Somerville had read Babbage's work, it was during the preparation of Laplace's translation that they started exchanging letters. There are twenty letters from Somerville in the Charles Babbage correspondence, housed at the British Library.[29] Their first extant letter, dated 16 June 1830, features Somerville requesting Babbage to lend her 'Mr Plana's theory of the moon' for a few days (MS 37185, f. 231). Babbage, along with Herschel, was consulted about the mathematics of Laplace's translation: 'I am in a difficulty and I would be truly obliged to you if you would look over the sheet I have sent from No 555, page 312 to No 558, page 314. I fear there is an error', she wrote to him on 26 February 1831 (MS 37189, f. 307). She further asked him to 'give me your opinion as to No 556' and most importantly, she requested a prompt reply: 'Do you think you can bring the answers tomorrow at 6' (f. 307)? Somerville's urgency stemmed from the fact that February 1831 was a critical period, with her book about to go to press, and she felt uncertain about the equations governing the motion of the planets, as Stenhouse has noted (2021, 105). Babbage was also consulted about both the preliminary dissertation and the title of the book, since the decision on the title proved challenging. However, there is no evidence that he contributed to the final choice (see Stenhouse 2021, 69).

The two mathematicians gradually became well acquainted, and by 1834, Somerville was frequently invited to Babbage's famous soirées at his Dorset Street home in Marylebone, where friends and acquaintances gathered to discuss science and view his experimental machines, as I will further discuss in the next section. Their correspondence reveals that they regularly exchanged ideas on mathematical work and publications: ‘I understand that Professor Powell of Baden has succeeded in bringing the dependence of light under the law of the undulatory theory; can you give me some account of it, or can you tell me if he has published anything on the subject and where’, she wrote to him on 6 November 1835 (MS 38189, f. 189). Babbage’s scientific library was a constant source for Somerville’s readings and literature research: ‘Will you have the kindness to lend me the Calculus of Functions by Sir John Herschel & you’ (f. 390).

However, their connection extended beyond science and mathematics. When her family faced financial difficulties in 1837, Babbage supported Somerville’s request for a £100 increase to her annual pension, which enabled them to live relatively comfortably in Italy, where they eventually relocated (see Patterson 1983, 160–1). In a letter of appreciation, dated 6 June 1837, she wrote: ‘I am proud of such a mark of your estimation, but ten times more delighted with your kindness and friendship’, signing off the letter as ‘your affect friend’ (MS 37190, ff. 204–5). Their correspondence became less frequent after the Somervilles moved to Italy, though she still found ways to share social and scientific news with him. ‘Let me introduce Professor Corridi of Pisa to your acquaintance’ (MS 37191, 369), she wrote from Florence on 2 May 1840. Corridi was being sent to England and France by the Grand Duke Leopold II, who was planning the establishment of the *Istituto Tecnico* in Florence,[30] with Corridi as its first director. Somerville was confident that the Italian mathematician would be ‘much pleased’ (f. 204) to meet Babbage and Herschel while in London, hence the epistolary introduction.

Throughout her long life, Somerville corresponded with leading mathematicians and scientists across Britain, Europe, and North America, some of whom she included in her autobiography to demonstrate her reputation and esteem. Her papers at the Bodleian Libraries feature a section titled ‘Letters from Celebrated People’,[31] which includes her correspondence with Lady Byron and Ada Lovelace. The exchange of letters between Somerville and Lovelace is particularly notable as it represents a unique case of two women mathematicians communicating within this feminist automathography, a topic I will explore further in the next section.

Ada Lovelace: ‘ever yours mathematically’

> I shall be very grateful if you will be kind enough the first time you have a few spare moments, to write me a letter about rainbows. I am very much interested on the subject just now, but I cannot make out one thing at all, viz: why a rainbow always appears to the spectator to be an arc of a circle. Why is it a curve at all, and why a circle rather than any other curve? I believe I clearly understand how it is that the colours are separated, and the different angles which the different colours must make with the original incident ray. I am not sure that I entirely understand the secondary rainbow—Is the spectator’s eye supposed to be in the centre of the circle of which the arc of the rainbow forms a portion?—Have you read Mrs Somerville’s new book, and what do you think of it? (LB 171, f. 127)[32]

On 15 March 1834, the young Ada Byron wrote a letter to her tutor, William Frend, with questions about rainbows. At the end of the letter, she mentioned Mary Somerville's 'new book', *The Connexion of the Physical Sciences*, which had just been published in 1834. Somerville, a highly regarded mathematician, was undoubtedly someone for the 19-year-old Lovelace to look up to and admire. Frend, recognising this, encouraged her to meet and develop a friendship with Somerville, whom he regarded as an ideal role model, 'not more distinguished by her scientific attainments than by her amiable qualities' (LB 71, f. 193). This indeed came to pass later that same month. Lovelace became Somerville's protégée, and the two embarked on a lifelong friendship and correspondence. 'Ada was much attached to me, and often came to stay with me. [...] She always wrote to me for an explanation when she met with any difficulty. Among my papers I lately found many of her notes, asking mathematical questions', Somerville wrote in her *Personal Recollections* (2001, 150).

Having been introduced to Somerville in February 1834, Lovelace wrote a letter on 19 March 1834, setting a date for them at Charles Babbage's famous soirées, where he invited friends and acquaintances to discuss science but also view his experimental machines, as already mentioned in the previous section. Her correspondence with Somerville often revolves around Babbage's gatherings, but also blends 'the personal and the scientific':

> My dear Mrs Somerville,
>
> I hope I may have the pleasure of meeting you on Saturday Evening at Mr Babbage's. I am going with Miss Montgomery to dine at Mrs Murchison's, and we shall be at Mr B [Babbage]'s for a short time in the evening. I have not seen him yet, but he sent a very respectful message through Mrs Murchison, inviting Miss Montgomery and me to his Saturday parties. I am afraid this may be my only chance of seeing you for the next three weeks. Mama intends going to Brighton for a fortnight on Tuesday and it is very doubtful whether I can get to Chelsea any day before that. Hoping therefore to see you, believe me,
>
> Most sincerely yours,
> A. Ada Byron
> (MSC, c.367, MSBY-2, f.40)

The first dinner party at Babbage's house in March 1834 was the beginning of many that Lovelace would attend, sometimes accompanied by her mother, but almost always in the company of Somerville. In the Victorian era, it was unthinkable for a young woman to go out in society without a chaperone. Lovelace's letters to Somerville often include arrangements for meeting and travelling together, as in the letter below dated 26 November 1834.

> My dear Mrs Somerville,
>
> I am very much obliged to you for your kindness, and shall be most happy to spend a few hours so delightfully on Friday. I shall be at No 10 Wimpole St by 1/2 past 11 o'clock, and as this is not out of the way from Chelsea to

Mr Babbage's I should be obliged to you if you would be so good as to call at the door as you pass, and then our carriages can go together, as I should wish to arrive at the same time with you. I shall therefore wait in Wimpole St till I am summoned.

Mama has never yet ventured out of the house, but she continues on the whole better, and hopes soon to have the pleasure of calling on you.

With my kind regards to your daughters, believe me,

Yours most sincerely obliged
A. Ada Byron
(MSC, c367, MSBY-2, f.45)

In addition to attending Babbage's dinner parties, Lovelace also went to a series of lectures on 'the Difference Engine', presented by Dionysius Lardner at the Mechanics Institute, this time accompanied by Mary King, her tutor's wife. When Lardner's book *The Economy of Machinery and Manufactures* was published in 1835, Lovelace was among its first readers. She was deeply fascinated by Babbage's inventions, writing to Somerville on 8 July 1834: 'I am afraid that when a machine, or a lecture, or anything of the kind, comes in my way, I have no regard for time, space, or any other ordinary obstacles' (MSC, C367, MSBY-2, f.38).

Additionally, Lovelace's relationship with Somerville had awakened her own inner teacher. While touring the Midlands to observe the new factories emerging across England during the Industrial Revolution, she visited her mother's friend Lady Gosford and her two daughters, Annabella and Olivia (Livi) Acheson, in Buxton, Derbyshire. In a letter to her teacher, William King, dated 1 September 1834, Lovelace described her new friends as 'amiable young women, with good natural abilities, but somewhat inclined toward indolence' (LB 172, f. 135). She thus decided 'to excite and rouse them to various objects of study and interest, and as much as possible to make my little talents, such as they are, of use to my young friends' (f. 135), and she particularly decided 'to induce one of them to take up mathematics'(f. 135). She soon found out that in teaching them 'I myself gain more perhaps than they do' (f. 136). Despite her newly found interests, Lovelace's fascination with Babbage's machines remained unwavering: 'This Machinery reminds me of Babbage and his gem of all mechanism' (f. 139), she wrote in the same letter cited above, recommending a recent article on Babbage's work in the latest issue of the *Edinburgh Review*.

Interestingly, upon returning to her home in Fordhook after the Midlands tour, Lovelace not only resumed her mathematical studies with renewed enthusiasm but also began a mathematical correspondence with her self-selected 'pupil', Annabella Acheson. 'So this you see is the commencement of "A Sentimental Mathematical Correspondence between two Young Ladies of Rank" to be hereinafter published no doubt for the edification of womankind' (LB 168, f. 23), she wrote to Annabella on 10 November 1834, ending her letter with the epistolary salutation 'Ever yours, mathematically' (f. 23).

Lovelace's relationship with Somerville flourished throughout the summer and autumn of 1834, and during her visits to Babbage's dinner parties, she had the opportunity to follow some notable and intellectually intense conversations between the two mathematicians. Seeing a woman debating important questions in science and mathematics with a man must have had a lasting intellectual and emotional effect upon a young aspiring woman. Eager to deepen her understanding, she borrowed the plans for Babbage's Engine and explored his notes and papers through the guidance of her mentor and teacher. In a letter dated 8 November 1834, she made the following request:

> May I trouble you the first time you see either Mr Babbage, or his son, to say how exceedingly obliged I am to the latter for his unexpected kindness in sending me the plates & account of the Machine, which is exactly what I was in want of; and is a very great help to me. I am very busy copying the Steam Engine paper I carried off from Mr Babbage's the other day. I have finished two and expect to complete the 3rd in a few days. I hope he is not in want of them and would have no scruples in claiming them immediately! (MSC, c367, MSBY-2, ff. 42–43)

But in February 1835, Somerville wrote a worrying letter to Lovelace's mother, wondering whether their lessons should stop, since Lovelace had shown signs of extreme tiredness and agitation during her last visit. Lovelace wrote back on 20 February 1835, admitting that she had felt 'shattered' and 'weak' when leaving her friend's house, to the point that 'at this moment [I] can hardly hold my pen from the shaking of my hand' (f.31). Still she could not 'complain of being what people call ill' (f.31) and was adamant that their lessons continue, insisting that 'in a few weeks I dare say I shall be quite strong (particularly if I see a good deal of you)' (f.31). Somerville must have been deeply concerned with her protégée's physical state, writing to her that 'riding on the downs and even in the riding school conduces more to health than reading mathematics' (LB 174, f. 23). Riding was indeed what Lovelace did in recovering from this bout of overexertion, as the following letter from Brighton, where her mother had taken her to recover in April, recounts:

> I am much stronger. I have been taking what has always been to me the finest of all medicines—horse exercise [...] I generally ride in the riding school everyday, and—best of all—leap to my heart's content. I assure you I think there is no pleasure in way of exercise equal to that of feeling one's horse flying under one. It is even better than waltzing (MSC, c.367, MSBY-2, ff. 35–36).

Lovelace's relationship with Somerville went beyond the realm of mathematics, strengthened by their shared interests and frequent visits. As their correspondence reveals, Somerville often invited Lovelace to her home, where they enjoyed activities such as attending concerts, a delight for Lovelace due to her deep passion for music. These personal exchanges helped to solidify their bond, combining intellectual engagement with a growing personal connection.

Lovelace's marriage to King, as well as the three children that followed soon after, temporarily disrupted her mathematical career, but not her perseverance in studying mathematics. During the early years after her marriage, Lovelace often turned to Somerville for help and advice, trying to reassure her mentor and perhaps herself that her marriage was not going to dampen her desire or tame her passion for mathematics. Her first letter to her tutor was penned on 1 November 1835, when Lovelace already knew that she was pregnant with her first child:

> My Dear Mrs Somerville,
>
> You must not think that I have forgotten you, although it is so long since I have written to you.—I am ashamed to think how long. We only came home last week.
>
> I now write, partly to ask news, partly to give news, principally and, to remind you of your promises to visit us [...]
>
> I now read Mathematics every day, and am occupied on Trigonometry and in preliminaries to Cubic and Biquadratic Equations. So you see that matrimony has by no means lessened my taste for those pursuits, nor my determination to carry them on, although it has necessarily diminished the time I have at command. But I suspect it is no bad thing to be limited in that respect.
>
> By the bye I have a mathematical question to ask you, which I hope is not too trifling to be beneath your notice. [...]
>
> I am sorry to find that your copy of Legendre's Geometry is still in my possession. I thought I had returned it and my only excuse is my marriage.
>
> (f.54)

This reassuring first letter was soon followed by more specific mathematical questions in a letter written only two weeks later: 'I have another trigonometrical question to ask you, and am encouraged by your kindness to trouble you with these things, which is a shame too, when you are so busy ...' (f.53).

Somerville replied on 28 November 1835, apologising for the delay due to her busy schedule. She provided Lovelace with detailed answers to her mathematical questions while also offering reassurance about her uncertainties. 'When you are more in the habit of using sines and cosines you will readily make out all the transformations but till that time I shall have great pleasure in giving you what aid I can' (LB 174, 31). She also expressed her joy at her friend's recovery from the discomfort of early pregnancy: 'I rejoice to hear you go on so well, indeed I never saw you so strong and I have no doubt that your health will now be perfectly established' (ff. 31–2).

The two women's correspondence continued through the very last days before Lovelace's confinement, and on 10 April 1836, she felt at ease to share with her tutor and friend the way she imagined her yet unborn child: 'How much I should like to have a mathematical child, and only think what pleasure I should have in teaching it, and how capable I might hope to be too by the time it was old enough' (MSC, c.367, MSBY-2, f. 71). After the Somervilles relocated to Italy, the correspondence between the two women became less frequent. However, Somerville continued to follow Lovelace's progress and sent a warm congratulatory letter following the publication of the work that solidified Lovelace's reputation as a mathematician—her translation and extensive notes on Luigi Menabrea's *Notions sur la Machine Analytique de M. Charles Babbage* (Lovelace 1843). This is what she wrote from Rome on 5 February 1844:

> A thousand thanks my dear Lady Lovelace for your kindness in sending me a copy of your publication on Babbage's machine the whole of which but especially the notes I have read with the greatest interest and most sincerely congratulate you on the proficiency you have made in the highest branches of mathematics and the clearness with which you have illustrated a very difficult subject (LB 174, f. 95).

Somerville's departure to Italy created a gap in Lovelace's mathematical education, which would take a new direction when Augustus De Morgan agreed to become her tutor, as we will explore in Chapter 4. However, her relationship with Babbage remained steady and uninterrupted. As we have already seen, Lovelace started frequenting Babbage's soirées around 1834 and she was soon immersed in what she considered the ingenuity of his mathematical ideas, as well as their applications in his Analytical and Difference Engines. It also seems that from the very beginning, Lovelace followed Babbage's guidance and advice in relation to her mathematics education: 'I have got on delightfully with Lardner's Trigonometry, which so far as I have yet gone I think *excellent*' (MS 37189, f. 281),[33] she wrote from Oakham on 18 January 1836, while also commenting that 'his road seems at all events to be the one suited to my mind' (f. 281). She did not hide the fact that her study was not easy, since 'I have had quantities of formulae to work out myself, & have destroyed a great deal of paper' (f. 281).

Their correspondence became infrequent in the years following her marriage and three pregnancies. Nevertheless, it appears that their intellectual relationship continued, despite her family obligations and health challenges during this time. 'I have just received amongst a number of packages from Town, the new edition of your work which you have so kindly sent me; & I will not delay another day thanking you for it, as I am much gratified by the kind recollection it shows' (MS 37190, f. 386), she wrote on 2 March 1838 from Oakham upon receiving the second edition of Babbage's *The Ninth Bridgewater Treatise*. She added that her health was improving after a 'very tedious & suffering illness which has occupied so many months, since a Miss *King* has been added to our family' (f. 386). Her letters frequently included invitations to their estate at Oakham: 'Dear Mr Babbage. Sat^dy^ next will suit us perfectly, but we hope you will stay on as far into the following week, as possible. Surely the machine allows you a holiday sometimes' (f. 386). Babbage was often invited to join them, sometimes in the company of her mother and the Somervilles. There were also cultural events to attend together whenever Lovelace spent time in London: 'My Dear Mr Babbage. What say you to taking a

Stall for the Italian Opera on Thurs[dy]' (MS 37191, f. 691), she wrote on a Tuesday night, in 1840 or 1841, inviting him as 'a fourth' in an opera night with two other women friends. In addition to exchanging ideas about readings and scientific news, and coordinating social and cultural visits, Lovelace's letters to Babbage would sometimes delve into metaphysical and existential reflections, as in the following letter dated 16 February 1840:

> You know I am by nature a bit of a philosopher, & a very great speculator,—so that I look on through a very immeasurable vista, and though I see nothing but vague & cloudy uncertainty in the foreground of our being, yet I fancy I discern a very bright light a good way further on, and this makes me care much less about the cloudiness & indistinctness which is near.—Am I too imaginative for you? I think not (MS 37192, ff. 333–4).

From their correspondence in late 1839, it appears that Lovelace initially sought Babbage's assistance in finding a tutor. However, it was through her family connections, rather than Babbage, that she ultimately found the right mentor to advance her mathematical knowledge. By the time she was well into her studies under De Morgan's tutelage, Lovelace continued to share her progress and enthusiasm for mathematics with Babbage. In a letter dated 5 January 1841, she wrote: 'Today, I have been working much at Mathematics. It has been bad for outdoors, & therefore I have got a *lift* at *in-doors* pursuits—I must show you a certain book called my Mathematical Scrap-Book…' (ff. 532–3). As her mathematical education progressed, she felt more confident in asking Babbage more detailed questions about his engines: 'Some day or other, you will have to put me in possession of the main points relating to your engine. I have more reasons than one for desiring this' (f. 533), she wrote in the same letter above. It appears that during this time the idea of deeply engaging with Babbage's work began to take shape in her mind, as suggested by the following epistolary exchange on 12 January 1841:

> I am very anxious to talk to you. I will give you a hint on *what*. It strikes me that at some future time, (it might be even within 3 or 4 years, or it might be *many* years hence), *my head* may be made by you subservient to some of *your* purposes & plans. If so, *if* ever I could be worthy or capable of being *used* by you, my head will be yours. And it is on this that I wish to speak most seriously to you. You have always been a kind and real & most invaluable friend *to me*; & I would that I could in any way repay it, though I scarcely dare so exalt myself as to hope however humbly, that I can ever be intellectually worthy to attempt serving *you* (ff. 543–4).

This initial idea came to fruition less than three years later, just as the letter had foreshadowed. Their correspondence indeed reveals Lovelace's dedication to the purpose and strategies of her mathematical studies with De Morgan, which revolved, however, around Babbage's work: 'I have been at work very strenuously since I saw you, & quite as successfully as heretofore' (ff. 566–7), she wrote on 22 February 1841, emphasising the connection between her studies and his work: 'I am now studying attentively the Finite Differences ... And in this I have more particular interest, because I know it bears directly on some of your business' (f. 567). In the same letter, she revisits the idea of 'serving his work', but this time it is reformulated as a collaborative effort, which she describes as 'a happy instinct' in her mind:

> I think much of the possible (I believe I may say the *probable)* future connexion between *us;* and it is an anticipation I increasingly like to dwell on. I think great good may be the result to *both* of us; and I suspect that the idea, (which by the bye is one that I believe I have *long* entertained, in a vague and crude form), was one of those happy instincts which do occur to one sometimes so unaccountably & fortunately. At least, in my opinion, the results *may* ultimately prove it such (ff. 567–8).

Lovelace was not only eager to engage with Babbage's work but also keen on expanding his circle of supporters: 'My Dear Babbage. Will you do me the favour of showing your Calculating Machine to an old friend of mine, who is on leave of absence from India, & is to return there immediately' (f. 75), she wrote on 2 May 1842. Although the publication of Menabrea's translation created ruptures in their intellectual relationship, Babbage and Lovelace remained friends to the point of appointing him executor of her will, as we have already seen in Chapter 1. Overall, 'the personal and the scientific' are effortlessly entangled throughout their correspondence, forming a consistent theme that links all the cases discussed in this chapter and leads us into Kovalevskaya's correspondence in the next section.

Deine Sonia: burned letters and other stories

> Dear Madam,
>
> Last evening, to my regret, I was prevented from visiting you, and at the same time, due to an accident, it was impossible for me to inform you of this. However, I hope that you will honour me with your visit tomorrow (IML/KW-SK/L1).[34]

On 11 March 1871, the renowned mathematician Karl Weierstrass sent a brief letter to Sofia Kovalevskaya, whose doctoral studies he had agreed to supervise in private. As already mentioned in previous chapters, at the time, it was impossible for her—or any woman—to formally enrol as a doctoral student at Berlin University, where Weierstrass was a professor, despite his efforts. A letter penned later that year, on 6 November 1871, shows that Weierstrass was at least successful in securing her access to the Library: 'From the enclosed letter from Professor Koner, you will see, dear Madam, that you have full access to use the University Library. Should any lower official give you any trouble again, you only need to refer to this letter' (L3). Despite the challenges of accessing the university library without being an enrolled student, Weierstrass went out of his way to support

Kovalevskaya's studies. In a letter dated 14 November 1871, he reassured her, writing, 'I think it is better if I write to [the librarian] directly. This will be done today' (L4). Along with the letter, he enclosed the necessary guarantee certificates and a copy of Schwarz's *Minimal Surface*, noting, 'As I mentioned before, you might find material in this paper for your work' (L4). He further encouraged her not to hesitate in seeking his help, stating, 'It will always give me great joy if you give me the opportunity to support you in your studies' (L4). Demonstrating his commitment, he informed her of his availability on most afternoons during the winter, underscoring his dedication to her academic progress.

There are four letters in the 1871 folder at the Mittag-Leffler Institute, marking the beginning of a lifelong correspondence, which began as a means of arranging their academic supervision but gradually evolved into an epistolary exchange where mathematical discussions were closely intertwined with a nurturing mentorship and a deep, intimate friendship. Unfortunately, we only have 150 extant letters that Weierstrass wrote to her, since Weierstrass burnt all her letters when he received the heartbreaking news of her untimely death in 1891. While studying their correspondence at the Mittag-Leffler Institute, where they are housed, Reinhard Bölling discovered the draft of a letter that Kovalevskaya wrote to Weierstrass in 1883, when she arrived in Stockholm for her first term as a private docent (Bölling 1992).

Although Kovalevskaya and Weierstrass typically met in person, there were occasions when letters had to bridge the gap of missed meetings. In one such exchange, Weierstrass, unwell due to a cold, regretted not being able to meet her and instead entrusted her with his lecture notes, remarking that they were 'so complete that you should be able to manage without my assistance' (L5). His request to have the notes returned in time for his next lecture underscored both his confidence in her abilities and the collaborative nature of their scholarly relationship.

One year after the initial letter, Kovalevskaya was addressed as 'dear friend', and Weierstrass felt comfortable enough to ask for her assistance: 'May I kindly ask you, dear friend, to copy the enclosed draft of the theorem we discussed yesterday onto a sheet of letter paper for me? I would like to share it with a friend' (L6), he wrote on 25 March 1872. Although Weierstrass admitted that 'I could do it myself', he was also 'reluctant to write something twice' and 'so I take the liberty of turning to you for help' (L6).

When later in the same year Kovalevskaya shared more personal details with Weierstrass, mostly concerning the difficulties of her fictitious marriage, their correspondence also became deeply personal. 'I have just found some older notes in my papers regarding the simplest case of the calculus of variations we have been discussing so far' (L7), he wrote in the morning of 26 October 1871, but the rest of the letter shows how much he had been affected by their personal discussions, at the same time having decided to offer her a topic for her doctoral dissertation:

> I have been preoccupied with thoughts of you throughout the night, as was inevitable. My thoughts wandered in various directions but always returned to one point that I must discuss with you today. Do not fear that I will touch upon subjects we have agreed, at least for now, not to discuss. What I have to

> say is closely related to your scientific endeavours, but I am unsure whether you, with the charming modesty with which you assess what you can already achieve, will be inclined to consider my proposal. However, all this is better discussed in person. Allow me, although only a few hours have passed since our last meeting, which brought us so close, to visit you again this morning for an hour to express myself (L7).

Their friendship deepened throughout 1872, with exchanges that extended beyond mathematics to more personal matters, such as the challenge of finding accommodation in Berlin. At the end of a lengthy letter dated 4 November 1872, filled with mathematical formulas, Weierstrass requested, 'If you or Miss L[ermontoff] can find out the current pension price for a single lady at the boarding house on Charlotten Street, please kindly let me know in a few words, as my sister would like to answer the inquiry directed to her soon' (L8). The letter's closing reveals the blending of personal affection with their shared intellectual pursuits: 'Farewell, my dearest heart, until I see you again, and when you immerse yourself in my formulas, let your thoughts occasionally wander to your faithful friend' (L8). Kovalevskaya was frequently addressed as 'dearest Sonia', and the formal epistolary ending of 'Yours sincerely' gradually gave way to the more familiar 'Your K.W.' (L9).

As Kovalevskaya and her friend Lermontova became closer to Weierstrass and his sisters, their correspondence began to include plans for social and cultural events. The two young women would spend Christmas at the Weierstrass home and participate in the season's festivities, as reflected in a letter dated 27 December 1872: 'The concert tonight, dearest Sonia, begins at 7:30 pm. If I'm not mistaken, you and Miss Lermontoff planned to join us. Please arrive before 7 pm so that we can leave on time. I'll arrange for a carriage' (L12).

Weierstrass's letters show deep concern about his student not just in matters of progress but also in matters of health, particularly given Kovalevskaya's frailty: 'How happy you have made me with the news that you are continuing to improve, and even feel inclined to "engage in mathematical matters" again' (L15), he wrote on 6 April 1873, while also strongly advising her to take it easy, care for herself, and dress warmly: 'Believe me, dearest, every hour you spend in the coming days, both here and later in Zurich, focusing on your physical recovery is time well spent. When you return to work with renewed strength, it will bear fruit' (L15). This letter was written shortly before Kovalevskaya's departure for Zurich, where she had gone to rest and visit her sister Anyuta, her husband Victor Jaclard, and their newborn child Yuri. They had emigrated to Switzerland as political exiles, following a turbulent period marked by his imprisonment and her exile after the fall of the Paris Commune in 1871 (see Kochina 1985, 73).

There are long extended personal letters between them during this period. In a letter written on 18 April 1873, Weierstrass expressed his great relief and joy upon receiving a letter from Kovalevskaya, in which she detailed her arrival in Zurich. He had been concerned that the journey, taken before she had fully recovered, would leave her too exhausted to write. His fears were alleviated, however, as she mentioned enjoying long daily walks, which suggested to him that she was feeling sufficiently strong. He wished for the weather to remain pleasant during her stay so that she could spend plenty of time outdoors, reminding her of a promise she had made to him before leaving:

> May the weather during your stay there remain as beautiful as it was upon your arrival, so that you can spend a lot of time outdoors and experience firsthand the truth of one of our medical authorities' statements—that besides chamomile tea, the only proven remedy is fresh, mild air. Don't forget what you promised me before your departure (L16).

Weierstrass's letter above continues to express his deep affection and concern for Kovalevskaya's well-being and future work. He urges her to take care of herself and return to him 'fresh and blooming', promising her that seeing her healthy and well upon her return would bring him great joy. He advises her to avoid any serious work until May, preserving her strength for the regular studies they would resume together. He also shares his own struggles with lingering illness, noting that he has been unable to engage in anything strenuous and has only managed light reading, although he had visits from several mathematical friends, with whom he has enjoyed discussions, particularly on topics like the geometry of finite spaces, which interests him greatly. Weierstrass reassures Kovalevskaya that his spirits remain intact despite his discomfort and that his interest in what matters to him most has not diminished: 'You write, dear Sonia, that you still have some questions to ask me. This gives me hope that I will receive at least one more detailed letter from you from Zurich. Isn't that right? I hope you won't wait so long that I won't be able to respond' (L16). His long letter concludes with a heartfelt reassurance: 'As I say goodbye for now, dear heart, I probably don't need to explicitly add that no matter where you are, I will always think of you with the deepest affection' (L16).

Kovalevskaya's friendship with Weierstrass was deeply intertwined with their pedagogical relationship, leading her to express concerns about the blending of these roles. However, Weierstrass remained steady and reassuring. The letter he wrote on 25 April 1873 reflects such discussions: 'As long as you can and want to learn something from me, you will remain my student in the best sense of the word, which I would not need to emphasize if a passage in your letter had not brought it to mind' (L17). Kovalevskaya must have expressed fears that 'if not the friend, then the student might become a burden' to him, to which Weierstrass firmly responded, rejecting the 'unpleasant word you used' (Kochina 1985) and making it clear how he viewed their relationship, declaring her his 'only true friend': 'speaking in all seriousness, dearest, most cherished Sonia, rest assured that I will never forget that it is the gratitude of my student to which I owe the possession—not of my best, but of my only true friend. Therefore, if you continue to show me the same sentiment you have demonstrated so far, you can be sure that I will always stand by you faithfully in your scientific endeavours' (L17).

Even when Weierstrass was appointed Rector of Berlin University in the summer of 1873, he wrote a reassuring letter to the rightfully concerned Kovaleskaya that their supervision would not be affected by his new responsibilities: 'If it seems to you that this winter you will see me very rarely, you misunderstood me: in any event we shall not give up our Sundays, and I shall still be able to find an hour that I can devote to my sweet friend' (L20), he wrote on 20 August 1873, while on summer vacation with his sisters in Sassnitz on the Island of Rügen. This is another lengthy letter from Weierstrass to Kovalevskaya, who was once again in Zurich. In it, Weierstrass reflects on the beauty of the landscape, using mathematical imagery to describe the scene at the time of writing:

> I am writing this letter at the open window of my room, and when I look up, I see the rolling sea about 1000 paces away, forming a complete semicircle, with its diameter not a sandy shore but a rather elevated, picturesque coastline adorned with fields, meadows, trees, and neat houses. Directly behind the hotel, one enters a forest that stretches for miles, filled with splendid deciduous trees, where a wonderfully refreshing breeze blows (L20).

Although Weierstrass was in good company with colleagues from Berlin, Leipzig, and Rostock, he longed for Kovalevskaya's presence: 'During my stay here, I have thought of you very often and imagined how wonderful it would be to spend a few weeks with you, my dear friend, in such magnificent nature' (L20). This passage captures the essence of their intellectual connection. 'How splendid we would dream and marvel here—you with your imaginative soul, and I stimulated and refreshed by your enthusiasm' (L20), Weierstrass writes, reflecting on the 'many puzzles left for us to solve, over the infinite and the infinitesimal, over the stability of the world system, and all the other great tasks of future mathematics and physics' (L20). For Weierstrass, such encounters were 'beautiful dreams', but he notes wistfully, 'I have long learnt to be content if not every beautiful dream is realized' (L20). Stanley has notably argued that a letter disrupts the perception of time and space and intensifies the present (2004, 208). What we read in the letter above is an intensification of feeling space and time through the imaginary presence of the addressee.

When later in October, however, the addressee had still not come back from Zurich, Weierstrass penned a letter to 'reprimand her' for neglecting to write for six weeks while also expressing his concerns about an upcoming official speech he was attempting to compose from his new position as rector of the university: 'Today, I am preparing for a speech I have to give in eight days, and I am still in the early stages of writing it' (L21), he wrote on 8 October 1873. We do not know the topic of the talk, but as Weierstrass explains, 'unless I choose a different topic, the content of my speech will be quite serious and will likely attract criticism from various sides' (L21). While acknowledging that confrontation 'cannot always be avoided', he was at the same time aware that 'I must carefully weigh each word to ensure I can stand by it' (L21). As a mathematician, he was good with numbers, but perhaps not so good with narratives. Yet, he still felt the need to write to his student and friend to share his worries about his own work. This theme of seeking support in his work recurs throughout their correspondence, reflecting the depth of their intellectual and personal bond.

During autumn, their correspondence went through difficulties since Kovalevskaya was still in Switzerland, and Weierstrass advised her to delay her return due to the outbreak of a cholera epidemic in Berlin: 'currently, we have around eight cases per day, and we hope that this unwelcome guest will leave us soon' (L22), he wrote on 12 October 1873. Complications with changing and forwarding addresses also arose, yet the desire to communicate and meet remained strong and persistent: 'I understand that you won't give me the exact date of your arrival, but if you wish to see me soon, please send me a short note by city post upon your arrival, so I can either come to you as soon as I can or let you know when you can certainly find me' (L22), he wrote in the same letter above. It appears that Kovalevskaya had offered him advice regarding the speech he was worried about, but in this letter, he reassures her that the issue has since been resolved: 'As for the rest of what I wrote, don't worry about it. My speech draft turned out so sharp that it's really not suitable for the occasion. I've already chosen another topic' (L22). It seems that his new administrative role was already taking a toll on his research, as the letter concludes with a sombre admission: 'Your faithful friend (even if he confesses that since the end of last semester, he hasn't had a single mathematical thought and you'll find him quite devoid of ideas)' (L22).

Despite his heavy administrative responsibilities, Weierstrass never lost sight of Kovalevskaya's doctoral research. Even amid his busiest days, her work remained a constant priority in his thoughts: 'I am writing this to you from the Senate Hall of the university during a profound examination on the fundamental properties of matter' (L26), he wrote on 26 November 1873, as he had just realised that in their last supervision 'we had overlooked one of the relations among the X' (L26), and the letter goes on hastily with dense mathematical formulas. Their supervision sessions provided Weierstrass with a welcome escape from the demands of university bureaucracy, a sentiment he conveys with great intensity in his letters: 'after several days of the most unpleasant work, I am especially longing to see my dear student and friend' (L27), he wrote on 29 November 1873. For Weierstrass, mathematics and poetry were deeply intertwined: 'I am sending you a complete series of "Songs Without Words", all of which relate to linearly bounded minimal surfaces' (L25), he wrote on 19 November 1873. He had famously claimed that one cannot be a great mathematician without possessing 'the soul of a poet', an imagery that Kovalevskaya embraced throughout her work, as I will further explore in Chapter 7.

As Kovalevskaya approached the completion of her work, her efforts intensified throughout December. On Christmas Eve morning of 1873, Weierstrass sent her two pages of mathematical formulas (L26). The urgency of this period is vividly captured in a letter Weierstrass wrote on 7 March 1874: 'I have been waiting for two days for a note from you indicating that you have achieved the desired result, and I now fear that my hopes might have been in vain. In any case, I need to be certain, so please expect me this afternoon around 5 o'clock (L32).

Following a prolonged period of frequent illnesses and ailments experienced by Weierstrass throughout the spring, the preparations for the doctoral work intensified over the summer. On 15 August 1874, Weierstrass wrote: 'Please send me your dissertation immediately via the delivery man, along with a packet of letters that you forgot yesterday. The printing office has already requested them, and we need to make some urgent corrections this afternoon' (L41). Three days later,

Weierstrass sent her his final letter while she was still in Berlin. Kovalevskaya had decided to return to Russia, having completed her doctoral studies. Despite her departure, Weierstrass continued to offer his support and guidance, extending his assistance not only to her but also to her friend Iulia, who stayed in Berlin to finish her own PhD, which she successfully completed later that year. This is what he wrote in a letter dated 18 August 1874: I forgot yesterday to explicitly tell you that if I can be of any help to you or Iulia today, you should feel free to send for me […] Also, tell your friend that I will certainly be back here from October 4th, and that she may feel free to contact me directly in any matter where I could assist her (L42).

Kovalevskaya was in Russia when she received the news that she had been awarded her doctoral degree *summa cum laude* (with the highest honours). In a letter dated 21 September 1874, Weierstrass informed her that he was overseeing the printing and distribution of her dissertation, shared stories from his summer holiday along the Rhine with his sisters, and expressed his support for her family's decision to 'lock away' her books during her stay in the countryside, as she had returned utterly exhausted. 'Whatever you miss in your studies now, you will easily make up for once you're fully recovered. From my own experience, I know how wretched it can feel to have a head full of problems but not enough physical strength to tackle them' (L43), he wrote and he also asked her to send him a photograph. The letter also mentioned that the news of the award of her doctoral degree had already become public in Heidelberg, the city where Kovalevskaya first began her mathematical studies before moving to Berlin.

> On the second evening of our stay, we spent the evening at Königsberger's house. Bunsen, who arrived a little later, brought a newspaper reporting that a learned lady in Moscow, Mrs. S. von Kowalewsky, had been awarded a doctorate by the Faculty of Philosophy in Göttingen—an honour previously bestowed on only one other woman, the famous Dorothee Schlözer. This caused quite a stir in the group—I had deliberately not mentioned it. Naturally, much was said about you and your two fellow students, and I learned some amusing details about your 'gypsy life' back then that entertained me greatly (L43).

Weierstrass was eager to hear about Kovalevskaya's time as a student in Heidelberg, a city that held fond memories for him as well. 'It felt strange revisiting a path I once walked with a friend many years ago—a friend who convinced me to finally act on my long-held decision to become a mathematician' (L43), he wrote in the same letter. He was particularly intrigued by the stir her presence had caused within the Heidelberg academic community and enquired whether the story behind Bunsen's description of her as 'a dangerous woman' was true: 'He [Bunsen] had vowed never to allow a woman, especially a Russian, into his laboratory. He refused to let Miss Lermontoff work or attend lectures there. Then you came to him and so sweetly asked that he couldn't resist and broke his vow. Is this true?' (L43).

This letter marks a transition between Kovalevskaya's graduate studies and the lengthy, challenging period before her first academic appointment in 1883. As I will elaborate in Chapter 4, Kovalevskaya's journey into academia was long, difficult, and unpredictable, yet Weierstrass' unwavering support remained a constant. Whether through letters or in person, his presence was the solid foundation amid the turbulent storms of her life and *agon* to become an academic.

Epistolary sensibility and the female self in science and mathematics

In this chapter, I have read and analysed the mathematical and scientific correspondences of women mathematicians through the lens of *epistolary sensibility*—an epistemological and methodological approach that examines letters as relational auto/biographical documents (Tamboukou 2020). This perspective attends not only to the content but also to the distinctive epistolary traits of their form. In doing so, I have particularly focused on correspondences, following Stanley's influential notion of 'the epistolarium' (2004).

In this context, mapping epistolary relationships was important in the analysis of all letters. A common denominator in all six cases was that women mathematicians were communicating with men—whether tutors, doctoral supervisors, fellow mathematicians, or even family members, lovers, and friends, when writing about mathematics. The only exception was the correspondence between Mary Somerville and Ada Lovelace. What also struck me in tracing epistolary relationships among women themselves is, for example, that Émilie Du Châtelet and Maria Gaetana Agnesi never communicated with each other, although they knew of each other's work. Agnesi had Du Châtelet's book, *Foundations of Physics*, in her library, Du Châtelet had praised Agnesi's work, and they had both become members of the Bologna Academy of Sciences around the same time through their connections with Laura Bassi, the first woman to hold a chair in physics in the early modern period.

In the same vein of glaring omissions, Mary Somerville never got to meet Sophie Germain when she visited Paris in 1817, just a year after Germain had become the first woman to be awarded the *grand prix des mathématiques* from the Paris Academy of Sciences. Although Somerville dined and wined with the members of this scientific community, as already pointed out in the Introduction, I often wonder how different things might have been if the few women mathematicians in eighteenth- and nineteenth-century Europe had been in communication with one another, especially when considering today's efforts to create and maintain similar networks of support.

Despite the absence of direct communication between women mathematicians, their mathematical and scientific correspondences became components of an *assemblage* of epistemological, philosophical, and literary practices that shaped the female self in science and mathematics within the intellectual landscape of the eighteenth and nineteenth centuries. Their letters both educated and connected them with the mathematical networks and circles of their time—introducing them, into the 'Republic of Mathematics', a transnational community sustained by correspondence, collaboration, and shared intellectual labour (see Oosterhoff 2016)—though this was never enough to bridge the gaps left by their exclusion from the formal academic institutions freely accessible to their male peers.

In light of the *epistolary sensibility*, women mathematicians' 'epistolaria' (Stanley 2004) carry traces of complex entanglements between science, culture, and affects in their dynamic 'intra-actions'. Here, the notion of 'intra-actions' comes from Karen Barad (2007), a feminist philosopher and atomic scientist who has coined the term as a theoretical juxtaposition to the usual notion of 'interactions'. The significant difference lies in the fact that while interactions occur between already established and separate entities, intra-actions emerge as relations between components. Entities, such as mathematics, culture, epistolary practices, and/or knowledge formations actually emerge as an effect of intra-actions between minor components, such as the materiality of writing, the darkness of a corridor, the subdued atmosphere of the library, a chat in the common room, a walk in the country, an enthusiastic salutation, a warm ending, or the delayed arrival of a letter.

For instance, Germain never attended the ceremony of her prestigious award because her notification letter arrived late and carried a condescending tone, as I will explore further in Chapter 4. This delayed and dismissive letter interacted with other practices and discourses through which she 'realized' herself as a woman mathematician in a man's world. In the same vein, Agnesi's ironic tone in a letter to her philosophy tutor, as we have seen in this chapter, offers the only glimpse we have into how becoming a public spectacle interrupted or perhaps disrupted her journey to becoming a mathematician.

In this context of epistolary 'intra-actions', I have thus considered 'the personal and the scientific' as a crucial dynamic in the process of becoming a woman mathematician, philosopher, and scientist in early and late modern Europe. What are the effects, then, of examining this personal-scientific interplay in the letters of women mathematicians? In the same way that the personal and the political cannot be separated, 'the personal and the scientific' create conditions of possibility for each other in their entanglement as components of the female self in science.

In this light, all six biographical cases reveal that personal circumstances played a significant role in shaping these women's journeys to becoming mathematicians, scientists, and philosophers within the conditions, constraints, and limitations of their geographical, socio-economical, and cultural contexts. For some, these personal conditions, along with their feelings and emotions, can be traced in their autobiographies—as in the case of Somerville and Kovalevskaya—personal letters—as in the case of Du Châtelet, Somerville, Lovelace, and Kovalevskaya—as well as in their philosophical and scientific writings. We don't have personal letters or autobiographies for Agnesi and Germain. Anything personal we can trace about them is either folded in the memoirs of people who knew them or is entangled in the pages of their scientific correspondence, as well as in their scientific and philosophical writings.

In thus considering the material and discursive conditions of possibility for the emergence of the female self within the entangled realms of gender, science, and mathematics, I have ultimately argued that unravelling the personal elements intertwined within scientific correspondences is both crucial and illuminating, especially in cases where no other autobiographical or biographical sources exist. In this context, the chapter contributes to the rich body of literature analysing women's letters and correspondences by shedding light on the often-overlooked intersection of gender, science, and mathematics.

Notes

1 See amongst others, the Darwin letters project on the importance of letters in Darwin's scientific work, https://www.darwinproject.ac.uk/ [Accessed 24 July 2022]

2 See George 2011 and Schurch 2019 for an overview of the literature.

3 For a discussion of her scientific correspondents see Bonnel 2000, 82–4. For her correspondence with Clairaut, see also Passeron 2001 and for her correspondence with Euler, see Suisky 2012; Musielak 2023.

4 Lettres de la marquise DU CHATELET à M. de Maupertuis (1734–1741), Bibliothèque Nationale de France, in the text and thereafter [BnF, fr.12269]. The collection is also available in Gallica, https://gallica.bnf.fr/ark:/12148/btv1b6000754n. All letters have been transcribed in Kölving and Brown 2018. Unless otherwise indicated, all translations are mine.

5 See Terrall 1995, 286, Bonnel 2000, 87.

6 See Hagengruber 1999, 198–9, for an analysis of Du Châtelet's and Maupertuis' discussions around the subtleties of the 'vis viva' controversy.

7 See, amongst others, Shank 2008, for an elaboration on these debates and controversies.

8 The vis-viva controversy in the eighteenth century was a significant scientific debate concerning the nature and measurement of motion and energy. This controversy goes back to the ideas of the German mathematician and philosopher Gottfried Wilhelm Leibniz. In the late seventeenth century, Leibniz proposed that the measure of a body's force should be proportional to its mass and the square of its velocity (mv^2), which he called vis viva. This was in contrast to the Cartesian notion, based on the work of René Descartes, which measured force as 'momentum', the product of mass and velocity (mv). See Terrall 2004 for an in-depth analysis of the vis-viva debate.

9 Maria Gaetana Agnesi to Michele Casati. Veneranda Biblioteca Ambrosiana, Agnesi, Maria Gaetana <1718–1799>, 0.180- 0.201. SUP., thereafter and in the text, [VBA] followed by codice number. Unless otherwise indicated all translations from Latin and Italian are mine.

10 For a discussion of the European culture of scientific spectacle, see Cavazza 2009, and Schurch 2019, 47.

11 The entire correspondence can be accessed on line, in Mazzone et al. 2010, http://bibdig.museogalileo.it/Teca/Viewer?an=000000990843. The correspondence has also been transcribed in Soppelsa 1985. Unless otherwise indicated, the letters cited in this chapter are translated in Roero 2015.

12 Niedersächsische Staats-und Universitätsbibliothek in Göttingen: Cod. ms. Gauss Briefe A: Germain, n. 9. Abteilung Handschriften und Selten Drucke. [Lower Saxony State and University Library in Göttingen: Cod. ms. Gauss Letters A: Germain, Department of Manuscripts and Rare Prints.] Thereafter and in the text [NS-UG/MS Gauss/Bref-Germain]. All letters have been transcribed in Del Centina and Fiocca 2012, 636–96 and

have been translated into English in Musielak 2020, 221–34. In the in-text references, I will only refer to their initial publication.

13 See Bucciarelli and Dworsky 1980, Del Centina and Fiocca 2012, 2018; Musielak 2020.

14 See Bucciarelli and Dworsky 1980, particularly Chapter 3; Musielak 2020, particularly Chapter 3.

15 Also published in Germain 1896, 271.

16 As his last and most popular play written in verse, Molière's play *Les Femmes Savantes* was a social satire of women's obsession with learning, as well as the effects of excessive pretension to knowledge.

17 This letter is included in Germain's papers at the Bibliothèque Nationale, MS. Fr. Nouvelle Acquisition, 4073, thereafter [BN 4073]. Translated in Bucciarelli and Dworsky 1980, 13.

18 See Musielak 2020, 219; Del Centina and Fiocca 2018, 672–4 and Bucciarelli and Dworsky 1980, particularly Chapter 3.

19 The paper was published in 1827 in the Mémoires de l'Académie des Sciences of 1825, but these results were also included in the Second Supplement of the *Essai sur la théorie des nombres* (Legendre 1825).

20 Also included in [BN 9118], translated in Bucciarelli and Dworsky 1980, 89.

21 Also included in [BN 4073].

22 Germain to Libri, Moreniana Biblioteca di Firenze, fondo Palazzi-Libri, filza 432, thereafter and in the text (MBF/FPL/) followed by folio number. Transcribed in Del Centina 2005, 66.

23 Draft of a letter, [BN 9118], translated in Bucciarelli and Dworsky 1980, 106–7.

24 Letter dated 8 February 1830, transcribed in Del Centina 2005, 68.

25 Also included in [BN 4073], translated in Bucciarelli and Dworsky 1980, 121–2.

26 Both Newton and Leibniz independently developed calculus, one of the most significant advancements in mathematics in the late seventeenth century, but then a dispute arose over who had done it first, known as 'the priority dispute'. Leibniz's version was called 'differential calculus', and his notation is the form still used today. It was different from Newton's, which was the one British mathematicians insisted on following at the time Somerville engaged in the study of mathematics. See Stenhouse 2020, 8 for a discussion of this debate with good references to the relevant literature.

27 The Royal Society, Herschel Papers, thereafter and in the text, HS/16/ followed by the number of the letter. Available online at https://makingscience.royalsociety.org/correspondence-explorer?first=Mary+Somerville&second=John+Frederick+William+Herschel [Accessed 22 August 2024] Unless otherwise indicated, all emphases are in the text.

28 Oxford Bodleian Libraries, Mary Somerville Collection, Dep. c.369-372; Dep. b. 233; Der. b. 206b; Dep.b. 157. Thereafter MSC followed by Department Folder and Folio Numbers.

29 British Library, Manuscript Collections, Correspondence of Charles Babbage, Add MS 37185-37189. In the text and thereafter, the MS number is followed by the folio number.

30 For more information about the history of the institute, see https://www.fstfirenze.it/istituto-tecnico-toscano/

31 OBL/MSC/'Letters from Celebrated Peple', 1791–1969 and n.d.

32 Oxford Bodleian Libraries, Archive of the Noel Byron and Lovelace Families, Dep. Lovelace Byron. In the text and thereafter, LB followed by the folder and folio number. Unless otherwise indicated, all letters cited in this chapter have been transcribed and published in Toole 1992.

33 British Library, Manuscripts Collection, Correspondence of Charles Babbage, Augusta Ada King, daughter of Lord Byron; wife of William, Lord King, 1st Earl of Lovelace :

Letters to C. Babbage: 1836–51, Add MS 37189-37192. In the text and thereafter, MS number followed by folio number.

34 Karl Weierstrass to Sofia Kovalevskaya, Library of the Mittag-Leffler Institute. Thereafter in the text, letter number only, since all letters are from Karl Weierstrass. All letters in this correspondence have been published in German in Bölling 1993 and have been translated into Russian in Kochina 1973. Unless otherwise indicated all translations are mine.

References

Agnesi, Maria Gaetana. 1748. *Instituzioni analitiche ad uso della gioventù italiana*, 2 vols. Milan: Nella Regia- Ducal Corte.

Altman, Janet. 1982. *Epistolarity: Approaches to a Form*. Ohio: Ohio State University Press.

Barad, Karen. 2007. *Meeting the Universe Halfway: Quantum Physics and the Entanglement of Matter and Meaning*. Durham: Duke University Press.

Bölling, Reinhard. 1992. '… Deine Sonia: A Reading from a Burned Letter.' Translated by David E. Rowe. *The Mathematical Intelligencer* 14 (3): 24–30.

Bölling, Reinhard, ed. 1993. *Briefwechsel Zwischen Karl* Weierstrass *Und Sofja Kowalewskaja*. Berlin: Academie Verlag.

Boncompagni, Baldassarre. 1879. *Lettera inedita di Carlo Federico Gauss a Sofia Germain* pubblicata da B. Boncompagni, Calcografia e autografia Achille. Paris. Firenze.

Boncompagni, Baldassarre. 1880. *Cinq lettres de Sophie Germain à Charles Frédéric Gauss,* publiées par B. Boncompagni d'après les originaux possédés par la Société Royale des Sciences de Göttingen, Berlin, Institut de photolithographie des frères Burchard. Berlin: Gustave Schade.

Bonnel, Roland. 2000. 'La Correspondance scientifique de la marquise Du Châtelet: La "Lettre-laboratoire"'. In *Femmes en toutes lettres: Les Epistolières du XVIIIe siècle*, edited by Marie-France Silver and Marie-Laure Girou Swiderski, 79–95. Oxford: Voltaire Foundation.

Bucciarelli, Luis, L., and Nancy Dworsky. 1980. *Sophie Germain: An Essay in the History of the Theory of Elasticity*. Dordrecht: Boston, MA.

Cavazza, Marta. 2009. 'Between Modesty and Spectacle: Women and Science in Eighteenth Century Italy'. In *Italy's Eighteenth Century: Gender and Culture in the Age of the Grand Tour,* edited by Paula Findlen, Catherine Sama, and Wendt Rowarth, 275–302. Stanford: Stanford University Press.

Cavazza, Marta. 2014. 'The Biographies of Laura Bassi'. In *Writing about Lives in Science: (Auto)Biography, Gender and Genre*, edited by Paola Govoni and Zelda Alice Franceschi, 67–86. Göttingen: V&R Unipress.

Del Centina, Andrea. 2005. 'Letters of Sophie Germain Preserved in Florence'. *Historia Mathematica* 32: 60–75.

Del Centina, Andrea, and Alessandra Fiocca. 2012. 'The Correspondence between Sophie Germain and Carl Frierich Gauss'. *Archive for History of Exact Sciences* 66: 585–700.

Del Centina, Andrea, and Alessandra Fiocca. 2018. 'On the Correspondence of Sophie Germain'. In *Mathematical Correspondences and Critical Editions*, edited by Maria Teresa Borgato, Erwin Neuenschwander, and Irène Passeron, 147–66. Cham: Birkhäuser.

Delambre, A., A. Choron, Jh. Fourier, Sophie Germain, De La Grange, Lalande, Guillaume Libri, and C. Henry. 1879. 'Les manuscrits de Sophie Germain et leur récent éditeur-documents nouveaux'. *Revue Philosophique de la France et de l'Étranger* 8: 619–41.

Deleuze, Gilles, and Felix Guattari. 1988 [1980]. *A Thousand Plateaus: Capitalism and Schizophrenia*. Translated by Brian Massumi. London: The Athlone Press.
Dubby, John. 1978. *The Mathematical Work of Charles Babbage*. Cambridge: Cambridge University Press.
Du Châtelet, Émilie. 1738. *Dissertation sur la nature et la propagation du feu*. Paris: Académie des sciences (concours).
Du Châtelet, Émilie. 1740. *Institutions de physique*. Paris: Prault.
Du Châtelet, Émilie. 1756. *Principes mathématiques de la philosophie naturelle*. Paris: Desaint & Saillant, Lambert.
George, Sam. 2011. 'Epistolary Exchange: the Familiar Letter and the Female Botanist 176–1820'. *Journal of Literature and Science* 4 (1): 12–29.
Germain, Sophie. 1821. *Recherches sur les théories des surfaces élastiques*. Paris: Vè Courcier.
Germain, Sophie. 1896 [1879]. *Œuvres philosophiques de Sophie Germain, suivies de pensées et de lettres inédites. Et précédées d'une notice sur sa vie et ses œuvres par Hte Stupuy (Nouvelle Édition)*. Paris: Firmin-Didot.
Goodman, Dena. 1994. *The Republic of Letters: A Cultural History of the French Enlightenment*. Ithaca, NY: Cornell University Press.
Grassi, Marie-Claire. 1998. *Lire l' épistolaire*. Paris: Dunod.
Hagengruber, Ruth. 1999. 'Émilie du Châtelet and Maupertuis: Eine Metaphysik in Briefen [Metaphysics in Letters]'. In *Pierre Louis Moreau de Maupertuis: Eine Bilanz nach 300 Jahren* [A Review after 300 years], edited by Hartmut Hecht, 187–206. Berlin: Spitz.
Hagengruber, Ruth, ed. 2022. *Époque Émilienne*. Cham: Springer.
Henry, Charles. 1879. 'Les manuscrits de Sophie Germain et leur récent éditeur, Documents nouveaux'. *Review of Philosophy* 8: 619–41.
Jordan, Charles-Etienne. 1735. *Histoire d'un voyage littéraire fait en MDCCXXXIII*. The Hague: Adoren Hoetiers.
Kochina, Pelageya, ed. 1973. *Письма Карла Вейерштрасса к Софье Ковалевской: 1871–1891* [Letters of Karl Weierstrass to Sofia Kovalevskaya]. Translated by P. Kochina. Moscow: Nauka.
Kochina, Pelageya. 1985. *Love and Mathematics: Sofia Kovalevskaya*. Translated by. Michael Burov. Moscow: Mir Publishers.
Kölving, Ulla, and Andrew Brown, eds. 2018. *La Correspondance d' Émilie Du Châtelet*. 2 vols. Ferney-Voltaire: Centre international d'étude du xviiie siècle (vol. 1, pp. 1–633; vol. 2, pp. 1–593).
Laubenbacher, Reinhard and Pengelley, David. 2010. 'Voici ce que j'ai trouvé': Sophie Germain's grand plan to prove Fermat's Last Theorem'. *Historia Mathematica* 37: 641–92.
Legendre, Adrien-Marie. 1825. *Essai sur la théorie des nombres, Second supplément*. Paris: Courcier.
Libri, Guglielmo. 1832. 'Sciences Mathématiques-Mlle Germain'. *Journal Des Débats*, 18 May 1832, 1–2.
Lovelace, Ada. 1843. 'Sketch of the Analytical Engine invented by Charles Babbage Esq. By L.F. Menabrea, of Turin, Officer of the Military Engineers, with Notes Upon the Memoir by the Translator'. *Taylor's Scientific Memoirs* 3: 666–731.
Margolin, Arianne Nicole. 2008. 'Le laboratoire épistolaire dans les oeuvres scientifiques de la marquise du Châtelet'. Graduate Student Theses, Dissertations, & Professional Papers, 614. ii. Accessed 23 February 2023. https://scholarworks. umt.edu/etd/614.
Maupertuis, Pierre-Louis Moreau. 1738. *La Figure de la Terre, déterminée par les Observations de Messieurs Maupertuis, Clairaut, Camus, Le Monnier & de M. l'Abbé Outhier, accompagnés de M. Celsius*. Paris: Imprimerie Royale.

Mazzone, Silvia, and Clara Silvia Roero, with the collaboration of Erika Luciano, eds. 2010. *L'epistolario di Jacopo, Vin-cenzo e Giordano Riccati con Ramiro Rampinelli e Maria Gaetana Agnesi*. Florence: Olschki. http://bibdig.museogalileo.it/Teca/Viewer?an=000000990843.

Mazzotti, Massimo. 2007. *The World of Maria Gaetana Agnesi, Mathematician of God*. Baltimore: Johns Hopkins University Press.

Mazzotti, Massimo. 2014. 'Rethinking Scientific Biography: The Enlightenement of Maria Gaetana Agnesi'. In *Writing about Lives in Science: (Auto)Biography, Gender and Genre*, edited by Paola Govoni and Alice Zelda Franceschi, 117–37. Göttingen: V&R Unipress.

Musielak, Dora. 2020. *Sophie Germain: Revolutionary Mathematician*. 2nd ed. Cham: Springer.

Musielak, Dora. 2023. 'Emilie du Châtelet and Euler: A Rare Convergence on the Hypotheses of Physics'. arXiv:2212.12665v2. Accessed 15 July 2024.

Oosterhoff, Richard, J. 2016. 'Lovers in paratexts: Oronce Fine's republic of mathematics', *Nuncius: Journal of the Material and Visual History of Science*, 31 (3): 549–583, https://doi.org/10.1163/18253911- 03103002

Passeron, Irène. 2001. 'Muse ou élève ? Sur les lettres de Clairaut à Madame du Châtelet'. In *Cirey dans la vie intellectuelle: la réception de Newton en France*, 187–97. Oxford: Studies on Voltaire and the Eighteenth Century.

Patterson, Elizabeth Chambers. 1983. *Mary Somerville and the Cultivation of Science 1815–1840*. The Hague: Martinus Nijhoff.

Plummer, Ken. 2001. *Documents of Life 2*. London: Sage.

Pomata, Gianna. 2013. 'Amateurs by Choice: Women and the Pursuit of Independent Scholarship in 20th Century Historical Writing'. *Centaurus* 55: 196–219.

Roero, Clara Silvia. 2015. 'M.G. Agnesi, R. Rampinelli and the Riccati Family: A Cultural Fellowship Formed for an Important Scientific Purpose, the *Instituzioni analitiche*'. *Historia Mathematica* 42: 296–314.

Schurch, Madeleine. 2019. 'Women, Empiricism and Epistolarity, 1740–1810'. PhD thesis, University of York.

Shank, John Bennett. 2008. *The Newton Wars and the Beginning of the French Enlightenment*. Chicago: Chicago University Press.

Somerville, Mary. 2001 [1873]. *Queen of Science: Personal Recollections of Mary Somerville*. Edited by Dorothy McMillan. Edinburgh: Canongate.

Soppelsa, Maria Laura. 1985. 'Jacopo Riccati-Maria Gaetana Agnesi: Carteggio (1745–1751)'. *Annali dell'Istituto e Museo di Storia della Scienza di Firenze* 10: 117–59.

Stanley, Liz. 2004. 'The Epistolarium: On Theorizing Letters and Correspondences'. *Auto/Biography* 12 (3): 201–35.

Stenhouse, Brigitte. 2020. 'Mary Somerville's Early Contributions to the Circulation of Differential Calculus'. *Historia Mathematica* 51: 1–25.

Stenhouse, Brigitte. 2021. 'Mary Somerville: Being and Becoming a Mathematician'. PhD thesis The Open University, 2021. https://doi.org/10.21954/ou.ro.0001346f

Suisky, Dieter. 2012. 'Leonhard Euler and Émilie du Châtelet. On the Post-Newtonian Development of Mechanics'. In *Émilie du Châtelet between Leibniz and Netwon*, edited by Ruth Hagengruber, 113–56. Dodrecht: Springer.

Tamboukou, Maria. 2020. 'Epistolary Lives: *Fragments, Sensibility, Assemblages in auto/biographical Research*'. In *The Palgrave Handbook of Auto/Biography*, edited by Anne Chapell and Julie M. Parsons, 157–64. Basingstoke: Palgrave.

Terrall, Mary. 1995. 'Émilie Du Châtelet and the Gendering of Science'. *History of Science* 33 (3): 283–310.

Terrall, Mary. 2004. 'Vis Viva Revisited'. *History of Science* 42 (2): 189–209, https://doi.org/10.1177/007327530404200202
Toole, Betty Alexandra. 1992. *Ada, the Enchantress of Numbers*. Moreton-in-Marsh: Strawberry Press.
Zinsser, Judith P. 2006. *Daring Genius of the Enlightenment, Émilie du Châtelet*. New York: Penguin.

3 The Dynamics and Politics of Translation

'It is certainly more valuable to give a good translation of an esteemed English or Italian book than to write a bad French one', Émilie Du Châtelet wrote in the Translator's Preface to Bernard Mandeville's *The Fable of the Bees* (2009, 45). This translation was her first attempt to actualise herself as an author, but it remained unpublished and hidden in Voltaire's St. Petersburg archive, as we have seen in Chapter 1.

According to the dates of the St. Petersburg manuscripts, Du Châtelet likely began her translation of *The Fable of the Bees* in 1735, the same year she took the critical decision to move to the Château de Cirey with Voltaire, immersing herself in the adventures and joys of intellectual life, as we have already seen in previous chapters. In a letter to her friend, the Duke de Richelieu, dated 30 June 1735, amidst her preparations, she wrote: 'the decision has been made [...] I am about to spend the three happiest months of my life. I leave in four days' (in Kölving and Brown 2018, vol. 1, 175). Her decision to join Voltaire reflected not only her deep affection for him but also marked a new beginning in her life: 'My mind is overwhelmed, but my heart swims in joy' (175).

In the 'Translator's Preface', Du Châtelet justified her decision to embark on a new intellectual journey alongside her personal transformation: 'we lose our own ideas when we neglect to cultivate them' (2009, 44), she remarked, outlining both the purpose of her study and her philosophy of translation. She viewed translators as vital contributors to the Republic of Letters, writing, 'Translators are the entrepreneurs of the Republic of Letters, and they should at least be praised for perceiving and knowing their limitations, for not attempting to produce original works themselves and thereby risking collapse under a weight too great to bear' (46). With this, she clearly defined both the scope and restrictions of translation, following a widespread trend of women translating culture in the early modern period in Europe.

In her important work on *Women and Early Modern Cultures of Translation* (2022), Hilary Brown notes that more women became involved in translation in sixteenth- and seventeenth-century Europe than ever before, making significant contributions to literature, philosophy, and science. Brown explores the link between gender and translation, highlighting how translation was historically perceived as one of the few intellectual pursuits deemed appropriate for women. She particularly refers to Lori Chamberlain's pathbreaking work (1988) on gender

DOI: 10.4324/9781003533993-4

and translation, which explores how translation enabled women to articulate their voices even in 'silent' forms of writing (2022, 1).

Brown, however, challenges the prevailing narrative that women primarily used translation as a means to assert agency, resist patriarchy, and express their voices through text choices or subtle interventions. Her central argument is that it is high time we reassessed traditional views of translation history, which have often been shaped by the assumption that translation is derivative and less significant than original writing. As Brown puts it, 'the cultural turn' in Translation Studies has emphasised that translation is not a marginal or silent act but a transformative and subversive one, positioning translators as agents of cultural change and elevating the importance of translation as a creative and significant literary practice (4).

Brown's approach aligns with a dynamic body of feminist scholarship that emphasises the central role of translation in literary and cultural history, where it shaped education, politics, religion, and the dissemination of ideas.[1] In this field, Paula Findlen's (1995) work has been particularly foundational in examining the connection between gender and translation during the early modern period, especially with regard to the transmission and circulation of scientific knowledge in the Italian states. Findlen has notably emphasised that the introductions and prefaces of significant scientific books and treatises translated by women in the seventeenth and eighteenth centuries served as key textual spaces for the translators' voices to emerge. These prefaces became the most common avenue for women scientists to assert their authority and influence within the Enlightenment Republic of Letters (173).

Situating my work within this body of scholarship, in this chapter, I explore the role of translation in the scientific, philosophical, and literary contributions of the six women mathematicians, scientists, and philosophers around whom this book revolves, while also drawing attention to the marginalisation of Wang Zhenyi's work. The chapter is structured thematically. The first section explores the critical role of translating science, culture, and mathematics not only as an original authorial practice but also as a vital means of creating and disseminating knowledge. The second section examines both historical and contemporary efforts, along with the challenges, of translating women's scientific, mathematical, and literary work. In the final section, I consider the impact of the politics of translation and its connection to gender dynamics, particularly in tracing the emergence of the female self in mathematics, science, and philosophy.

Gendered cultures of translation

As discussed in the previous section, alongside her scientific experiments at the Château de Cirey and her dedicated study of mathematics, Du Châtelet's first literary endeavour was translating Mandeville's *The Fable of the Bees*. The choice of this particular book was not accidental, as Du Châtelet clearly explained in her Preface, which engages *avant la lettre* with Peter Burke's six critical questions regarding the 'cultures' of translation in early modern Europe: 'Who translates?

With what intentions? What? For whom? In what manner? With what consequences?' (2007, 11)

Du Châtelet begins by justifying her choice of genre, stating: 'of all works, those of discursive thought seem to me the most suited to good translation' (2009, 45). In her view, treatises and essays dealing with reason and morals are more easily translatable because they rely less on the specific genius of a language and more on the ideas conveyed. In contrast, works of imagination, such as poetry, are much harder to translate well, as they require the translator to match the original author's poetic skill: 'to translate a good poet, one must be almost as good a poet as the author' (45). Thus, while 'poetic imagination' may be difficult to faithfully render in multiple languages, reason can be effectively translated, argued Du Châtelet. Her example was the translation of John Locke's philosophical works into French, which made Locke's complex ideas accessible to non-English speakers who would otherwise be hindered by both language and the intricacies of the subject matter (45).

In Du Châtelet's view, the translator's agency emerges not through 'the genius of creativity' (46), but from their essential role in disseminating knowledge, for which they deserve greater appreciation, even if they receive less recognition: 'it calls for an application for which they must be even more grateful, as they can expect less glory from it' (46). As already mentioned above, Du Châtelet described translators as the 'entrepreneurs of the Republic of Letters'. As 'entrepreneurs', they had the freedom to make significant alterations to existing texts 'in the process of transposing the words and phrases from one language and one culture to another', Judith Zinsser has aptly commented (2002, 609).

In underscoring the importance of 'application', Du Châtelet therefore highlights the crucial role of translation in the circulation of knowledge, portraying translators as educators and facilitators of transnational cultural exchange. It is at this point that Du Châtelet makes the connection between gender and translation: 'However mediocre this kind of work, it may be thought that it is audacious for a woman to aspire to do it', she poignantly notes (2009, 48). Feeling the weight of prejudice that often excludes women from the sciences, Du Châtelet reflects on the absence of significant contributions from women in various artistic and intellectual fields, raising the question of why women, whose understanding she believes to be equal to men's, seem to face an insurmountable barrier to achieving recognition:

> Let us reflect briefly on why for so many centuries, not one good tragedy, one good poem, one esteemed history, one beautiful painting, one good book of physics, has come from the hands of women. Why do these creatures whose understanding appears in all things equal to that of men, seem, for all that, to be stopped by an invincible force on this side of a barrier; let someone give me some explanation, if there is one (48).

The Translator's Preface thus becomes a manifesto for human rights, including women's right to education. Du Châtelet argues that until a physical explanation is provided by naturalists, women have every right to protest their limited education. She envisions a reform that would grant women equal rights, particularly in intellectual pursuits, suggesting that this change would benefit humanity as a whole. She believes that current educational practices diminish women's potential and that a new system would enhance their value and foster healthy competition among men (48–9).

Du Châtelet reflects on her own experiences, noting that many women may remain unaware of their talents due to inadequate education or societal prejudice. She expresses regret that her realisation of her intellectual capabilities came too late for her to fully develop them. At the same time, however, she highlights her determination to use her remaining time to cultivate her intellect. Acknowledging that she may lack the creative genius to discover new truths, she finds fulfilment in clearly translating and conveying the ideas of others, making them accessible to a wider audience hindered by language barriers. In doing so, she firmly situates herself as a translator in the Republic of Letters (49). Having clarified why she had chosen to become the translator of 'a discursive book', Du Châtelet delves into specifics about the text, framed by her admiration for the clarity and strength of English philosophy:

> I have chosen this book titled The Fable of the Bees, among all those I could have translated, because it appears to be, of all the books in the world, one of those most designed for humanity in general. It is, I believe, the best book of ethics ever written, that is to say, the one that most leads men to the true source of the feelings to which they abandon themselves almost all without examining them. Mandeville, who is the author of it, could be called the Montaigne of the English if it were not for the fact that he has more method and healthier ideas about things than Montaigne (49–50).

Although Du Châtelet boldly compares Mandeville to Montaigne, noting that the author of *The Fable of the Bees* exhibits a greater sense of method and healthier ideas, she does not consider him without flaws, thus castigating 'the idolatrous respect of all translators' (50). She has read enough to think that Mandeville's work is poorly written in English, with parts that are too lengthy and ideas that 'sometimes go too far' (50). She also points out certain dangerous or untrue claims. To mitigate this, she does not hesitate to add corrective statements and edits the text by refining the style and removing content too specific to English customs. Additionally, she 'takes the liberty' (50) to include her own reflections, although she highlights them with quotation marks to distinguish them from the original text. Finally, Du Châtelet cautions her readers that the text contains bold and controversial ideas, urging them to approach these statements with care, subtly anticipating the discerning nature of her intended audience.

> The reader will find in this book thoughts that may seem a little bold, but before making this judgment it is only a matter, I believe, of examining whether they are accurate. For, if they are true, and if they teach men how to know themselves, they cannot fail to be useful to those thinking men, and it is for those only that this book is destined. *Odi prophanum vulgus et arceo*[2] (50).

Since the translator's role as an 'entrepreneur of the Republic of Letters' involves transposing words and meanings across languages and cultures, Du Châtelet is clear in her Preface about her flexible approach to translation. In exercising such liberties, she questions the very concept of 'faithfulness to the text'. She criticises 'unfaithful' translators, noting that some adhere too rigidly to word-for-word translations out of fear of straying from the original, while others, unable to fully grasp the author's meaning, obscure brilliant ideas with their incomplete understanding (47–8).

The conclusion to the Preface returns to the link between gender and translation. Du Châtelet acknowledges the boldness of her undertaking and expresses a strong desire to succeed. She believes that, given society's unfair exclusion of women from the sciences, women should use this disadvantage to avoid producing poor work. Since this is her first writing project, she hopes to achieve excellence, 'so that this tyranny will be a happy necessity for us, leaving nothing for them [men] to condemn in our works but our names' (50).

Du Châtelet's translation eventually became very much 'her own creation, [...] her 'fable', not Mandeville's, her own work of morality reflecting her privileged, aristocratic, French world, not his Anglo-Dutch, dissenting, mercantile perspective', as Zinsser has commented (2002, 610–1). Not only did she make important interventions in the text, as noted above, but she also notably altered the structure of the book. The *Fable of the Bees* originally consisted of three parts: a poem, an essay, and a series of 'Remarks'. Du Châtelet made deliberate choices about what to include, notably omitting Mandeville's allegorical verse fable, 'The Grumbling Hive'. Given her previous criticism of the impossibility of translating poetry, it is unsurprising that she left the poem out entirely. She also altered or omitted significant portions of the accompanying essay, 'An Enquiry into the Origins of Moral Virtue', instead focusing on the explanatory 'Remarks'. Even within the text she retained, she made substantial revisions, changing words, sentences, and paragraphs, and substituting her own examples to better suit a French audience, as already mentioned above.

Du Châtelet's translation project is thus a serious deviation from a wider trend in the history of translation—the idea that translation simply transfers the essence of a source text to create an equivalent target text, often leading to judgements of inadequacy or loss. But, as Brown has commented, this perception of translation has been critiqued as overly simplistic in the literature of 'the cultural trend' (2022, 5). Brown draws attention to the work of translation scholars like Karen Emmerich and Lawrence Venuti, who argue that source texts are not stable entities, as they are interpreted differently by various readers and exist in multiple versions (5). As Brown notes, Emmerich (2017) has actually questioned the concept of a single 'source', highlighting that texts are continuously remade in various formats. She proposes that translations are not inferior copies but part of a continuum of editions, contributing to the growth of a work in new interpretive forms (5). Venuti (2019) has further expanded on this argument by advocating for a 'hermeneutic' approach to translation, viewing it as an interpretive act that varies based on the receiving culture's context. As Brown observes in her critical commentary, this model recognises the translator's agency and the importance of translation in shaping cultural and social institutions (2022, 5).

Taking a hermeneutic approach to Du Châtelet's translation within the context of eighteenth-century debates on human-animal distinctions, Elisabeth Wallmann (2020) has argued that Du Châtelet's substantial modifications to Mandeville's text should be seen as a response to his vitalistic views on physical and political bodies as well as the consequences of these views for his political economy. By reading both versions closely, the author demonstrates that Du Châtelet's revisions aimed to bypass Mandeville's notion that all living bodies, human or animal, are interconnected systems animated by vital 'spirits'. Thus, these two texts, 'the source' and 'the translation', serve as a lens for examining differing perspectives on the human-animal hierarchy. In her interpretive reading of various sections of both 'the source' and 'the translation', Zinsser argued that Du Châtelet shortened, summarised, or even omitted Mandeville's views with which she disagreed, while translating verbatim—or even enhancing and clarifying—passages that not only aligned with her views but also resonated with her personal experiences (2002, 615–17).

There are four drafts of the Translator's Preface in Voltaire's Library at St. Petersburg and two complete versions of the manuscript of the translation itself, although it remains unclear how or why these manuscripts ended up in Voltaire's papers (see Zinsser 2002, 612). These multiple drafts, along with the numerous revisions to the translation, reflect not only the seriousness of the project, at least for Du Châtelet, but also her awareness of the potential controversies it could provoke. In this context, it is hardly surprising that the translation was never published, as it likely would not have received royal approval—an essential requirement for all publications under the Ancien Régime. Or perhaps Du Châtelet continued revising the text because she had originally intended to publish it and sought to 'avoid Voltaire's experiences with French government censors', as Zinsser has speculated (611).

In any case, different copies of the translation circulated privately among trusted friends and intellectual circles. These various versions were also 'texts in becoming', as I would call them: they reveal how Du Châtelet's own reading and understanding kept changing while she was actualising herself as an author, philosopher, mathematician, and scientist. Reflecting on the enigmas surrounding the manuscripts, Zinsser insightfully observed: 'This is the luck of the historian: to have the manuscripts, to be able to trace changes in successive drafts and the progression of a woman's thought in another time and place, and still to have no explanation of why and how such an extraordinary work was preserved' (622).

The *Fable of the Bees* was Du Châtelet's first translation, but it was not the one that brought her lasting fame. That recognition came with her translation of Newton's *Principia*, which remains the authoritative French edition to this day. In addition to the translation, she wrote an extensive commentary in which she gave her own description of Newton's *System of the World* and analytical solutions to key disputed aspects of Newton's theory of universal gravitation. She also included summaries of two mathematical essays that clarified and confirmed Newton's application of his theory to observed phenomena: Aléxis–Claude Clairaut's work on the shape of the Earth and Daniel Bernoulli's work on the effects of the Sun and Moon on the tides (see Zinsser 2001, 227).

The importance of this work in the history of science and mathematics was cemented in the 1765 entry on Newtonianism in Diderot and d'Alembert's renowned *Encyclopédie*, where Du Châtelet was listed among seven physicists and mathematicians who made Newton's ideas accessible to their contemporaries.[3] This work has also been thoroughly examined within the extensive body of scholarship dedicated to Du Châtelet's contributions to philosophy, science, and mathematics.[4]

Du Châtelet's translation and commentary on Newton's *Principia* was not an isolated occurrence. As feminist historians of science and mathematics have demonstrated, women were instrumental in disseminating scientific knowledge during the Enlightenment. Despite being barred from formal institutions of learning, they participated in the European Republic of Letters mainly through their work as translators and synthesisers of important scientific texts.[5] Thus, translation became a widely favoured path for women who sought to publish their work, allowing them not only to contribute to scientific literature but also to establish themselves as public figures. Despite being viewed as a 'humble task' within the rigid hierarchies of intellectual work, translation allowed women to express their views in print without the direct authority tied to original authorship. But while translating often increased their visibility by linking them to renowned authors, it also cast doubt on the authenticity of their voices, as their contributions were usually confined to prefaces, footnotes, commentaries, and word choices rather than original ideas.

Du Châtelet's project of translating Newton's *Principia,* however, extended far beyond the mere desire to become an 'entrepreneur of the Republic of Letters' or the ambition to publish. She was already a renowned author, with her *Institutions de Physique* [Foundations of Physics] first published in 1740, earning her widespread fame across Europe. Her work had been translated into Italian and German and had even led to her election to the Bologna Academy of Sciences. Du Châtelet's translation project rather aligns with a broader conceptualisation of translation as a dynamic field of knowledge production and dissemination in scientific, literary, and cultural history (Brown 2022), as discussed in the previous section.

Du Châtelet understood that Newton's Latin text posed a challenge for many readers. Translating it into French, the common language of Europe at the time, appeared to be the most effective means of disseminating new knowledge. Here it is important to note that prior to Du Châtelet's work, only one major translation of the *Principia* had been completed: Andrew Motte's 1729 translation from Latin to English of Newton's third edition. But this was admittedly written in a complex and obscure language, even for English readers, let alone French audiences. Zinsser has specifically highlighted Bernard Cohen, who, in his preface to the new English translation of Newton's *Principia* (1999), had praised the clarity and conciseness of Du Châtelet's translation, noting that he referred to it when encountering difficulties with wording or meaning (in Zinsser 2001, 239 n2).

As Zinsser has noted, the manuscript of Du Châtelet's translation, which she deposited at the Bibliothèque Nationale shortly before her death, shows that she worked methodically, making detailed notes in the margins about specific Latin terms and calculations. She often drafted multiple versions of prose sections and carefully chose between scholarly interpretations of contested passages. Zinsser argues that these extensive revisions highlight her clarity and precision, underscoring her exclusive authorship. This challenges claims that Clairaut contributed significantly, as he was heavily engaged in his own mathematical work during the same period, as discussed later (231–2).

Du Châtelet's commentary on Newton's *Principia* went beyond a literal translation, with the goal 'to show Newton's system in all its glory and to perfect it' (233). This extensive commentary, comprising two-thirds of the second volume, played a crucial role in disseminating Newton's ideas on the Continent. It offered a clear synthesis of his work alongside the latest scientific discoveries, explaining complex concepts through progressively detailed chapters, using accessible language, and mathematical demonstrations. Additionally, it provided solutions to key propositions and unresolved problems from Newton's original text and addressed the inadequacies of previous geometric approaches in the face of complex phenomena.

And yet, the reviews that followed Du Châtelet's posthumous publication of Newton's *Principia* reflect both stances vis-à-vis women's translations of scientific texts. Zinsser has poignantly observed that despite the recognition of her contributions by contemporaries, including her entry in the *Encyclopédie*, the most prominent scholars in the history and philosophy of science, with few exceptions, have largely ignored her, attributed her writings to others, or dismissed her work as irrelevant to the broader narrative of modern science (27). Even Cohen, who had praised the quality of her translation, as we have seen above, attributed sole authorship of the explanatory sections, her *Commentaire*, to her tutor and final editor of the posthumous publication, Clairaut (239, n2).

Recent scholarship on Du Châtelet's *Commentaire* has convincingly challenged claims that diminish her authorship and the originality of her work. Contributing to the expanding body of literature seeking to restore Du Châtelet's rightful place in the history of philosophy, science, and mathematics—the *Époque Émilienne* (Hagengruber 2022)—George Smith (2022) has shown that her commentary provides an up-to-date overview of Newtonian theory as it stood in the late 1740s, while also identifying discrepancies between Newton's theories and the research of that time. However, in the middle of 1749, while Du Châtelet was dealing with personal and family matters, both Clairaut and d'Alembert made significant breakthroughs concerning two of these discrepancies. These advancements fundamentally altered the status of Newtonian theory, rendering her commentary outdated by the time it was finally published in 1756. Nonetheless, according to Smith, her commentary remains the most comprehensive 'snapshot' of Newtonian theory just before these pivotal developments, while also ruling out the possibility of any intervention by Clairaut in the text (255).

While Du Châtelet's multifaceted translation and commentary on Newton's *Principia* significantly contributed to the dissemination of his ideas to a wider audience in the Continent, including young mathematicians, and reinforced his prominence in Enlightenment thought, the calculus priority dispute between Newton and Leibniz had led to a stagnation in British mathematics during the nineteenth century, as discussed in Chapter 2. It was thus in the context of reviving British mathematics that Mary Somerville was approached to translate Laplace's *Mécanique Céleste*, and it was through this translation that she gained recognition as an accomplished mathematician throughout Europe.

As we have already seen in previous chapters, Somerville and her husband William visited Paris in 1817. They were welcomed into the Parisian academic community, met Pierre-Simon Laplace and his wife Marie-Charlotte, and paid a visit to their country house at Arcœuil. Somerville had a particular rapport with Laplace, and in her autobiography, she recalls that 'we had a great deal of conversation about astronomy and the calculus, and he gave me a copy of his Système du Monde with his inscription, which pleased me exceedingly' (2001, 140). After this visit, Somerville maintained a correspondence with Laplace until his death in 1827.

Writing to her in August 1824 about the completion of his *Mécanique Céleste*, Laplace expressed his admiration for Newton's *Principia* but acknowledged the necessity of modern algebraic calculus, which had been crucial for French mathematicians like Clairaut, as we have seen above. In closing his letter, Laplace expressed optimism about the gradual adoption of differential calculus in Britain, stating, 'I view with great pleasure your mathematicians now embracing the analytical method, and I have no doubt that by following it with the wisdom characteristic of your nation, they will be led to significant discoveries' (inSomerville 2001, 204 n66). Indeed, notable mathematicians in Somerville's times, including her friends and correspondents John Herschel and Charles Babbage, advocated for reform in British scientific education, favouring the Continental approach to calculus, as we have seen in Chapter 2.

It was in this context that, in early 1827, the year of Laplace's death and the centenary of Newton's, Lord Brougham approached William Somerville, requesting that his wife write a popular account of Laplace's *Mécanique Céleste,* emphasising the need for wider public understanding. 'This letter surprised me beyond expression', Somerville wrote in her autobiography (2001, 187). She believed that Lord Brougham 'must have been mistaken' about her abilities and concluded that her self-taught knowledge was far inferior to that of men educated at universities. She felt it would be highly presumptuous to attempt writing on such a subject, or any other, given her background.

Despite her initial self-doubt, Somerville eventually agreed to this translation project under the condition that she would write a more serious work and keep it secret, burning the manuscript if it did not meet her standards. As discussed in Chapter 2, Somerville worked diligently on her expanded version, consistently drawing on the support and advice of Herschel and, later, Babbage, as evidenced by their mathematical correspondence.

The book was released in 1831 and received widespread acclaim for making *Mécanique Céleste* accessible, with some scholars ranking it just below Newton's *Principia* in significance, although some criticism arose as well. As Robyn

Arianrhod (2012, 194) has noted, Somerville's translation and her seventy-page 'Preliminary Dissertation' clarified many mathematical concepts that Laplace had assumed were obvious and included additional background material on gravitational theory based on Newton's laws. While the dissertation was accessible to non-specialists, the main text was complex and primarily comprehensible only to those trained in higher mathematics. It inadvertently became suitable for advanced university courses, leading professors like Peacock and Whewell to adopt it as a textbook for their higher astronomy classes at Cambridge, marking a significant achievement in her career. Arianrhod has further speculated that Somerville 'must have been thrilled to know that many reviewers commented on how extraordinary it was that a self-taught woman wrote a book that few men could understand' (194). Its reception was equally enthusiastic in France, with the Paris Academy of Sciences commissioning a review that highlighted her exceptional blend of charm and rigorous knowledge.

Translation was thus Somerville's entry into the world of publishing and public recognition, paving the way for her career as a professional writer. Apart from establishing herself as a prominent mathematician and scientist, Somerville's scientific writings became a crucial source of income that supported her entire family, especially during a difficult period of economic hardship, as discussed in Chapter 2. Like Du Châtelet, Ada Lovelace did not engage with translation for financial gain but for the glory of science and personal challenge. As in Somerville's case, however, her translation and extended notes of Luigi Menabrea's memoir on Babbage's Analytical Engine cemented her reputation as a mathematician, as I will discuss next.

In 1840, Charles Babbage travelled to Turin, Italy, to deliver a presentation on the Analytical Engine. His lecture, accompanied by charts, drawings, models, and mechanical notations, highlighted the Engine's key feature: its capacity to direct its own operations. Among the attendees was a young Italian mathematician, Luigi Federico Menabrea (who would later become Italy's prime minister). Menabrea meticulously documented Babbage's ideas, corresponded with him regarding the planned exposition of his work, and subsequently authored a paper detailing the principles of the Analytical Engine. Although he did not publish it until two years after Babbage's presentation, in 1842, the paper, written in French, appeared in the *Bibliothèque universelle de Genève* and became the first comprehensive account of the Analytical Engine's design. The following year, Lovelace translated the twenty-three-page memoir into English.

In his autobiography, Babbage recalls that when Lovelace first informed him of her translation of Menabrea's memoir, he asked why she hadn't authored an original paper on a subject she knew so well. Lovelace replied that this idea had simply not occurred to her. It seems that she was still in the frame of mind of 'serving' his work, as we have seen in Chapter 2. However, when Babbage suggested she add some notes expanding on the design and operation of the Analytical Engine, she eagerly embraced the idea (1864, 136).

Lovelace's task of composing the *Notes* to accompany Menabrea's translation was demanding, as it required her to synthesise Babbage's extensive collection of ideas, writings, and plans amassed over years of experimentation with the Engines. Her role went beyond demonstrating the academic value of his work; she also aimed to support Babbage in his repeated, yet ultimately unsuccessful efforts to secure funding from the British government to build the machine—a goal that, unfortunately, was never realised during his lifetime.

As we have seen in Chapter 1, some biographers and commentators have questioned Lovelace's contribution to the *Notes*, suggesting she may have been merely following Babbage's instructions. However, Babbage himself confirmed in his autobiography that Lovelace authored the *Notes* based on his provided material. The collaboration was substantial, and even when Babbage tried to edit her work, she did not hesitate to critique his editing skills: 'I am much annoyed at your having altered my Note', she wrote on a Saturday afternoon in late July 1843 (MS 37192, f. 360). While she did not oppose revisions, she could not tolerate interference with her actual writing: 'You know I am always willing to make any necessary alterations myself, but I cannot endure another person meddling with my sentences', she insisted (f. 360).

Despite such occasional tensions and disagreements, their correspondence from the summer of 1843 reveals just how closely they collaborated. 'My dear Babbage. I am working *very* hard for you; like the Devil in fact; (which perhaps I *am*)', she wrote at the end of June 1843 (f. 362). She was assured of the progress of the Notes and unconcerned about any possible delays in publication, declaring, 'I will have it well and fully done, or not at all' (f. 363). Despite her confidence, she still needed Babbage's help: 'I want to put something about Bernoulli's Numbers [...] Give me the necessary data and formulae' (f. 363). According to Babbage's autobiography, completing the table for the Bernoulli numbers was the only help he gave to Lovelace in the project:

> We discussed together the various illustrations that might be introduced: I suggested several, but the selection was entirely her own. So also was the algebraic working out of the different problems, except, indeed, that relating to the numbers of Bernouilli[sic], which I had offered to do to save Lady Lovelace the trouble. This she sent back to me for an amendment, having detected a grave mistake which I had made in the process (Babbage 1864, 136).

It is not unexpected for her to identify 'a grave mistake', whether it stemmed from Babbage's table or her own calculations. Lovelace had a keen eye for detail. In an undated letter, likely from early July, she wrote to Babbage once more to explain the reasons for the delay in sending him the work she had been preparing. 'Think of my horror then at just discovering that the Table & Diagram, (over which I have been spending infinite patience & pains) are *seriously* wrong in one or two points' (MS 37192, f. 401). The very next day, on a Sunday evening, the corrections were completed: 'I have worked incessantly, & most successfully all day. You will admire the *Table & Diagram extremely*. They have been made out with extreme care, & all the *indices* most minutely & scrupulously attended to' (f. 337).

The exchange of letters between them during July and August 1843 was remarkably intense, with correspondence being sent at all hours of the day—morning, afternoon, and evening. Although most of these letters are undated, with dates added later in pencil or red ink, the day of the week and the time are consistently noted in the correspondence.[6] In addition to their correspondence, they also held several in-person meetings, primarily when Lovelace was in London, as well as during Babbage's visits to the Lovelace estate in Ockham. 'I find that I do *not* come up on Friday next. But if you will give me Monday Evening (I shall stay the night), it will be desirable',[7] she wrote from her London residence in St James's Square on a Tuesday afternoon in early July (f. 364).

As emerging from their early correspondence, imaginary figures soon entered Lovelace's epistolary discourse: 'Think of my having to *walk*, (or rather *run*), to the Station, in half an hour last evening; while I suppose you were feasting & flirting in luxury & ease at your dinner' (f. 347), she wrote on a Tuesday morning in early July, sketching herself as a Fairy: 'It must be a very pleasant merry sort of thing to have a *Fairy* in one's service, mind & limbs!—I envy you!—I, poor little Fairy, can only get dull heavy *mortals*, to wait on *me*!' (347)

The playful tone of their correspondence carries traces of how literary metaphors, philosophical ideas, and metaphysical reflections were entangled in their mathematical discussions and scientific exchanges, as we have already seen in Chapter 2. Responding to Babbage's epistolary question: 'Why does my friend prefer *imaginary* roots for our friendship?' (f. 350), Lovelace offers a whimsical reply: 'Just because she happens to have some of that very imagination which *you* would deny her to possess; & therefore she enjoys a little *play & scope* for it now& then' (f. 350). This playful response then shifts into a more metaphysical reflection: 'Besides I deny the *Fairyism* to be entirely imaginary' (f. 350).

In late July, Lovelace openly expressed to Babbage her satisfaction with the nearly completed Notes, boldly admitting that she believed them to surpass the quality of the original text of her translation and commentary: 'To say the truth, I am rather *amazed* at them & cannot help being struck *malgré moi* with the really masterly nature of the style, & its superiority to that of the Memoir itself', she wrote to Babbage (f. 393). Her *Notes* meticulously contextualise Babbage's work by underscoring the key distinctions between the Difference Engine and the Analytical Engine. To further illuminate the originality and capabilities of the Analytical Engine, she not only detailed its technical aspects and mathematical operations but also constructed a metaphysical framework to articulate its overarching conceptual design, mapping out both its strengths and limitations. In doing so, she employed poetic metaphors to illustrate its conception and functioning.

Lovelace speculated that the Analytical Engine could be used to compose music or create art, noting its ability to process symbols beyond numerical values. 'Supposing, for instance' she wrote, 'that the fundamental relations of pitched sounds in the science of harmony and of musical composition were susceptible of such expression and adaptations, the engine might compose elaborate and scientific pieces of music of any degree of complexity or extent' (Lovelace 1843, 21). She famously wrote that 'the Analytical Engine weaves algebraic patterns just as the

Jacquard loom weaves flowers and leaves', drawing a metaphor between weaving and programming (23).

When Lovelace's *Sketch of the Analytical Engine* was published in Taylor's *Scientific Memoirs* in August 1843, Menabrea's original article had nearly quadrupled in length, with Lovelace's *Notes* comprising forty-one of its sixty-six pages. Although the full work was credited to her under her complete name, her Notes themselves were signed using only her initials. Lovelace's correspondence with Babbage reveals the careful consideration she gave to how she would sign it: 'Lord L—suggests my signing the translation & the Notes; by which he means, simply putting at the end of the former: "translated ~ A. A. L." & adding to each note the initials A. A. L.', she wrote on a Tuesday morning in early July 1843 (MS 37192, f. 344). However, as she explained to Babbage, she viewed this translation as potentially the first in a series of future works she hoped to publish: 'It is not my wish to proclaim who has written it; at the same time, I rather wish to append anything that may tend hereafter to individualise, & identify it, with other productions of the said A. A. L.' (f. 344).

Overall, Lovelace's extensive *Notes* blend mathematical formulas and technical details with philosophical and poetic reflections. She envisions a holistic approach to science and creativity, arguing that analytical thought and imagination should work in tandem, as I will further discuss in Chapter 6. Her emphasis on understanding the Analytical Engine's broader implications suggests a unique vision of computing that goes beyond mechanics to consider the machine's potential impact on society and human thought.[8] As Christopher Hollings, Ursula Martin, and Adrian Rice succinctly note in their pivotal study on Ada Lovelace:

> Lovelace's paper is an extraordinary accomplishment, probably understood and recognised by very few in its time, yet still perfectly understandable nearly two centuries later. It covers algebra, mathematics, logic, and even philosophy; a presentation of the unchanging principles of the general-purpose computer; a comprehensive and detailed account of the so-called 'first computer program'; and an overview of the practical engineering of data, cards, memory, and programming (2018, 86).

Translation was thus Lovelace's pathway to asserting herself as a published author in the scientific world, but as we have seen, translation was just the trigger of a wider creative project in science, philosophy, and mathematics. Babbage called her 'the Enchantress of Numbers' in a letter sent on 9 September 1843, which ended with the salutation: 'Farewell my dear and much admirer Interpreter' (LB 168, ff. 49–50). Lovelace was obviously much more than an interpreter, as a glimpse into the paper that made her famous has shown.

What is important to highlight here, however, is that Lovelace, as well as Somerville and Du Châtelet before her, did not necessarily engage with translation because there was no other alternative or choice for them. As Brown (2022, 48) compellingly argues, there is no intrinsic association between translation, secondary status, and femininity. While some early modern thinkers, especially in England and France, may have regarded translation as subordinate to 'original' writing, this does not suggest that translation was seen as, or became, a distinctly feminine pursuit. In contrast, in regions such as the German states, translation was frequently viewed as equal to, or even surpassing, 'original' work, with no evidence to suggest it was gendered as a feminine activity.

Instead of slipping into essentialist and ahistorical generalisations, Brown has argued that the perception of translation at any given time and place is often shaped by the specific context in which a text is produced (49). In highlighting this, she has raised several critical questions that have shaped the chapter's analysis and discussion, including a consideration of the status of the translator, the genre and significance of the source texts used, the languages involved, and the objectives behind any translation project as well as their temporal framework. Brown has suggested that to grasp the uniqueness and specificities of any translation project, 'we should read evidence from contemporary sources such as egodocuments and paratexts with caution' (49). I have adopted this approach by reading translations and commentaries in conjunction with relevant letters and correspondences.

In this same context of charting translation projects, I will now consider the relatively uncommon phenomenon of how some scientific, philosophical, and literary works produced by these women mathematicians were translated into other languages by their contemporaries, despite the fact that most of them were ultimately forgotten over the course of the twentieth century, with some still remaining untranslated to this day.

Women's scientific and literary work as source texts for translations

In 1740, Du Châtelet anonymously published *Institutions de Physique* in Paris. Written initially as a guide for her son's education, the work synthesises complex ideas from various philosophical traditions, bridging the gap between Leibnizian metaphysics and Newtonian science. As I will further discuss in Chapter 6, this work is a substantial philosophical and scientific treatise.

The treatise not only solidified her reputation as a serious scholar but also contributed significantly to the dissemination and interpretation of Newtonian science in continental Europe, culminating with the translation and commentary of Newton's *Principia,* as we have seen above and in Chapter 2. It is no wonder then that just a year after the release of the second edition in 1742—published under her name and featuring her portrait in the frontispiece—the *Institutions* were translated into both German and Italian. The German translation was published in Halle and Leipzig, but the title was changed to: *Der Frau Marquisinn von Chastellet Naturlehre an ihren Sohn*, viz. [The doctrine of nature that the Marquise Du Châtelet gave to her son]. The Italian translation was published in Venice, and its title was closer to the original than the German: *Istituzioni di Fisica di Madama la Marchesa du Chastelet Indiritte a suofi Gliuolo*. [Foundations of Physics by Madame the Marquise Du Châtelet. Addressed to her son]. Despite their differences, both titles are noteworthy for highlighting that this scientific work was crafted by 'a mother' specifically for her son, as women's engagement with science was sometimes justified and permitted on the grounds that it would contribute to the education and upbringing of their children.

As discussed in the previous section, Brown has shown that in the German states during the early modern period, translation was not considered inferior to 'original' writing; instead, it was viewed as integral to a broader national reform programme that spanned language, culture, religion, and politics (2022, 48). This programme was envisioned as requiring the active participation of both men and women. In her detailed study of Luise Gottsched's work as a translator, Brown traces Gottsched's early interest in science to her involvement in the scholarly circles of Leipzig, which were fascinated by Leibniz's and Wolff's contributions to mathematics. Since Gottsched's scholarly focus was primarily on translating works from various fields—including physics, metaphysics, ancient and modern history, classical studies, and archaeology—her translations include essays and studies by prominent figures in the European Republic of Letters, such as Newton and French Newtonians like Du Châtelet and Voltaire (Brown 2012, 161–2).

In this context, Gottsched was drawn to Du Châtelet's *Institutions de Physique*, a work that offered a critical overview of contemporary research in the field. Gottsched was particularly captivated by the bold and innovative manner in which Du Châtelet sought to navigate and articulate the ideas of both Leibniz and Newton, whose rivalry defined one of the central scientific debates of the period (see Hagengruber 2012). Although her original intention was to translate the *Institutions*, she discovered that the project had already been undertaken. Instead, she chose to translate and publish the public epistolary debate between Du Châtelet and Jean-Jacques Dortous de Mairan, the Perpetual Secretary of the Paris Academy of Sciences (Gottsched 1741), which I will discuss further in Chapter 4.

As Brown has commented (2012, 164), Gottsched did not choose to translate this polemical correspondence as a discreet means of engaging with scientific discourse. She did not view translation as a secondary intellectual endeavour typically assigned to women. On the contrary, she was well-versed in a wide range of scholarly disciplines and occasionally tackled highly complex material. Brown has observed that the Châtelet-Mairan letters were filled with pages of intricate mathematical calculations, yet Gottsched navigated the long, dense sentences and technical terminology with confidence. In carefully reading the German text, Brown has noted that Gottsched even took the liberty of refining Châtelet's convoluted syntax by strategically adjusting the punctuation to improve clarity (164).

Brown has further remarked that Gottsched's translation reveals her thorough command of the complex relationships between bodies, forces, and velocities, which were central to the *vis viva* debates, as we have seen in Chapter 2. Moreover, her introduction to the translation was clear and illuminating, providing readers with a detailed context for the controversy, and she confidently asserted her authorship on the title page (165). Her translation was part of a broader effort to keep German readers informed about the latest controversies in the European Republic of Letters, which was predominantly shaped by French and English scientists and intellectuals, while also promoting the work of German scholars, thereby fostering scholarship and stimulating debate within the German states (168–72).

The Italian translation of the *Institutions*, including her correspondence with Mairan, was most likely overseen by François Jacquier, a mathematician and co-editor of the renowned Geneva 'Jesuit' edition of Newton's *Principia*. (see Zinsser 2006, 209). His intellectual pursuits and connections within the broader Republic of Letters helped bridge scientific communities across Europe. Du Châtelet had hosted Jacquier at her 'private academy' in Cirey (116), and they maintained correspondence, especially during the years she was engaged in her own work with Newton's *Principia*. According to Zinsser, Jacquier must have also arranged her election at the Bologna Academy of Sciences in 1746, following the translation and publication of the *Institutions* (209). Laura Bassi was definitely using the Italian translation in her lectures at Bologna University, while a copy of the Venetian translation was also in Maria Gaetana Agnesi's rich scientific library at the Palazzo Agnesi in Milan (see Mazzotti 2007, 102).

Agnesi likely read Du Châtelet's *Instituzioni di fisica* while preparing her own *Instituzioni analitiche*, published in 1748 to widespread acclaim in Italy and across Europe, as discussed in Chapter 2. Among the many congratulatory letters Agnesi received upon the publication of her magnum opus was one from Jean-Paul Grandjean De Fouchy, the secretary of the *Académie Royale des Sciences*. In his letter to Agnesi, dated 13 December 1749, he forwarded a highly favourable report on the book, presented to the Academy on 6 December 1749: 'We regard it as the most complete and well-crafted treatise of its kind, and we believe that the Academy will agree with us when we say that it is highly deserving of its approval and praise' (VBA. 0 201. SUP. 44).[9] The report was signed by its authors, Etienne Mignot de Montigny and Mairan, but there was no controversy surrounding Agnesi's work. On the contrary, in a personal letter to Agnesi, dated 8 December 1749, De Montigny described her book as 'the most clear, methodical, and complete' presentation of calculus published to date, particularly praising its unprecedented methodological uniformity (VBA. 0 201. SUP. 45).

It appears that French academicians were far more generous in their recognition of women mathematicians who were geographically distant, as we have also seen with the contrasting treatment of Germain and Somerville. Such discrepancies highlight a tendency to offer more praise to foreign women scholars than to those closer to home. The fact remains that this book was the first on calculus to be published in Europe. Grounded in a predominantly geometrical approach to algebra and calculus, it devoted significant attention to the methods and objectives of Cartesian geometry (see Mazzotti 2007, 116). This focus provided a boost to the French scientific nationalism that Du Châtelet had criticised, as discussed in the previous chapter.

The second volume of the book was translated into French in 1775 under the title *Traités élémentaires de calcul différentiel et de calcul intégral* by Pierre-Thomas Antelmy, with additional notes by Charles Bossut.[10] In his brief preface, the translator noted: 'The principles of differential and integral calculus are explained in a clear and precise manner, making this work particularly suited for those wishing to acquire the necessary knowledge to delve into certain branches of mechanics, hydrodynamics, etc.' (Agnesi 1775, iii). When the translation was submitted to the

Academy for approval, D'Alembert and Vandermonde were appointed to review it. After their report, the Academy determined that the translation deserved to be published under its official 'privilege', as recorded in its minutes dated 30 August 1775, and signed by its perpetual secretary, Grandjean de Fouchy—the same academician who had sent the congratulatory letter to Agnesi twenty-six years earlier.

Beyond the French academic circles, Agnesi utilised her connections at the Bologna Academy to send her work to the Royal Society in London, another major hub of the European Republic of Letters. Her book was enthusiastically received among British mathematicians, and it was from John Colson, the Lucasian professor of mathematics at Cambridge, that it was translated into English. Although the translation was ultimately released in 1801, Colson must have translated it almost immediately after its Italian publication in 1748, since he died in 1760. According to the editor's preface of the posthumous publication of Colson's translation, 'he found after all the Analytical Institutions of Agnesi to be so excellent, that he was at the pains of learning the Italian Language, at an advanced age, for the sole purpose of translating that work into English; that the British Youth might have the benefit of it as well as the youth of Italy' (Agnesi 1801, v). The historical context of this translation is very interesting in terms of its cultural politics in the geographies of science and mathematics. As Mazzotti has commented (2007, 119), in her presentation of calculus, Agnesi incorporated approaches from the northern Italian Leibnizian tradition, while also emphasising themes that were increasingly becoming peripheral to it. Although she used the Continental differential notation, she maintained that the 'differential' was essentially equivalent to Newton's 'fluxion'. Additionally, she prioritised geometrical reasoning and overlooked the mechanical applications that were more prevalent during her time.

The publication of the English translation at the start of the nineteenth century occurred within the context of the British debate on the nature of calculus and the ongoing priority dispute, as discussed in Chapter 2. More than fifty years after the original text's publication and forty years after the translator's death, Francis Maseres and John Hellins championed its publication (see Agnesi 1801, vi–vii). As Mazzotti has commented, they viewed the book as an important introduction to algebra and calculus, particularly aligned with the Newtonian 'geometrical' tradition (2007, 119). Since Agnesi had argued that the Continental and Newtonian notations were equivalent, the English translation converted her differential notation into fluxional notation (see Huffman 2023). What we also learn from the editor's preface is that Colson had designed and begun a popular account of the *Institutions*, 'under the title of *The Plan of the Lady's system of Analyticks,* explaining article by article what was contained in it' (vi). It seems that Colson had envisioned an English version of Algarotti's *Newtonianismo per le dame*, since his unfinished draft of *The Plan* was intended to render Agnesi's work 'more easy and useful for the Ladies of this country' (vi).

The conditions surrounding the translations of Agnesi's *Instituzioni* support Brown's (2022) argument that translations, particularly in the realm of science, were not inherently tied to femininity or viewed as secondary intellectual activities. Eminent academicians and scholars of mathematics were responsible for translating and reviewing her work in both its French and English editions. The effort to popularise her work was in line with similar initiatives undertaken by male scholars for male mathematicians, such as the well-known case of Francesco Algarotti's widely read book, *Il Newtonianismo per le dame* (1737). It is therefore crucial to recognise the role of translation in shaping the power dynamics among different schools of thought in philosophy, science, and mathematics—an influence that extends into the present day.

Agnesi's translations further emphasise what Stephen Muecke describes as 'the political labour of translation', asserting that translation is inherently about 'movement and transformations', while challenging 'the old dream' of transparent communication and the myth of linguistic unity (2024, 3). Indeed, the numerous translations of Kovalevskaya's autobiographical and literary works illustrate how translation functions to create textual palimpsests, layering and reconfiguring discontinuities, gaps, and differences. These translations continuously navigate and negotiate the inevitable disjunctions both within and across languages and societies, as I will discuss next.

As noted in Chapter 1, Kovalevskaya wrote two versions of her autobiography: one in Swedish, narrated in the third person and concealing her authorship (Kovalevskaya 1889), and another in Russian, written entirely in the first person. There were several publications of her *Memories of Childhood* in Russian with differences between the early versions and the later ones (Kovalevskaya 1945, 1951, , 1974). Her chapter 'Palibino', for example, was not included in the early Russian versions of her autobiography, as it had not yet been discovered in the archives of the then USSR Academy of Sciences (see Kovalevskaya 1978, xi).

During her lifetime, Kovalevskaya was aware that her autobiography had already been translated into French, and she intended to include sections omitted from the Russian version. She shared these plans in a letter to her friend Marie Mendelson from Stockholm on 7 October 1890.

> My 'Memories,' which you praised so highly, and some new chapters, which I did not publish in the Russian language (partly out of consideration for censorship), have already been translated into French by a young lady here. The translation has been reviewed by the French ambassador in Stockholm, Mr. Millet, who is an excellent stylist. [...] Mr. Millet is a nephew of Mr. Gaston Paris, who has excellent connections to literary circles, and has promised to take an interest in my book. Currently, he is in France, and I have just received a letter from him informing me that he has good prospects of placing my 'Memories' in the *Nouvelle Revue* (in Mendelson 1897, 611).

Kovalevskaya's autobiography was indeed published in *La Revue de Paris* in three parts, 1 August, 15 August, and 1 September 1894 (Kovalevskaya 1894), and one year later, it appeared as a book together with Leffler's biography, also translated into French (Kovalevskaya 1895c). Following her death, her autobiography was further translated into several languages including Czech, Danish, Dutch, German, Japanese, and Polish. Two English translations appeared in 1895: one in London, *Life in Russia: The Sisters Rajevski* (Kovalevskaya 1895a), and one in New York,

Her Recollections of Childhood (Kovalevskaya 1895b). Both versions, however, presented challenges that led to a third translation in 1978 by Beatrice Stillman (Kovalevskaya 1978). This edition includes a translation of her short autobiographical essay (Kovalevskaya 1891), which was posthumously published in *Ruskaya Starina*. In addition, Stillman's translation used as its source text the 1974 Russian edition of Kovalevskaya's *Vospominaniya* detstva [Memories of Childhood] contained in *Vospominaniya, povesti* [Recollections, Novellas], which is, according to Stillman, the most complete collection of Kovalevskaya's fictional and critical writings (Kovalevskaya 1974). This version also incorporates the first translation into English of the chapter 'Palibino', although its order has been changed from its source text, 'to fit best chronologically' in the translator's view (Kovalevskaya 1978, xi).

Other changes made in this translation include titles for the initial memoir's chapters. The first Russian editions did not feature any titles, while the editions of 1951 and 1961, edited by Solomon Yakovlevich Shtraikh, provided titles that Stillman felt needed to be changed as 'more appropriate to the content of the chapters' (1978, xi). Finally, the very title of the book was changed from *Memories of Childhood* to *A Russian Childhood*, since the translator felt that this title would be more evocative in the minds of English-speaking readers. This title change in a twentieth-century translation reflects the earlier title modifications of Du Châtelet's *Institutions* when translated into Italian and German in the eighteenth century. Such shifts support the understanding of translation as a culturally adaptive process, aligning with broader patterns in translation projects across historical and linguistic contexts.

It is notable that despite the extensive literature surrounding Kovalevskaya's life and work, the Russian collections of her memories, letters, diaries, literary essays, poetry, and plays have yet to be translated into English. Only selected excerpts from these collections have been translated and incorporated into various biographies accessible to a wider audience, as already noted above. An exception is her autofictional novella, which has appeared in English under two titles: *Vera Barantzova* (1895) and *Nihilist Girl* (2001). To the best of my knowledge, the latter has also been translated into French (2004), Hungarian (1904), Italian (2005), and Spanish (2014), but there might be more translations in other languages. Additionally, her two theatrical plays, co-written with Carlotta Leffler, along with a selection of her poetry, were only recently translated into English by Sandra DeLozier Coleman in 2021. In her preface, Coleman describes her fascination with Kovalevskaya's literary writings, detailing her efforts to learn Russian solely to translate Kovalevskaya's work. She also shares the challenges she faced with the Cyrillic alphabet and the intricacies of the poetic metre and rhyme in Kovalevskaya's verses: 'I begun trying to translate the first poem one word at a time. This proved to be a particularly slow and difficult process for someone who was not yet even familiar with the order or sounds of the letters of the Russian alphabet' (2021, 12).

What is intriguing about reading the various versions and translations of Kovalevskaya's auto/biographical and literary writings is how the subtle variations in the texts create a rich interplay of differences and repetitions, reinforcing Muecke's argument above that translation is about 'movement and transformations' (2024, 3). Even the various spellings of her name and surname change across languages create a semiotic puzzle that both dazzles and problematises the reader and the researcher who excavates the bibliographical archive of different editions and translations. These nuances, both semantic and semiotic, captivate the reader, drawing attention to the rhythm of Kovalevskaya's prose and poetry, as I will further discuss in Chapter 7. It is important to remember, however, that some of these differences may simply be the result of linguistic shifts arising from the translation of her work from Swedish or Russian into other languages. In any case, variations in the several publications and translations might seem insignificant for the general reader, yet for the reader/researcher who studies these texts to excavate details of how Kovalevskaya constitutes herself in the very act of writing, such 'little differences' can be hugely important.

Despite gaps in the translation and accessibility of Kovalevskaya's writings, there is no doubt that a substantial body of literature surrounds her, as discussed earlier and in previous chapters. This is not the case, however, for Wang Zhenyi. As noted in the Introduction, Zhenyi wrote six books on astronomy and mathematics, but none of these have survived. What remains today are the *Collection of the Defeng Pavilion* and the *Jingling Series*, which include her nine volumes of prose, a single volume of poetry, and three volumes of *ci* poetry. Her surviving work has not yet been translated into English or any other European language, making her writings an archival translation project par excellence. It is thus with some thoughts on the politics of translation that I will conclude this chapter.

The cultural politics of translation

In this chapter, I have considered historical and contemporary issues of translation in relation to women mathematicians' scientific, philosophical, and literary work. I have studied them both as translators of important scientific works of their time as well as scientists, philosophers, and literary authors whose own work has, in turn, been translated across different eras. In this context, I agree with Brown's (2022) important argument that translation in the early and late modern periods in Europe was not necessarily taken as a feminised secondary activity, a substitute for 'real creation' as well as an outlet for women's suppressed desire to write and become published authors. Women translated foundational works by prominent male mathematicians, and notably, their own work was translated by male contemporaries in both scientific and literary circles. However, a historical discontinuity emerges: despite the celebrated and widely translated contributions of women mathematicians in the eighteenth and nineteenth centuries, much of this legacy was marginalised in the twentieth century. Nevertheless, we now see encouraging signs of a dynamic revival as we approach the second quarter of the new millennium.

What is striking to our own days is the fact that a large part of women mathematicians' auto/biographical writings, and particularly their letters, remain within the linguistic barriers of their geographical context, and consequently the secondary literature that has been revolving around them has not travelled well.

As noted earlier and throughout this work, it is striking that key writings by pioneering women in science and mathematics remain untranslated into English, limiting their accessibility to global audiences. Du Châtelet's and Kovalevskaya's extensive correspondences, Germain's philosophical explorations, Agnesi's biographies, Kovalevskaya's diaries and literary works, and Wang Zhenyi's surviving texts all hold immense historical and intellectual value. Yet, without translation, these vital contributions remain confined to their original languages, hindering broader engagement and appreciation within international scholarly and public circles. Furthermore, the scholarship surrounding these women remains largely confined by national and linguistic boundaries, with few transnational comparative studies available.

As I will further discuss in the Conclusion, when I first undertook the project of writing a feminist genealogy of *automathographies*, I had not anticipated the scarcity of comparative studies focusing on these women—particularly those who share similar ethnic and linguistic backgrounds, like Du Châtelet and Germain, or Somerville and Lovelace. Comparative analyses on a bilateral level are even rarer, with the notable exception of Arianrhod's work (2012), which provides an excellent study contrasting Du Châtelet's translation and critique of Newton with Somerville's work on Laplace.

As shown in this chapter, the 'cultural turn in translation studies' (Brown 2022) plays a pivotal role in transcending linguistic boundaries. Equally critical to current trends, however, is addressing the politics of translation. Drawing connections between the concepts of 'bordering' in politics and the 'translational turn' in linguistics, Naoki Sakai argues that, like borders, translation involves more than delineating linguistic boundaries, and it engages instead with complex social and political dimensions (2009, 71).

Languages are neither coherent nor continuous argues Sakai. Accepting that difference is intrinsic to language, translation emerges as a form of 'political labour' that proposes—and actively constructs—the unity of language. He further argues that without assuming that language is a unified or singular entity, it would be difficult to discuss translation either as a movement from one distinct language to another—inter-lingual translation—or as occurring within a language, where variations are perceived as internal distinctions—intra-lingual translations (72).

Sakai offers the concept of 'many in one' as an alternative to the assumption of linguistic unity, suggesting that the notion of a singular national or ethnic language is always inherently linked with the idea of other languages coexisting in similar frameworks of unity. This element of 'many in one' thus implies that the unity of a language only emerges within a context of plurality, where different languages coexist and interact, challenging the notion that any language can exist as a purely isolated, monolithic entity (75). In this framework, translation is understood not merely as a conduit between languages but, more critically, as 'an ambiguous act of constructing continuity from discontinuity' (71).

Transferring these critical insights into the realm of scientific and philosophical languages, we can see that the translation projects undertaken by women mathematicians in the eighteenth and nineteenth centuries were deeply involved in the 'politics of translation'. This is evident in both cross-linguistic and intra-linguistic translation efforts. For instance, translating Newton's works into French or Laplace's into English demonstrates the transnational dissemination of knowledge. Similarly, intra-lingual cases highlight the reinterpretation of ideas within a single linguistic tradition, such as Lovelace's distinctive rewriting of Babbage's works to make them more accessible to English audiences. Another striking example is Zhenyi's adaptation of Mei Wending's *Principles of Calculation*. Believing the text required simplification for broader comprehension, she reimagined it under the title *The Musts of Calculation*, ensuring its accessibility to a wider readership (see Peterson 2000, 344). These instances underscore the intellectual and cultural labour of translation involved in making complex scientific ideas more universally understood.

Moreover, women mathematicians' works were adapted within various cultural contexts, as evidenced by the title changes in Du Châtelet's translations into German and Italian, or Kovalevskaya's autobiography into English. Their writings were also translated amid ongoing scientific wars and debates, as was the case with Agnesi's translation into English, which included the transposition of the differential notation she used, into the Newtonian fluxion. Across all cases examined in this chapter, Sakai's (2009) element of 'many in one' emerges, highlighting the complex dynamics of translation. This is true whether these translations involved scientific texts or literary and autobiographical writings, as seen in Kovalevskaya's varied translations both across languages and within the Russian language itself.

As the assumption of linguistic unity recedes, languages increasingly permeate and cross boundaries, and transnational feminist studies continue to expand (see Hundle, Szeman, and Hoare 2019), there are strong reasons to anticipate that the gendered landscapes of science will similarly broaden and deepen. Within this context, both this chapter and the book as a whole seek to address an important gap in scholarship on gender, translation, and the production and circulation of scientific knowledge. This work contributes to a richer understanding of how gender and translation are entangled in shaping scientific discourse across different languages and cultures.

Notes

1 See Brown (2022) for an excellent review of the literature in this field.

2 [I hate the vulgar rabble and drive them away," Horace Odes, III, 1,1.] translated and cited in Du Châtelet 2009, 50n46.

3 See D'Alembert 2015, originally published as 'Newtonianisme, ou Philosophie Newtonienne', in *Encyclopédie* ou Dictionnaire raisonné des sciences, des arts et des métiers, 11:122 (Paris, 1765).

4 For the most recent overview of this literature, see Hagengruber 2022, particularly Chapters 7; (Hecht 2022), Chapter 10; (Toulmonde 2022) and Chapter 11 (Smith 2022).

5 As we have seen in previous chapters, in the Italian states, some women also had the opportunity to work as educators in academies and universities, as was the case with Laura Bassi, who became the first woman professor of physics in Europe in 1776.
6 These later annotations by unidentified individuals, according to the British Library archivists, have created conceptual confusion and temporal inconsistencies in the correspondence, which have further been exacerbated by biographers and researchers attempting to reorder them and make corrections.
7 Unless otherwise indicated, all emphases marked in italics are in the epistolary texts as underlined words.
8 For a more detailed exposition and in-depth discussion of Lovelace's Memoir, see Hollings, Martin, and Rice 2018, particularly Chapter 7.
9 A copy of the report is in Agnesi's papers at the Veneranda Biblioteca Ambrosiana in Milan, *Extrait des registres de l'Academie Royale des Sciences (sul merito delle "Instituzioni analitiche" di Maria Gaetana Agnesi), 6 dicembre 1749* ; thereafter [VBA] followed by codice number.
10 See the book at: https://www.e-rara.ch/zut/content/titleinfo/1017133 [Accessed 12 October 2024]

References

Agnesi, Maria Gaetana. 1775. Traités élémentaires de calcul différentiel et de calcul intégral, trans. Pierre-Thomas Antelmy. Paris: Claude-Antoine Jombert.

Agnesi, Maria Gaetana. 1801. *Analytical Institutions, in Four Books*: Originally Written in Italian. Translated by John Colson. Edited by John Hellins. London: Taylor & Wilks.

Algarotti, Francesco. 1737. Il Newtonianismo per le dame, ovvero dialoghi sopra la luce e i colori. Napoli [i.e., Milano]: n.p.

Arianrhod, Robyn. 2012. *Seduced by Logic: Émilie Du Châtelet, Mary Somerville and the Newtonian Revolution*. Oxford: Oxford University Press.

Babbage, Charles. 1864. *Passages from the Life of a Philosopher*. Longman, Green, Longman, Roberts, & Green.

Brading, Katherine. 2019. *Émilie Du Châtelet and the Foundations of Physical Science*. New York and London: Routledge.

Brown, Hilary. 2012. Luise Gottsched the Translator. Studies in German Literature, Linguistics, and Culture 118. Rochester, NY: Camden House.

Brown, Hilary. 2022. *Women and Early Modern Cultures of Translation*. Oxford: Oxford University Press.

Burke, Peter. 2007. 'Cultures of Translation in Early Modern Europe'. In *Cultural Translation in Early Modern Europe*, edited by Peter Burke and R. Po-chia Hsia, 7–38. Cambridge: Cambridge University Press.

Chamberlain, Lori. 1988. 'Gender and the Metaphorics of Translation'. *Signs: Journal of Women in Culture Society* 13 (3): 454–72.

Cohen, Bernard I. 1999. 'A Guide to Newton's Principia'. In *The Principia: Mathematical Principles of Natural Philosophy*. Translated by I. Bernard Cohen and Anne Whitman, assisted by Julia Budenz. Berkeley: University of California Press.

Coleman, Sandra DeLozier. 2021. *Mathematician with the Soul of a Poet: Poems and Plays of Sofia Kovalevskaya*. NC: Bohannon Hall Press.

D'Alembert, Jean Le Rond. 2015. 'Newtonianism or Newtonian Philosophy'. In *The Encyclopedia of Diderot & d' Alembert, Collaborative Translation Project*, translated by Terry Stancliffe. Ann Arbor: Michigan Publishing, University of Michigan Library.

Du Châtelet, Émilie. 1759. *Principes Mathématiques de la Philosophie Naturelle*. Edited by Isaac Newton, E. Du Châtelet, translator. 2 vols. Paris: Desaint & Saillant.

Du Châtelet, Émilie. 2009. *Selected Philosophical and Scientific Writings*. Edited with an Introduction by Judith Zinsser, translated by Isabelle Bour and J. Zinsser. Chicago and London: The University of Chicago Press.

Emmerich, Karen. 2017. *Literary Translation and the Making of Originals*. New York: Bloomsbury.

Findlen, Paula. 1995. 'Translating the New Science: Women and the Circulation of Knowledge in Enlightenment Italy'. *Configurations* 3 (2): 167–206.

Gottsched, Luise. 1741. *Trans. Zwo Schriften, welche von der Frau Marquis von Chat- elet, gebohrner Baronessinn von Breteuil, und dem Herrn von Mairan, beständigem Sekretär bey der französischen Akademie der Wissenschaften, Das Maaß der lebendigen Kräften betreffend, gewechselt worden* [Two Writings Exchanged between the Marquise du Châtelet, née Baroness of Breteuil, and Monsieur de Mairan, Permanent Secretary of the French Academy of Sciences, Concerning the Measure of Living Forces]. Leipzig: Breitkopf.

Hagengruber Ruth, ed. 2012. *Emilie du Châtelet: Between Leibniz and Newton*. London and New York: Springer

Hagengruber Ruth, ed. 2022. *Époque Émilienne*. Cham: Springer.

Hecht, Hartmut. 2022. 'Three French Newtonians and Their Leibnizian Background'. In *Époque Émilienne*, edited by Ruth Hagengruber, 175–96. Cham: Springer.

Hollings, Christopher, Ursula Martin, and Adrian Rice. 2018. *Ada Lovelace: The Making of a Computer Scientist*. Oxford: Bodleian Library.

Huffman, Cynthia, J. 2023. 'Agnesi vs Colson: Did Location Matter?' In *Research in History and Philosophy of Mathematics, Annals of the Canadian Society for History and Philosophy of Mathematics/ Société canadienne d'histoire et de philosophie des mathématiques*, edited by Zack, Maria, Waszek, David, 103–14. Cham: Springer.

Hundle, Anneeth Kaur, Ioana Szeman, and Joanna Pares Hoare. 2019. 'What Is the Transnational in Transnational Feminist Research?' *Feminist Review* 121 (1): 3–8.

Kölving, Ulla, et Andrew Brown, eds. 2018. *La Correspondance d' Émilie Du Châtelet*. 2 vols. Ferney-Voltaire: Centre international d'étude du xviiie siècle (vol. 1, pp. 1–633; vol. 2, pp. 1–593).

Kovalevskaya, Sofia. 1891. 'An Autobiographical Sketch' in *Russkaya starina, No1, 1891, 450–63*, translated by Beatrice Stillman in Kovalevskaya, Sofia, 1978. A Russian Childhood, 213–29. New York: Springer-Verlag.

Kovalevskaya, Sofia. 1893. *Литературные сочинения Софьи Васильевны Ковалевской С портр. авт.* [Literary Works of S.V. Kovalevskaya with a portrait of the Author]. St. Petersburg: M. Stasyulevich's Printing House.

Kovalevskaya, Sofia. 1894. 'Sophie Kovalevsky, Souvenirs d'Enfance', La Revue de Paris, (1 August, 1er partie, 567–611), (15 August, 2de partie, 839–69),1 September, 3^{e} partie, 164–95) Available in Gallica. https://gallica.bnf.fr/ark:/12148/bpt6k2079741/f565.item; https://gallica.bnf.fr/ark:/12148/bpt6k2079741/f837.item; https://gallica.bnf.fr/ark:/12148/bpt6k207975d/f162.item

Kovalevskaya, Sofia. 1895a. 'Reminiscences of Childhood [Life in Russia: The Sisters Rajevski]'. In *Sonia Kovalevsky. Biography and Autobiography*, translated by Louise Von Cossel, 167–317. London: Walter Scott.

Kovalevskaya, Sofia. 1895b. 'Recollections of Childhood'. In *Sónya Kovalévsky, Her Recollections of Childhood. With a Biography*, translated by Isabel F. Hapgood, 1–151. New York: The Century Co.

Kovalevskaya, Sofia. 1895c. *Souvenirs d'enfance: écrits par elle-même et suivis de sa biographie par Mme A. Ch. Leffler, Duchesse de Cajanello*. Paris: Librairie Hachette.

Kovalevskaya, Sofia. 1895d [1892]. *Vera Barantzova*. Translated by Sergius Stepnak and William Westall. London: Ward and Downey.

Kovalevskaya, Sofia. 1904. *A nihilista nő* [Nihilist Girl]. Budapest: Sachs Frigyes.

Kovalevskaya, Sofia. 1945. *Vospominaniya detstva i biograficheskie ocherki* [Childhood Memories and Autobiographical sketches]. Edited and annotated by Solomon Yakovlevich Shtraikh. Moscow: USSR Academy of Sciences. Accessed 24 May 2024. https://www.prlib.ru/item/404617

Kovalevskaya, Sofia. 1951. *Воспоминания и письма* [Memories and Letters]. Edited and annotated by Solomon Yakovlevich Shtraikh. Moscow: USSR Academy of Sciences. [2nd ed. 1961]

Kovalevskaya, Sofia. 1974. *Воспоминания повести* [Recollections, Novellas]. Edited by Pelageya Kochina. Moscow: Nauka.

Kovalevskaya, Sofia. 1978. *A Russian Childhood.* Translated by Beatrice Stillman. New York: Springer-Verlag.

Kovalevskaya, Sofia. 2001 [1892]. *Nihilist Girl.* Translated by Natasha Kolchevska with Mary Zirin. New York: The Modern Language Association of America.

Kovalévskaya, Sofia. 2014. *Una Nihilista.* Translated by Natalia Litvinova. Buenos Aires: Mardulce.

Kovalevsky Sonja. 1889. *Ur ryska lifvet: Systrarna Rajevski* [A Russian Life. Sisters Rajevski]. Stockholm: Hæggström.

Kowalewsky Sonja. 1896a. *I. Teil. Kindheitserinnerungen (Von ihr selbst erzählt) [Part I: Childhood Memories (Told by Herself)]; II. Teil: Was ich mit ihr zusammen erlebt und was sie mir von sich erzählt hat. Von Charlotte Leffler* [Part II: What I Experienced with Her and What She Told Me About Herself by Charlotte Leffler]. Translated into German by M. Kurella. Halle: Otto Hendel.

Kowalewsky, Sonja. 1896b. *Die Nihilistin.* Translated by Louise Flachs-Fokschaneanu. Vienna: Vienner Mode.

Kovalevskaïa, Sophie. 2004. *Une Nihiliste.* Translated by Michel Niqueux. Paris: Phébus.

Kovalevskajia, Sofiya. 2005. *Una ragazza nichilista.* Translated by Alessandro Sfrecola. Trieste: Asterios Editore.

Lovelace, Ada. 1843. 'Sketch of the Analytical Engine invented by Charles Babbage Esq. By L.F. Menabrea, of Turin, officer of the Military Engineers, with notes upon the memoir by the translator'. Taylor's Scientific Memoirs 3, 666–731.

Mandeville, Bernard. 1724. *The Fable of the Bees: or, Private Vices, Public Benefits. With an Essay on Charity and Charity-Schools. And a Search into the Nature of Society.* 3rd ed. London: J. Tonson.

Mazzotti, Massimo. 2007. The World of Maria Gaetana Agnesi, Mathematician of God. Baltimore: Johns Hopkins University Press.

Menabrea, Luigi, F. 1842. 'Notions sur la machine analytique de M. Charles Babbage'. *Bibliothèque universelle de Genève, nouvelle série* 41: 352–76.

Mendelson, Marie. 1897. 'Briefe von Sophie Kowalewska' [Letters from Sofia Kovalevskaya]. *Neue Deutsche Rundschau* 6: 589–614 (in German).

Muecke, Stephen. 2024. 'Bruno Latour and Translation'. *Theory, Culture & Society* 41 (5). https://doi.org/10.1177/02632764241275572

Peterson, Bennet, Barbara. 2000. 'Wang Wang Zhenyi' in Notable Women Of China: Shang Dynasty to the Early Twentieth Century, edited by Barbara Bennett Peterson with He Hong Fei, Han Tie, Wang Ziyu and Zhang Guangyu, 341-346. New York: M. E. Sharp.

Sakai, Naoki. 2009. 'How do We Count a Language? Translation and Discontinuity'. *Translation Studies* 2 (1): 71–88.

Smith, George, E. 2022. 'Du Châtelet's Commentary on Newton's Principia': An Assessment'.In *Époque Émilienne*, edited by Ruth Hagengruber, 255–310. Cham: Springer.

Somerville, Mary. 2001. [1873] *Queen of Science: Personal Recollections of Mary Somerville*, edited by Dorothy McMillan, Edinburgh: Canongate.

Toulmonde, Michel. 2022. 'Émilie Du Châtelet and Newton's Principia'. In *Époque Émilienne*, edited by Ruth Hagengruber, 235–54. Cham: Springer.

Venuti, Lawrence. 2019. *Contra Instrumentalism: A Translation Polemic.* Lincoln: University of Nebraska Press.

Wallmann, Elisabeth. 2020. 'The Human-Animal Debate and the Enlightenment Body Politic: Emilie Du Châtelet's Reading of Mandeville's Fable of the Bees'. *Early Modern French Studies* 42 (1): 86–103.

Zinsser, Judith P. 2001. 'Translating Newton's "Principia": The Marquise du Châtelet's Revisions and Additions for a French Audience'. *Notes and Records of the Royal Society of London* 55 (2) (May): 227–45.
Zinsser, Judith P. 2002. 'Entrepreneur of the "Republic of Letters": Emilie de Breteuil, Marquise Du Châtelet, and Bernard Mandeville's The Fable of the Bees'. *French Historical Studies* 24 (4): 595–624.
Zinsser, Judith, P. 2006. *Daring Genius of the Enlightenment, Émilie du Châtelet.* New York: Penguin.

4 Exceptional Women? The *Agonistics* of Knowledge

'I love studying with more passion than I ever loved the world, but I realized it too late' (BnF, fr.12269. f. 116), Du Châtelet wrote to her tutor Maupertuis on 24 October 1738, her words reflecting a 'frenzy to learn', as Robert Mauzi described it in his introduction to her philosophical treatise *Discourse on Happiness* (1961, xii). Mauzi has further commented that Du Châtelet's passion for study was undeniably a core element of her life: 'For her, study is not what it was for her beloved Cicero—a refined pastime or a restful harbour for the soul after life's tempests—but rather a profound possession' (xxiii).

Du Châtelet's life thus unfolded as a deep, consuming commitment to the pursuit of knowledge—both as the focus of her education as well as the intellectual legacy she would leave as a mathematician and scientist. As we have seen in Chapter 2, all of her scientific correspondence reflects an *agonal* stance and passion for study and reveals 'an insatiable spirit for truth' (Brunet 1929, 28). Building on Du Châtelet's rigorous and passionate pursuit of knowledge, in this chapter, I deploy the concept of *agonism* as a lens through which women's archival documents including their autobiographies, notebooks as well as their scientific correspondences and writings are read. By focusing on agonism, I aim to illuminate the challenges these women faced in their quest for knowledge and recognition within diverse scientific communities. Here, however, *agonism* is not solely understood in terms of conflict, contestation, or struggle, often emphasised in several theories of the political (see Mouffe 2005; Rancière 2010) Instead, and perhaps more significantly, I adopt the Arendtian perspective on *agonism*, emphasising its role as affirmative action towards achievement, creation, and public recognition.

This understanding of *agonism* as a dynamic, constructive force will be a central notion in this chapter, which unfolds in four sections. Following this introduction, I outline and utilise the notion of *agonism* in mapping women mathematicians' discourses and practices as inscribed in their philosophical, scientific, and literary writings, as well as their archival documents, including their autobiographies, notebooks, and letters. In the third section, I propose the notion of the *event* as a modality for understanding how women were historically constituted as mathematicians, scientists, and philosophers in the early and late modern periods in Europe. Finally, in the conclusion, I reflect on the effects of the discourse of 'the exceptional woman' on the formations and problematics of the present.

DOI: 10.4324/9781003533993-5

In the *agonal* spirit

Agonism is an important notion in the philosophical thought of Hannah Arendt, underpinning her configuration of action and politics. As its etymological root in the ancient Greek word ἀγών [struggle] indicates, *agon* presupposes appearance in the public sphere, where human beings actively engage in political discourses, debates, and collective decision-making. In the spirit of *agonism,* they enter processes of constructive contestation. Arendt's notion of *agonism* is informed by her readings of classical Greek political life, where citizens engaged in the *agon*, or contest, in the public sphere not to destroy or overpower one another, but to achieve excellence and recognition within the collective: 'the urge toward self-disclosure [...] became the prototype of action for Greek antiquity and influenced, in the form of the so-called *agonal* spirit, the passionate drive to show one's self in measuring up against others that underlies the concept of politics prevalent in the city-states' (1998, 194, emphasis added).

Central to Arendt's agonism is the idea of action as the means by which individuals disclose their identities and enter the web of human relations. In The *Human Condition*, action is the modality par excellence for appearing in the public realm, being seen and heard by others, and is intrinsically linked to plurality: 'Action, the only activity that goes on directly between men [sic] without the intermediary of things or matter, corresponds to the human condition of plurality, to the fact that men, not Man, live on the earth and inhabit the world' (7). Regardless of its particular aim, action for Arendt creates connections among individuals and has 'an inherent tendency to force open all limitations and cut across all boundaries' (190).

In this context, Arendt's image of *agonism* is affirmative and generative, rather than destructive, as it underscores the role of discourse, recognition, and shared space in building a vibrant political life. Arendt believes that through *agonistic* action, a space for freedom is created, allowing for new beginnings, the potential of each human to bring something unprecedented into the world. As already noted in the previous section, Arendt's idea of *agonism* is not driven by personal victory or conquest but by a shared commitment to political life. She contrasts this with more antagonistic or adversarial conceptions of politics, in which power is often sought through dominance. In her view, authentic political engagement occurs when individuals join together in plurality, working through differences to pursue collective goals while maintaining individual uniqueness. This concept of *agonism* reframes political engagement as a generative and creative process rather than solely an arena for power struggle, but it can also be transferred beyond politics into different domains of the public sphere, including education and science, in the case of this chapter.

Here, it is essential to recognise that while *agonism* is a well-established, though debated, concept within political philosophy, its application within education is a developing yet vital area. However, current educational explorations of *agonism* primarily address the sociological, political, and philosophical dimensions of institutions, discourses, and practices, rather than focusing directly on the processes of learning or the dissemination of knowledge and understanding.[1] In thus transposing the Arendtian conceptualisation of agonism as affirmative contestation in the field of gender and science, I particularly consider the importance of plurality in a shared world of appearances, the realm of what Karen Detlefsen (2022) has configured as 'public science', as we have already seen in the Introduction. Building on Adriana Cavarero's concept of 'surging democracy' (2021), which envisions politics as a generative and affirmative force, an agonistic approach to the realm of 'public science' reveals an analogous image of *surging knowledges*, as I have argued elsewhere (Tamboukou 2024a). These *surging knowledges* emerge from the joy of creative engagement, where women mathematicians explore innovative perspectives, experiment with new ways of seeing, and develop novel approaches to understanding and solving problems.

Thinking with Arendt's take on *agonism* in the sphere of 'public science', it is also important to consider that action is always unpredictable, contingent, boundless, and difficult to prefigure or foreseen, as Bonnie Honig (1993) has aptly observed. It is precisely 'the unruliness of action, its excess, its resistance to being captured—tamed—by any perspective, interpretation, or story' (529) that I have found crucial when using the *agonistic* lens in understanding women's engagement with science and mathematics in the eighteenth and nineteenth centuries, as I will further discuss.

On fire

In introducing this chapter, I have considered Du Châtelet's passion for learning as a starting point for exploring the concept of *agonism,* noting that an *agonal* spirit permeates not only her education but also her scientific contributions, both in published and unpublished works. In this context, her essay on fire (1739) was her first successful attempt to appear in the world of public science and thus constitute herself as a scholarly persona. The opportunity arose in 1737 when the Royal Academy of Sciences in Paris announced a prize competition on the nature and propagation of fire. Voltaire decided to enter the contest, and together with Du Châtelet, they undertook a series of experiments in the laboratory they had established at the Château de Cirey. As Zinsser has recounted, their research included observing the melting and cooling of various substances as well as measuring temperatures of boiling liquids, in an effort to understand the properties of heat and fire as well as their relationship (2006, 152). However, their interpretations of the results diverged, and this led to a scientific rift between them. Voltaire's view was that the experiments confirmed *Newton's* idea that fire was a form of matter, thereby subject to gravitational forces, while Du Châtelet remained unconvinced: 'Fire has no weight, and it could very well be a particular being that is neither spirit nor matter, just as the existence of space, which is proven, is neither matter nor spirit', she wrote to Maupertuis on 21 June 1738 (BnF, fr.12269, f. 92). As Maupertuis was on the assessment panel of the Academy's prize competition, she also felt the need to explain the conditions that had made her decide to write her own essay and submit it to the Academy—the first woman to do so:

> I believe you were quite astonished that I had the audacity to compose a paper for the Academy. I wanted to test my abilities under the protection of

> anonymity, for I was quite hopeful of never being known. M. Du Châtelet was the only one who was in my confidence, and he kept the secret so well that he didn't tell you anything about it in Paris (f. 91).

Du Châtelet's admission that she decided to enter the competition 'to test' her abilities is an Arendtian *agonistic* action par excellence, despite the fact that she did it anonymously to adhere to the patriarchal protocols of the time while also ensuring her husband's support, which she needed for public exposure. In her letter, she also confided to Maupertuis that the process was slow and ambivalent: 'I started working without knowing if I would send my paper and I did not tell M. de Voltaire because I did not want to feel embarrassed in his eyes over an endeavour that I feared he might disapprove of'(f.91), she wrote. Her struggle with her own existential uncertainties and her lack of self-confidence were accompanied by material and practical difficulties: 'I could not conduct any experiments because I was working without the knowledge of M. de Voltaire, and I could not hide them from him [...] I could only work at night, and I was still quite inexperienced in these subjects' (f. 91). Her letter is a fine balance between respecting Voltaire's views while also firmly stating her objections and standing her ground by going for an independent submission. 'M. de Voltaire's work was almost finished before I began mine, which inspired me with ideas and the desire to pursue the same path' (f. 91), she wrote, but she also added: 'I opposed nearly all his ideas in my work' (f. 91).

When the competition results were released in April 1738, neither Voltaire nor Du Châtelet won, and it was at this point that she revealed her secret to Voltaire, as being rejected alongside a famous philosopher did not feel awkward for her anymore: 'I only confessed it to him when I saw in the Gazette that neither he nor I had received a prize. It seemed to me that a rejection shared with him became honourable' (f. 91). The letter thus stages a fine scene of contest wherein scientists compete with each other in the pursuit of truth and knowledge creation. In the Arendtian plane of *agonism*, Du Châtelet and Voltaire worked together in plurality and through differences, in search of scientific truth, while maintaining their individual uniqueness.

Moreover, the outcome of the competition was indeed transformed into an honour rather than failure, since both of their essays were recognised as significant and were included in the Academy's publication of the top five submissions, alongside the three prize-winning essays, one of which was authored by the renowned mathematician Leonhard Euler.[2] In her letter to Maupertuis, Du Châtelet had actually acknowledged Voltaire's support in having her essay printed, a sign of what the Ancient Greeks called *εὐγενὴς ἅμιλλα* [eugenís ámilla], that is noble rivalry, a form of competition that emphasises honour, mutual respect, and the shared goal of improvement or excellence, rather than mere victory over others. 'Instead of resenting my discretion, M. de Voltaire only thought of helping me, and being quite pleased with my work, kindly undertook to request its publication', she wrote (f. 91).

At the time of writing this letter to Maupertuis, Du Châtelet was still waiting for the Academy's response, but she was bold enough to require her tutor's assistance towards the final decision: 'I am hopeful of obtaining it, especially if you would be so kind as to write a word to M. Du Fay and M. de Réaumur, M. de Voltaire has written to both of them' (ff. 91–2). In this context, *the agonal* spirit *in* Du Châtelet's epistolary discourse subtly shifts to a struggle against entrenched patriarchal norms, as the publication of a paper written by a woman was breaking new ground in the Academy's tradition. This *agon* for recognition took on added significance, as it would be over two centuries before the Academy began admitting women members (see Petrovich 1999).

Having cleared the grounds of her participation in the prize competition, Du Châtelet's letter went on to defend her position vis-à-vis the nature of fire, entering the discursive milieu of the 'laboratory letter' that we have discussed in Chapter 2. Although she acknowledged that the idea of fire having no weight and being neither spirit nor matter might seem 'strange', Du Châtelet insisted that it was not 'untenable' and urged Maupertuis to study her essay carefully: 'if you can bear it, I would passionately wish for you to read it' (BnF, fr.12269, f. 92). As Ruth Hagengruber (1999) has observed, Du Châtelet and Maupertuis had lengthy epistolary exchanges on the dangerous liaisons between physics and metaphysics and certainly the idea that fire is 'no spirit, nor matter' falls well into the realm of metaphysical speculations. Such an approach to the nature of fire as 'neither spirit, nor matter' is also an instance par excellence of *surging knowledges*, an idea anticipating 'the modern concept of "energy", an entity animating all substances', as Judith Zinsser has commented (in Du Châtelet 2009, 54).

In his book on *The Psychoanalysis of Fire*, Gaston Bachelard has particularly referred to Du Châtelet's essay, observing that her perspective suggested a metaphysical view where fire defies gravity, creating an oscillation in nature between fire's expansive properties and matter's weight and cohesion. This stance, according to Bachelard, challenged scientists of her time who attributed weight to fire, as we have seen in her disagreement with Voltaire (Bachelard 1964 [1938], 81). Bachelard's engagement with Du Châtelet's essay, 200 years after its first publication, is a forceful sign of the long-lasting effects of her *agonistic* intervention in the scientific debates of her time, which further reaches our own days. As Michael Veldman (2023) has succinctly pointed out, the essay on fire reveals Du Châtelet's commitment to experimentation, rigorous argument, mathematical reasoning, and hypothesis formation. It also traces the development of her ideas, clearly reflecting her intellectual independence in bringing together Newton and Leibniz. In this light, it is not surprising that Du Châtelet was keen to have Maupertuis' view on her first work under publication. Moreover, in anticipating the Academy's positive decision, she wished to improve the essay, hoping that the Academy would accept 'a few corrections' (BnF, fr.12269, f. 92). This seemingly minor point was important, as 'the errata' Du Châtelet insisted on including in the 1739 publication were related to her turn to Leibniz's views on living forces [vis viva] in general and the nature of collisions in particular, thus dismissing her previous endorsement of Mairan's views, as Zinsser has pointed out (in Du Châtelet 2009, 55).

Du Châtelet's turn to Leibniz, which was further elaborated in the *Institutions*, first published only a year later after the dissertation on fire, sparked an intense scientific dispute between her and the Academy's perpetual secretary, Jean-Jacques Dortous de Mairan, as already mentioned in Chapter 2. What began as an instance of 'noble rivalry' in the overall *agonal* spirit quickly escalated into a fierce intellectual battle conducted through an exchange of letters, known as *La Correspondence*, which was included in the second publication of the essay on fire in 1744, that Du Châtelet herself oversaw. 'Read, I implore you, Madame, & reread', Mairan emphatically urged her (Mairan 1741, 9), while sharply accusing her of misunderstanding the mathematics, misrepresenting his arguments, and interjecting her own interpretations in place of his (11). Du Châtelet's reply was equally pointed and laced with irony: 'You repeat to me here once again, Sir, that I have not read your memoir; from hearing it so often, I fear that in the end, you may have succeeded in persuading me. So I have just reread it once again, to be absolutely certain I have read it, but I must admit that I found none of the things you had led me to expect', she concluded with unmistakable wit in her rebuttal (1741, 40).

Although in agreement with Du Châtelet's position, Euler, the winner of the prize competition, expressed his objection to the acrimonious way that the *vis viva* debate was held in his correspondence with her (see Reichenberger 2012, 168). Nevertheless, Du Châtelet's clash with Mairan gained significant traction in the 1740s, with their arguments reprinted in various French publications and translated into German and Italian, as we have seen in Chapter 3. It also captured the interest of contemporary thinkers and was widely discussed in philosophical circles (See Zinsser 2006, 195). Notably, the young Immanuel Kant cited the debate in 1747 in his first published work, *Thoughts on the True Estimation of Living Forces*.[3]

Whether expressed through 'noble rivalries' or being transposed into polemical disputes and confrontations, the *agonal* spirit was a creative force that permeated the experiences of women mathematicians throughout various phases of their lives and work. I will further discuss this below, in the context of Maria Gaetana Agnesi's *agonistic* relationship with science and mathematics.

Visual archives of agonistic affects

In the summer of 1727, the nine-year-old Maria Gaetana Agnesi delivered a spirited address at Milan's Palazzo Agnesi, challenging the notion that women should be excluded from studying the liberal arts: 'I consider it worthwhile today to reveal to my illustrious listeners, no matter how many surround me, how contrary to the truth is the opinion of those who suppose that the studies of the liberal arts are judged unsuitable in women' (Agnesi 2005, 130). To fully understand the significance of this speech, it is essential to consider its historical, political, and cultural context. This event was one of many gatherings showcasing the intellectual and artistic talents of Maria Gaetana and her sister, Maria Teresa. As detailed in previous chapters, these evenings were structured as *conversazioni*—dialogues and debates on hot topics in natural philosophy, including light, colour, gravity, celestial movement, the earth's shape, and the nature of tides.[4] Agnesi's oration transcended its immediate setting in Milanese intellectual circles, however, to align with broader debates about women's rights to education, such as those initiated by the Academia de Ricovrati in Padua in 1723.[5] Framed as both a scholarly project and a performative display, the oration highlighted women's contributions to science and philosophy, showcasing Agnesi's alignment with an intellectual movement advocating for women's academic participation.

Agnesi opened with a caution against reducing intellectual work to spectacle, fearing her speech might be dismissed as the 'frivolous opinion of a clever girl' (Agnesi 2005, 129). However, she celebrated women who had become 'marvels of their sex' (129) through education and achievement, emphasising the critical role men—fathers, brothers, or husbands—played in fostering women's intellectual development. She traced this tradition back to Pythagoras, who educated his daughter Damo and sister Themistoclea, and highlighted Hypatia as the pinnacle of this legacy, a woman who surpassed societal expectations to achieve scholarly distinction (136). Agnesi extended this lineage into her own era, citing figures like Cornelia Piscopia, the first woman to receive a doctorate in philosophy, and Daceria, a scholar of Homeric texts (139). Through these examples, she constructed a genealogical path of learned women to legitimise women's intellectual contributions and envision a future of greater inclusion. However, Agnesi limited this vision to 'exceptional women', leaving broader systemic barriers to women's education largely unaddressed.

While Agnesi's oration was a pivotal moment in her intellectual development, she did not write the text herself. According to her first biographer, Antonio Francesco Frisi, the oration was originally composed in Italian by her tutor, Gemelli, which Agnesi then translated into Latin and memorised (1799, 10). This raises questions about her authorship—whether she simply translated the text or contributed her own insights. Paula Findlen suggests that her tutors likely structured the oration and provided material, with Agnesi having minimal input (inAgnesi 2005, 120). Evidence of grammatical inconsistencies and awkward syntax in the text suggests her struggles with advanced concepts, possibly reflecting her attempt to articulate ideas drawn from her tutors or father, Findlen has observed (121).

The oration was later published as part of a 1729 collection edited by the Academia de' Ricovrati, which included debates on women's education (Volpi 1729). Her oration stood out as one of only two essays by women in the volume, highlighting its significance within the ongoing intellectual debate. Agnesi's dedication of her first published work to the Theatine priest, Augustino Tolotae, reveals her humility and acknowledgement of collaboration, describing the oration as 'a girl's little gift' and noting that she supplemented her own ideas with those of others (Agnesi 2005, 92). This experience not only introduced her to the *agonistic* spirit of intellectual discourse but also laid the foundation for her future scholarly pursuits, as seen in her notebooks, preserved at the Ambrosiana manuscripts, which reveal her developing engagement with complex ideas and *surging knowledges*—a theme I will explore next.

Agnesi's notebooks have been analysed extensively by her biographers, particularly Maria Luisa Anzoletti, who conducted a detailed examination of the twenty-five codices for her 1900 biography. Anzoletti sought to uncover overlooked insights, describing her process as a meticulous, page-by-page review to ensure accuracy. However, she concluded that the manuscripts revealed no trace of Agnesi's personal emotions or inner life, characterising them as devoid of sentiment or intimacy (1900, 123). Contrary to Anzoletti's interpretation, my reading of Agnesi's notebooks revealed a different affective dimension (see Tamboukou 2024b). While they may not include personal reflections, they are imbued with emotion expressed visually and rhythmically, rather than through explicit discourse. This archival rhythm deeply influenced my understanding of Agnesi's approach to learning, shedding light on the *agonistic* dynamics and intellectual energy that shaped her scholarly work as well as the *surging knowledges* that derived from it.

Overall, Agnesi's notebooks demonstrate that she did not merely absorb the knowledge presented to her; rather, she actively engaged with it in the *agonal* spirit of affirmative action. As Anzoletti has importantly underscored, Agnesi's lessons were far from being pre-packaged, ready-to-digest material presented to the student without effort: 'They were not didactic dishes already cooked and served on the table, so that the student had nothing else to do but put them comfortably in his mouth' (1900, 149). Instead, she was required to actively engage with them, 'cook them herself' (149), thus constructing her own understanding. Indeed, Agnesi had to critically analyse each subject from its most basic principles, exercising her dialectical judgment at every step. This process demanded that she arrive at the conclusion of a given question only after carefully weighing all arguments for and against it. Agnesi was trained to navigate the challenging path of discovery, learning through trial and error to discern what seemed true—or was widely accepted as true—before such notions could be considered axiomatic (149).

Agonism in this context leaps from the pages of these documents in a striking visual manner: continuous underlinings, frequent corrections, and numerous additions, alongside margin scribbles and rough sketches of physical phenomena and geometric figures. Furthermore, the notebooks exhibit a distinct spatio-temporal rhythm in their textual signs: while they often begin with neat, firm, calligraphic script, the handwriting becomes increasingly unstable, and the pages grow uneven and convoluted with smudges, scribbles, jottings, and other textual marks. This evolution reflects both her intellectual struggle and her dynamic engagement with the material, marking her journey as a learner and thinker.

What I suggest, therefore, is that although the Ambrosiana notebooks have been rightly discredited as scholarly texts by Agnesi's twentieth-century biographers, their value lies elsewhere: they vividly reveal intense processes of *agonistic* learning and become archives of *surging knowledges*. Viewed from this perspective, the notebooks show that Agnesi became a mathematician, philosopher, and scientist not through the often-exaggerated exceptionality of her mind but through hard and meticulous work as well as her desire to participate in the world of public science, compete with her peers, and excel. This was achieved in an era marked by harsh gendered exclusions and limitations.

Most importantly, far from being dry and devoid of Agnesi's feelings—as Anzoletti has suggested—the documents burst with emotions emanating from the struggle of learning: anticipation, impatience, frustration, and disappointment, but also hope and a resolution to persevere in the process of learning and understanding. Agnesi's notebooks unleash visual forces that momentarily illuminate a young girl's struggle to make sense of a world of knowledge that had unexpectedly opened to her.

Never giving up, thinking the impossible

While Du Châtelet's and Agnesi's *agonistic* spirit culminated in their published work, Sophie Germain's path to authorship emerged through the framework of intellectual competition, culminating in her participation in the 1809 prize contest organised by the Institute of France. The competition sought a mathematical theory on the vibration of elastic surfaces, offering a gold medal and public recognition to the winner. Despite the renowned mathematician Joseph-Louis Lagrange's assertion that existing mathematical tools were insufficient to solve the problem, Germain entered the contest, motivated not by financial need or professional ambition but by a desire to engage with a historical tradition and assert her place within the scientific community (see Bucciarelli and Dworski 1980, 38).

Indeed, the contest represented an opportunity for Germain to step into the public sphere, engage in intellectual rivalry, and challenge herself within the broader tradition of mathematical contests, which have historically driven innovation and participation in the sciences. Though the competition was not tailored for her, Germain's decision to participate reflected her growing confidence and ambition, nurtured through her intellectual exchanges with Lagrange and Adrian-Marie Legendre, who provided her with advice, support, and encouragement (see Musielak 2020, 73). While Siméon Denis Poisson, a protégé of Laplace, emerged as her rival in the competition,[6] Germain's informal mentorships took on new significance, offering her both guidance and validation. The contest became a pivotal moment for Germain, marking her transition from the fringes of academic circles to active participation in the *agonistic* dynamics of scientific inquiry.

Between 1809 and 1811, Sophie Germain devoted herself to tackling the problem of vibrating plates, drawing heavily on Euler's principles while also referencing the work of Lagrange, the Bernoullis, and d'Alembert. In the preface to her prize-winning paper, Germain described her motivation as stemming from scientific curiosity rather than competitive ambition, stating that she studied Euler's memoir 'not with the intention of competing … but with the sole desire to appreciate the difficulties' presented by the problem (1821, v). This intellectual pursuit, sparked by the competition, allowed her to engage deeply with mathematical theory and establish herself within the Parisian mathematical community.

Her dedication is evident in her extensive study of Euler's work and her correspondence with Legendre in early 1811, where she sought guidance on Euler's analytical methods. Over three letters, Legendre provided clarifications, explained complex derivations, and expanded on algebraic challenges she faced, particularly in interpreting sections of Euler's Latin-texted memoir.[7] This brief yet significant mentorship underscores Germain's *agonistic* drive to overcome obstacles and refine her methods, ultimately paving the way for her breakthrough in the field.

Germain submitted her memoir on 21 September 1811, confident she had achieved something significant but seeking validation from the academic community. In the preface to her published paper, she expressed her reliance on external judgment to dispel her doubts and refine her understanding: 'By expressing my feeling, I also present my reasons, so that the enlightened public may weigh them and teach me to judge better' (1821, v). Despite her limited formal training and the challenges she faced, she felt her hypothesis was triggering *surging knowledges*, and eagerly submitted it for evaluation: 'Neither my sense of inadequacy, nor the lack of experience with calculus, nor the little time I had left until the competition [...] could prevent me from submitting a Memoir to the Institute, in which I proposed the hypothesis I had imagined' (vi).

The Institute's competition required submissions to be anonymous, identified only by a motto or epigraph, with the option to include a sealed envelope containing the same phrase along with the entrant's name and address, and to be revealed only if the entry won (Maindron 1881, 54). Germain selected an epigraph from Newton's *Principia*, *Regula II*: *"Effectum naturalium ejusdem generis eædem sunt causæ"* [The effects of natural things of the same kind have the same causes]. This choice reflected her alignment with classical physics and underscored her deep engagement with foundational scientific principles—a theme that would resonate throughout her broader philosophical work, as explored further in Chapter 5.

After submitting her memoir in September 1811, Sophie Germain sought reassurance about its status. On 22 October, Legendre confirmed its receipt, noting it was the only submission. He informed her that the Academy had appointed a review panel, including himself, Laplace, Lagrange, Sylvestre Lacroix, and Étienne-Louis Malus. Aware of the potential conflict of interest due to their correspondence, Legendre discreetly advised her to maintain confidentiality: 'I have said nothing; I advise you as well, to keep silent until a definite judgment is made' (BnF 9118, f. 46).

During the review process, Germain and Legendre continued their exchange. On 10 November, Legendre updated her, explaining that her memoir was actively circulating among the panel members and that a supplemental piece she had submitted later would be reviewed as well: 'Your memoir is in circulation. M. Lacroix had it in his hands last Monday. I will find out tomorrow to whom he handed it over and will attach the supplement to it. The commissioners will then decide whether or not to take this supplement into account' (f. 44). While such interventions technically breached anonymity rules, this informal approach was not uncommon among the Academy's tight-knit intellectual circles, as Bucciarelli and Dworski have noted (1980, 52–3).

Despite her efforts, Germain's memoir was ultimately deemed flawed. On 4 December, Legendre wrote with regret: 'I do not have good news to give you regarding the examination of your memoir. It is found that your main equation is not correct' (BnF 9118, f. 48). He explained her errors in detail, offering alternative methods to refine her approach. Encouragingly, he suggested the Academy might reopen the competition, urging her to continue: 'I imagine that the same question will be asked again—with a new deadline; thus, hope is not lost. On the contrary, it is more important than ever to think about winning the prize' (f. 49).

In her response, Germain expressed gratitude for Legendre's guidance and acknowledged her limitations: 'You have highlighted a flaw in my work, one for which I was somewhat unprepared. Realizing how much I have deceived myself makes me question my abilities. Still, I am driven to pursue the truth to ensure the certainty of knowledge, even when it is not in my favour' (BnF 5166, f. 22). She also appreciated Legendre's efforts to protect her reputation, writing: 'Your kindness truly encourages me, and I deeply appreciate the care you take to shield me from the potential ridicule my work might otherwise attract' (f. 22).

After submitting her first memoir, Sophie Germain paused her correspondence with Legendre to focus on a second entry, prepared for the reopened competition of 1812, with submissions due by October 1813. During this time, Germain demonstrated a growing confidence in her mathematical abilities, despite her ongoing self-doubt. In an undated draft letter to an unidentified recipient—likely a high-ranking scientist at the Institute—she reflected: 'I believe that my theory is supported by sufficient evidence and that it is more developed'. Yet she admitted, 'Even with all the reasons I see in favour of my ideas, I have so little faith in my judgment that I still question their value' (BnF 5166, f. 23).

This letter illustrates Germain's intellectual humility and her awareness of the biases her work might face. She feared that her memoir might be unfairly judged, particularly given Lagrange's earlier assertion that the problem she tackled was nearly unsolvable. 'Without doubt, the problem has been abandoned only because this grand geometer judged it difficult. Possibly this same prejudgment will mean a condemnation of my work without a reflective examination' (f. 24). Her concerns were reinforced by her choice of an epigram for the second memoir, quoting Francis Bacon: 'But by far the greatest obstacle to the progress of science and to the undertaking of new tasks and provinces therein is found in this: that men despair and think things impossible'.[8] This epigram encapsulated her resolve to challenge the boundaries of mathematical knowledge despite the limitations imposed by her gender and the scepticism of her contemporaries.

The submission process for her second memoir was even more complex than the first. Once again, hers was the only entry. The new judging panel included Legendre, Lacroix, Poisson, and Sadi Carnot, with Poisson's involvement posing a clear conflict of interest due to his own work on vibrating plates. Despite these complications, Germain submitted corrections during the review process, but this time Legendre's response was less accommodating. On 4 December 1813, he wrote with evident frustration: 'I do not understand the analysis you have sent me at all; there is certainly an error either in the writing or in the reasoning' (BnF 9118, f. 50). He declined further assistance, noting: 'The die is cast, there is nothing more to change in the report, and with all my goodwill, I could do nothing about it' (f. 50). However, his tone softened later in the letter, acknowledging that while Germain's analysis was flawed, her explanation of the phenomena held merit: 'The rest may be correct in terms of the explanation of the phenomena' (f. 50). He encouraged her to consider independent publication: 'In any case, you have the option to publish your research by either restoring the correct analysis or omitting it, and your work will still bring you honour' (f. 50).

The Academy ultimately recognised Germain's contribution with an honourable mention, reflecting the *surging knowledges* her work had inspired, even without full mathematical rigour. Yet, this phase of her career underscored the precariousness of her position as a woman working on the margins of the mathematical community. Legendre's support, while invaluable, was conditional and often strained, as evidenced by his references to discretion and an unspecified prior incident: 'I always promise you the utmost secrecy, and if you haven't committed some other indiscretion, the matter will be as if it never happened' (f. 50) .

Germain's isolation, reliance on correspondence, and lack of institutional support highlight the broader risks faced by those working on the periphery of their fields, then and now. Her work unfolded against the backdrop of significant social and political upheaval in France, including the fall of Napoleon and the restoration of the monarchy (see Chandler 1980). These challenges, coupled with her determination to contribute to the progress of mathematical knowledge despite systemic barriers, underscore her *agonistic* stance and the lasting impact of her intellectual endeavours.

Despite facing setbacks and working amidst political turmoil, Germain demonstrated extraordinary resilience by entering the Academy's competition for a third time, with the deadline extended to 1 October 1815. Her persistence reflects her commitment to solving the problem of vibrating surfaces and her determination to persevere despite prior criticisms and her exclusion from formal academic circles. For this submission, she chose a motto from Virgil: 'Felix qui potuit rerum cognoscere causas' [Happy is he who was able to know the causes of things][9]—a testament to her passion for uncovering the truths of mathematical phenomena.

This time, Germain's persistence was rewarded, and she became the first woman to win the Academy's prestigious prize. Although the judges expressed reservations about the rigour of her proof—describing her methods as 'awkward and clumsy' compared to the mathematical tools of her time (Bucciarelli and Dworsky, 1980, p. 7)—her achievement was groundbreaking. However, the moment was marred by an administrative mishap. The letter notifying her of the award was delayed, and her tickets for the ceremony were lost in the mail. Jean-Baptiste Delambre, the Academy's perpetual secretary for mathematical sciences, sent her a curt note explaining the oversight and offering handwritten tickets for her guests as a substitute. His tone, strikingly formal and devoid of congratulations, reflected the Academy's ambivalence towards Germain's accomplishment:

> M. Delambre has the honour of presenting his respects to Mademoiselle Germain and sending her two tickets for the Institute ... However, having heard from Mr. Sédillot that Miss Germain had not received any tickets as of last evening, he fears there may have been some oversight and, in that case, asks her to contact him. Since the printed tickets are depleted, he can substitute them with handwritten tickets for as many guests as Miss Germain wishes to bring (BnF 9118, ff. 8–9).

This lukewarm acknowledgement of her historic win may have dissuaded Germain from attending the ceremony. Contemporary accounts, such as the *Journal de Débats*, noted her absence, observing that the public had eagerly anticipated seeing a 'virtuoso of a new kind'. The report remarked: 'The expectation of the public was deceived: the young lady did not go to take the trophy that no one of her gender has ever received in France'.[10] Her absence, while partially due to the logistical delay, also reflects the broader isolation and challenges she faced as a woman mathematician.

Germain's award marked a historic moment but was neither easily attained nor fully celebrated.[11] The delayed recognition and administrative mishaps were emblematic of the systemic biases and obstacles she encountered throughout her career. Yet, the award was the culmination of years of perseverance and intellectual work—the Arendtian *agon* transposed into real struggle. Germain's *agonistic* spirit was a critical component in the process of becoming a woman mathematician and scientist in nineteenth-century Europe.

As previously noted, a year after Germain's award, Somerville visited Paris, yet no record exists of them meeting. This missed opportunity for a connection between two accomplished women mathematicians gains further significance when viewed in the context of Germain's ongoing tensions with the French academic establishment, particularly with Poisson. As a rising star in the academic community and as Laplace's protégé, Poisson was among the guests at a dinner in his patron's house in honour of the Somervilles. Somerville was most probably unaware of the underlying tensions between Germain and the French scientific community. However, she had undoubtedly immersed herself in the *agonistic* process of becoming a mathematician, as I will discuss next.

Mathematical puzzles and medals

As discussed in previous chapters, Somerville was a self-taught mathematician whose initial interest in the field was sparked by a ladies' fashion magazine. However, in describing how she became deeply engaged in the serious study of mathematics, Somerville also mentions her connection and correspondence with fellow Scots William and John Wallace. At the time of their acquaintance, William Wallace was serving as Professor of Mathematics at the Royal Military College and editor of *The New Series of the Mathematical Repository*, later moving on to become Professor of Mathematics at the University of Edinburgh (see Craik, 1999). Somerville recounts their encounter as follows:

> I became acquainted with Mr. Wallace, who was, if I am not mistaken, mathematical teacher of the Military College at Marlow, and editor of a mathematical journal published there. I had solved some of the problems contained in it and sent them to him, which led to a correspondence, as Mr. Wallace sent me his own solutions in return (Somerville 2001, 114).

In her autobiography, Somerville describes returning home from London as a widow with two children and fully dedicating herself to mathematical studies.

During this time, she appears to have been a regular reader of the *Mathematical Repository*, submitting solutions to the mathematical puzzles featured in the journal. What was the context of these exchanges? The *Mathematical Repository* was a six-volume mathematical journal published intermittently from 1806 to 1835. Each volume was organised into sections that included original mathematical essays, mathematical extracts from notable works, and a large collection of mathematics problems with reader-submitted solutions. In the third volume of the journal, a 'Cambridge Problems' section was also added, featuring questions from Cambridge University's Senate-House exams. Each volume included up to 120 questions, distributed across four 'Numbers' of around 30 questions each, which were likely circulated separately before compilation into bound volumes. The final question of each 'Number' was designated as a 'Prize Question', for which the editors awarded a silver medal to the best solution. Prize Questions were similar in content and difficulty to the other problems in the journal and were submitted by a range of contributors, including professors and amateur mathematicians, often using pseudonyms. Notable contributions included a puzzle from a memoir by Gauss and solutions sourced from the *Annales de Mathématiques pures et appliquées*. Each issue also concluded with mathematical notices (Stenhouse 2021, 29–30).

Mathematical periodicals with questions-and-answers sections, like the *Mathematical Repository*, actively engaged readers in the creation and circulation of mathematical and scientific knowledge. This type of mathematical journalism was part of a wider pattern in Britain, Europe, India, and the USA, since the seventeenth century (see Despeaux 2014, 47–50). In studying British mathematical journals in the eighteenth and nineteenth centuries, Sloan Evans Despeaux (2014) has particularly noted that the enduring popularity and prevalence of the 'questions and answers' genre in British journals during this period can be understood as the merging of two distinct mathematical traditions: (1) the recreational problem-solving tradition, which originated from the vibrant almanac market of the eighteenth century, and (2) the problem-solving preferences of Cambridge graduates and their students. The latter was nurtured by the transformation of the Cambridge Tripos into a written examination at the start of the nineteenth century and by the spread of written exams across Britain in the latter half of the century. According to Despeaux, this blend of practices fostered a unique approach to mathematical communication and discovery that managed to transcend both class distinctions and, eventually, the pressures of professionalisation (5).

As Stenhouse has noted, these periodicals provided access to cutting-edge and pioneering mathematics (2020, 8) and played a vital role in the education and career paths of those unable to access formal schooling (7). This group included women of the minor gentry, like Mary Somerville, and men such as her correspondent William Wallace, who, lacking formal education after age eleven, began as a bookbinder's apprentice before eventually moving to Edinburgh to attend university classes (Craik 1999, 240).

Viewed through an Arendtian lens, Somerville's engagement with mathematical journals was thus an act of appearance—her entry into the public sphere, where she exposed her ideas to scrutiny, engagement, and critique from others.

This step into public engagement was fundamental to her intellectual growth, transforming her from a solitary learner into a participant in the shared, *agonistic* dialogue of mathematics. Through her contributions and solutions, Somerville gained recognition, gradually building confidence as a scholar within a vibrant mathematical community. Her mentor's role was crucial in this journey, as he encouraged her to take part in this shared intellectual space. Given Wallace's own unconventional path to academia, it is no surprise that he was eager to support Somerville as she ventured into mathematics, a journey she reflects upon in her autobiography:

> Mr Wallace was elected Professor of Mathematics in the University of Edinburgh, and was very kind to me. When I told him that I earnestly desired to go through a regular course of mathematical and astronomical science, even including the highest branches, he gave me a list of the requisite books, which were in French, and consisted of Francœur's *Pure Mathematics,* and his *Elements of Mechanics*, Lacroix's *Algebra*, and his large work on the *Differential and Integral Calculus*, together with his work on *Finite Differences and Series*, Biot's *Analytical Geometry and Astronomy*, Poisson's *Treatise on Mechanics*, Lagrange's *Theory of Analytical Functions*, Euler's *Algebra*, Euler's *Iso-perimetrical Problems* (in Latin), Clairault's *Figure of the Earth*, Monge's *Application of Analysis to Geometry*, Callet's *Logarithms*, Laplace's *Mécanique Céleste*, and his *Analytical Theory of Probabilities*, &c., &c., &c. (2001, 115).

As discussed in Chapter 2, Somerville's mathematical studies followed the Continental tradition, and we have encountered the authors referenced in her mathematical bibliography earlier, including those who were mentors, tutors, and competitors to Du Châtelet and Germain. Stenhouse further notes (2020, 12) that beyond recommending readings, William Wallace provided Somerville with guidance on solving mathematical problems and shared insights into his preferred approaches to mathematical work. On 12 May 1816, he wrote to her: 'I have been led to consider the problem and have found a solution which I intended to send in this letter' (c. 375/MSDIP–1),[12] referring to her solution to the 14th Question in Issue 14 of the *Mathematical Repository*. However, he held off on sending it, wanting 'to polish it and revise the calculations' (c. 375/MSDIP–1). Wallace believed in taking time to refine a solution, as the initial approach to a problem was often rough. As a mathematician who valued the beauty and elegance of his formulas and proofs, he aimed for a concise and polished result: 'I hardly ever resolved a problem in the most direct manner possible at first: In general I find that a first solution may be improved and shortened, hence it always happens that a short and simple solution is the result of long meditation' (c. 375/MSDIP–1), he explained in the same letter. As I will discuss further in Chapter 6, simplicity and beauty are essential elements of mathematical aesthetics, closely connected to clarity and deeper understanding. Somerville's correspondence with John Wallace was also important but for different reasons, as we read in her autobiography:

> I had now read a good deal on the higher branches of mathematics and physical astronomy, but as I never had been taught, I was afraid that I might imagine that I understood the subjects when I really did not; so by Professor Wallace's advice I engaged his brother to read with me, and the book I chose to study with him was the *Mécanique Céleste*. Mr John Wallace was a good mathematician, but I soon found that I understood the subject as well as he did. I was glad, however, to have taken this resolution, as it gave me confidence in myself and consequently courage to persevere (2001, 117–18).

Somerville's relationship with John Wallace was thus one of mutual respect and intellectual equality, especially in their understanding of advanced mathematics. Working alongside someone she regarded as a peer at that time likely boosted her self-confidence. In this context, their interactions can be seen as a form of intellectual collaboration that embodies Arendt's *agonistic* spirit—a space where participants challenge one another, not to dominate but to elevate the discourse through rigorous, equal exchange. Although there is no evidence that John Wallace formally tutored Somerville, he refers to her as 'his pupil' in a letter dated 12 July 1811, congratulating her on winning a silver medal for her solution to prize-question 310 in Volume 3 of the *Mathematical Repository* (c. 375/MSDIP–1).[13] John Wallace himself was also submitting solutions to the journal's problems, and by 1817, his involvement with the *Mathematical Repository* contributed to his appointment as Master of Arithmetic at the Royal Military College (Stenhouse 2021, 29).

Although only a few letters between Mary Somerville and John and William Wallace survive in the Somerville collection at the Bodleian Libraries, Stenhouse's (2020, 2021) in-depth analysis of their correspondence offers a revealing window into Somerville's engagement with French mathematical analysis and her intellectual development in differential calculus. These exchanges are particularly significant when viewed through the lens of Arendt's image of *agonism*—the idea that true political and intellectual life thrives through contestation and the active participation of individuals in a shared space of discourse and debate.

Somerville's correspondence with the Wallaces, as well as her contributions to the *Mathematical Repository*, highlights her role in an intellectual *agon*, where ideas were not merely exchanged but tested, refined, and often contested in a dynamic, evolving dialogue. Far from being a passive recipient of knowledge, Somerville actively participated in this intellectual struggle, engaging deeply with the cutting-edge mathematical theories of the time and pushing the boundaries of the discipline, thus facilitating the emergence of *surging knowledges*. In this way, her contributions were integral to the reform of British mathematics in the early nineteenth century, a reform that began long before the publication of her celebrated *Mechanism of the Heavens* in 1831 (see Stenhouse 2020).

For Germain, entering the prestigious contests of the French Institute was a way to stake her place in the world of mathematics, using the grandeur of the *Grand Prix des Mathématiques* to establish herself as a serious mathematician in a highly visible way. By contrast, Somerville's approach was quieter but no less significant: sending solutions to problems in mathematical journals like the *Mathematical*

Repository became her initiation into the world of mathematics beyond her own solitary study. For Ada Lovelace, however, the process of becoming a mathematician unfolded differently: her aristocratic status, well-connected networks, and financial resources created different conditions for her *agonistic* engagement with mathematics, as I will explore in more depth in the next section.

The infinite that there is to be done

As previously mentioned in Chapter 1, Dorothy Stein, in her assessment of the Lovelace-Somerville mathematical correspondence, argued that Lovelace was mathematically weak and that her abilities were limited (1985, 90). In transcribing and publishing a selection of Lovelace's letters, Betty Toole argued that it is misleading to judge Lovelace's abilities solely based on the questions she asked her teachers when she encountered difficulties (1992, 2). Christopher Hollings, Ursula Martin, and Adrian Rice have highlighted Toole's observation in considering Lovelace's early education (2017b, 225). They have further suggested that while Somerville played an important role in Lovelace's mathematical development, it was under Augustus De Morgan, the first professor of mathematics at London University, that her mathematical education was truly solidified.

In the early 1840s, Lovelace sought to strengthen her mathematical education, and De Morgan agreed to become her tutor. A renowned mathematician from Cambridge, De Morgan was a prolific writer across various fields of mathematics. He was also married to Sophia Frend, the daughter of Lovelace's tutor, William Frend, as we have seen in Chapter 2. It is important to note that their tutoring relationship was conducted primarily through correspondence. De Morgan guided Lovelace's independent reading, supplemented by a series of mathematical problems and exercises (see Hollings, Martin, and Rice 2017a, 206). The Lovelace-De Morgan correspondence covers the period between July 1840 and January 1842 and consists of 63 surviving letters—43 from Lovelace and 20 from De Morgan—housed in the Lovelace-Byron collection at the Bodleian Libraries.[14] The collection also includes mathematical notes that Lovelace most probably sent to de Morgan.[15] This correspondence has also been digitised and is accessible online through the Clay Mathematics Institute, together with its transcription by Christopher Hollings.[16]

Lovelace seems to be aware of the gaps in her mathematical knowledge from the very beginning of the correspondence: 'I am very much obliged by your remarks & additions. I believe I now understand as much of the points in question as I am intended to understand at *present*' (LB 170, f. 48),[17] she wrote to him on 13 September 1840. Apart from exchanging mathematical ideas and problems, she also felt at ease to confide to her tutor the difficulties of mathematical work: When I sit down to study, I generally feel as if I could *never* be tired;—as if I could go on for ever. I say to myself constantly, 'Now today I will get through so & so'; and it is very disappointing to find oneself after an hour or two quite wearied, & having accomplished perhaps about one twentieth part of one's intentions, perhaps not that (ff. 48–9).

In this epistolary extract, we see an open acknowledgement of the profound challenges within Lovelace's mathematical endeavour—a struggle that both resonates with Arendt's concept of the *agon* but also bends it. Lovelace's words reveal the force of her desire to continue working through intellectual and embodied difficulties. While for Arendt, the *agon* is manifested in the realm of public action, Lovelace's *agon* is largely solitary and private. And while Arendt's *agon* emphasises the visible, performative space where action and speech unfold before others, here the 'disappointment' is one that may go unrecognised and hidden. Yet in this act of confiding in her letter, she sheds light on her *agonistic* engagement with mathematics. Her recipient becomes the singular you, the *tu* in Adriana Cavarero's image of relational narratives (2000). Lovelace's *agon* is shared but does not necessarily become public.

In the context of Lovelace's esoteric experience of the mathematical *agon*, metaphysical reflections inevitably emerge: 'When I compare the *very* little I *do*, with the *very* much—the infinite I may say that there is to *be done*; I can only hope that hereafter in some future state, we shall be cleverer than we are now' (LB 170, f. 49). There is 'work to be done', *the œuvre à faire,* in Étienne Sourieau's philosophical configuration (2009), creative work in process, that keeps Lovelace's existential forces alive and kicking. De Morgan seems to tune into this epistolary rhythm. In addition to addressing her mathematical questions, his response provides reassurance. In a letter dated 15 September 1840, he not only offers her guidance on the appropriate pace for her studies but also connects her reflections to the ideas and writings of figures like Newton and Laplace:

> [...] above all, never estimate progress by the number of pages. You can hardly be a judge of the progress you make, and I should say that it is more likely you progress rapidly upon a point that makes you think for an hour, than upon an hour's quick reading, even when you feel satisfied. That which you say about the comparison of what you do with what you see can be done was equally said by Newton when he compared himself to a boy who had picked up a few pebbles from the shore; and the last words of Laplace were 'Ce que nous connaissons est peu de chose; ce que nous ignorons est immense' [What we know is a small thing; what we do not know is immense.] So that you have respectable authority for supposing that you will never get rid of that feeling; and it is no use trying to catch the horizon (LB 170, f. 14).

Beyond philosophical reflections, Lovelace's letters to De Morgan provide detailed accounts of her readings and the challenges she faced with her tasks. When there were long time gaps between letters, Lovelace's epistolary style would take on a tone of self-irony—a literary way of acknowledging her struggles while also diffusing any potential frustration from her correspondent. 'I hope you have not by this time come to the conclusion that I have drowned the Differential Calculus at least (if not myself with it also) in the Seine or the Channel' (f. 50), she wrote on a Saturday morning in June 1841 from her London residence in St James Square. This letter was written after a four-month hiatus, seemingly due to the emotional turmoil, most probably prompted by the discovery of her late father's purported relationship with his half-sister, Augusta Leigh (see Hollings, Martin, and Rice 2017a, 206). In her letters, Lovelace recognised that disruptions in the continuity of her mathematical work could significantly impact her overall progress:

> I am setting about work again now in good earnest. To say the truth I find myself a little bit rusty & awkward. I have been some days getting my head in again, & it is not yet even at all what it was. However this will soon mend, I doubt not. I very easily lose a habit, but then I very easily re-form it. But I am a little vexed at this interruption. I was going on so nicely (f. 50).

Geographical distance and reliance on written correspondence posed additional challenges in Lovelace's mathematical *agon*. Being in London, therefore, offered valuable opportunities for face-to-face meetings with her tutor, providing a vital supplement to her epistolary education. 'I should prefer *seeing* you to writing, on this occasion' (f. 51), she wrote in the same letter above. Despite her busy social life, 'I have at present no *evening*' (f. 51), she would carve out time for their lessons, either in her house or at the De Morgan family home at 69 Gower Street in Bloomsbury: I would suggest, (if it is not trespassing too much on your time & kindness) that I will send my carriage to bring you here on Tuesday Morning at 1/2 past eleven o'clock. I propose sending for you thus, in order to facilitate your coming. Or would you prefer instead that I should go to Gower St at that hour on Tues^{dy}? (f. 51)

Even when residing in the family estates in the countryside, Lovelace was aware that inertia was to be avoided: 'The last fortnight has been spent in total idleness, *mathematically* at least; for we have had company & been as they say *gadding* about' (f. 62), she wrote on 10 November 1840. And yet, the slow passing of time had revealed a new dimension in her educational philosophy: 'I fancy great proficiency in Mathematical Studies is best attained by *time*;—constantly & continually doing a *little*' (f. 66), she noted in the same letter above. This insight is particularly relevant today, offering a critique of the hurried pace of higher education and emphasising the value of what feminist philosopher Isabelle Stengers (2018) has theorised as 'slow science'.

> If this is so, surely then the University cramming system must be very prejudicial to a real progress in the long run, particularly when one considers how *very very* little School-boys are generally prepared on first going to the Universities, with anything like distinct mathematical or even arithmetical notions of the most elementary kind (ff. 66–7).

The cyclical rhythms of the seasons also found their way into her correspondence as she adapted to the winter weather, study, and life in the countryside. Writing from the estate at Oakham Park on 17 December, she remarked 'This is very mathematical weather' (f. 68), underscoring the pleasure she derived from her studies. 'When one cannot exercise one's muscles *out* of doors, one is peculiarly inclined to exercise one's brains indoors. Accordingly, I have been setting vigorously to work again, with much satisfaction' (f. 68). Attuning herself to the rhythm of 'slow science', Lovelace gradually came to appreciate the pedagogical value of errors.

'I used once to regret these sort of errors, & to speak of time lost over them. But I have materially altered my mind on this subject. I often gain more from the discovery of a mistake of this sort, than from 10 acquisitions made at once & without any kind of difficulty' (f. 70).

In a careful study of the mathematical aspects of her correspondence with De Morgan, Hollings, Martin, and Rice have noted Lovelace's limited background at the outset of her studies, her persistence in tackling challenging concepts, and her growing ability to recognise and learn from her mistakes (2017a, 208). In doing so, they have come to the conclusion that this correspondence provides a window into Lovelace's development over approximately eighteen months, during which she advanced from a relatively elementary understanding to acquiring 'a working knowledge of several major branches of what was then university-level mathematics' (215). In their approach, which centres on the mathematical aspects of the correspondence, the use of epistolary references to contemporary events aids in more accurately dating the letters, especially when dates are absent or misleading—issues compounded by Lovelace's early biographers who sometimes speculatively pencilled in incorrect dates (207).

While recognising the importance of situating the correspondence within its proper time frame, my own analysis has focused on content that delves into philosophical reflections, pedagogical relationships, and the intertwining of the personal and scientific, a crucial analytical theme in my study, as we have seen in Chapter 2. Furthermore, the sheer volume and complexity of Lovelace's letters to De Morgan, as well as her mathematical notes preserved in the Bodleian Libraries, form a vivid, intricate map of her intellectual struggle—a visual cartography that reveals her *agonistic* journey into the world of mathematics.

While her correspondence with De Morgan has been digitised and is accessible online, physically opening Box 170 of the Lovelace-Byron family papers in the reading room of the Weston Library in Oxford offered me a tangible, almost visceral connection to the intense intellectual energy of her mind at work. Indeed, as the opening web page of the Clay Mathematics Institute notes, the documents in this box 'present the most coherent picture of Lovelace's mathematical abilities'[18], revealing her dedication and progress as she grappled with increasingly complex mathematical problems.

Every page, filled with mathematical formulas, diagrams, geometric shapes, and visual representations of equations, stands as a testament to her hard work and intense dedication. These documents further trace her journey from early studies focused on elementary geometry—a foundation that shaped her initial understanding—to her advanced explorations, particularly her engagement with calculus. As the Clay Mathematics Institute website highlights, this collection vividly illustrates the progression of her mathematical thinking and abilities.[19]

Her notes reflect this *agonistic* process visually, with folded A3 pages densely packed with calculations, annotations, and visual aids—arrows connecting related concepts, figures inscribed with layers of geometric transformations, and occasional side notes. It is in her sketches and equations that we see Lovelace experimenting with ways to represent mathematical ideas and problems visually, as if the physical act of writing and drawing were integral to her learning process. These notes offer glimpses into her mathematical thinking, a blend of rigorous structure and exploratory curiosity, as she worked to synthesise mathematical theories with her own understanding and leave her own trace in the field, always in the form of *surging knowledges*. The process of becoming a mathematician unfolds as an embodied, tactile experience; thus, Lovelace's letters and notes create a rich visual cartography of her *agon* in entering the world of mathematics.

Overall, Lovelace's correspondence with De Morgan shows that her position as a mere tutee evolved during this period, as she demonstrated 'a keen eye for detail' (Hollings, Martin, and Rice 2017a, 216), but also provides ample evidence of her independent thinking. While studying De Morgan's *Elements of Algebra*, she felt compelled to question one of his assumptions, writing to him on 10 November 1840: 'Then I am not at all sure that I like the assumption in the last paragraph of page 212. It seems to me somewhat a large one, and much more in need of proof than many things which in Mathematics are rigorously & scrupulously demonstrated' (LB 170, f. 65). Yet, she diplomatically acknowledged that her critique might stem from her own inexperience: 'But these inconsistencies have always struck me occasionally, and are perhaps only in reality the inconsistencies in a beginner's mind' (ff. 65–6). This exchange reflects Lovelace's growing confidence in analysing mathematical reasoning, even as she maintained a modest awareness of her own limitations.

Three years later and in the wake of her published paper on Babbage's *Analytical Engine,* Lovelace decided that she was now a recognised and acknowledged scientist: 'I am now a completely professional person, to speak plainly; and am engaged in studies and in literary and scientific avocations', she wrote to Noel Byron on 9 August 1843 (LB 173, f. 155). As I have shown in Chapter 3, her correspondence with Babbage during the period she worked on Menabrea's translation serves as a testament to her maturation as a mathematician and scientist. Most of all it is deeply embedded in the *agonal* spirit of research and learning. Lovelace's *agonism* was thus marked by an inwards, almost metaphysical striving—a drive to bridge imagination with mathematical insight. In this way, it resonates with Kovalevskaya's own poetic approach to mathematics, which was entangled with her *agonal* spirit, as I will discuss next.

Say what you know, do what you must

As discussed in previous chapters, in November 1883, Sofia Kovalevskaya arrived in Stockholm to take up a visiting lecturer position at the newly established Department of Mathematics at Stockholm University. This temporary role came without a salary, so Kovalevskaya knew she would need to work hard to prove her value and, ideally, secure a permanent position. Alongside adjusting to life in Stockholm—finding an apartment, meeting the academic community, and forming new friendships—Kovalevskaya meticulously prepared her lectures. She was thrilled to have the chance to present her research and contributions to the field (see Kochina 1985, 129). Yet, she still sought guidance from her former doctoral advisor, Karl Weierstrass, as a draft of her only surviving letter to him from December 1883 reveals: 'It is true that I regret somewhat that I did not choose from the first to lecture on the calculus of variations [...] But please, be so kind, my dear best friend, and help me by giving me your advice in my distress' (in Bölling 1992, 27). This letter highlights both her dedication to her teaching and her continued reliance on Weierstrass's mentorship as she navigated this new stage of her career.

Kovalevskaya's need for mentorship extended beyond teaching to her research and publications as well. In the same letter to Weierstrass, she wrote, 'I wish to turn to the detailed work [Ausarbeitung] on my last study, as it is most necessary that it appear this winter in *Acta Mathematica*, and without your help I cannot take a step forward' (27). Her determination to publish while awaiting a decision on her tenure led to the eventual appearance of her paper in the sixth volume of *Acta Mathematica* in 1885 (Kovalevskaya 1885). This publication was a strategic move, showcasing her research contributions and reinforcing her presence in the mathematical community. Apart from publishing work that was already completed, Kovalevskaya also started a new research project on the problem of the rotation of a solid body about a fixed point, which would eventually become the highlight of her academic career. In the mood of *surging knowledges*, this is what she wrote to her friend Maria Jankowska-Mendelson on 19 January 1884:

> I am currently very busy and completely wrapped up in my worries about consolidating my position at the university so as to open this path for women in this way. The new mathematical work I have recently embarked upon fascinates me now, and I would not like to die without having discovered what it is I am looking for. If I manage to solve the problem I am currently working on, then my name will be entered among the names of the most eminent mathematicians (Kovalevskaya 1951, 279).

Being in the fever of preparing for her first lectures, Kovalevskaya was also entering the process of establishing herself in the field not as a 'woman professor', but as a distinguished mathematician, an Arendtian *agonistic* act par excellence. She knew that the process would be long: 'By my reckoning, I need another five years to achieve good results' (279), she wrote in the same letter, 'but I hope that in five years, more than one woman will be capable of replacing me here and I will then surrender myself to the other ambitions of my gypsy nature' (279). Her words reflect both her ambition and her desire to pave the way for future women mathematicians. It is this *agonistic* process that I want to trace through her correspondence with Mittag-Leffler below.

In the summer of 1886, Kovalevskaya wrote to Mittag-Leffler from Paris, sharing news of her interactions with leading French mathematicians, including Henri Poincaré, Joseph Bertrand, and Charles Hermite. These eminent scholars were reportedly impressed by her recent findings on the rotation of a solid body around a fixed point. 'Hermite came to see me soon after', she wrote, 'expressing in the most flattering terms his *admiration* for my new discovery, which I explained to him in great detail' (IML/SK-ML/L115, emphasis in the text).[20]

In a follow-up letter dated 26 June 1886, Kovalevskaya conveyed more exciting news: Bertrand had informed her that the Paris Academy of Sciences was planning a meeting to discuss prize problems, and he intended to propose her rotation problem as a candidate for the prestigious academic prize of 1888: 'Bertrand has the idea of proposing the problem of rotation as a subject for judgment' (L116). To support this proposal, she discussed her findings in detail with Hermite, Bertrand, Camille Jordan, and Jean Gaston Darboux—all members of the committee that would determine the prize topic (L116).

Indeed, later that year, the Academy announced the question for the 1888 Prix Bordin: 'to perfect an important aspect of the theory of the movement of a solid body'. The prize would consist of a gold medal valued at 3,000 francs, with a submission deadline set for 1 June 1888.[21] Kovalevskaya had two and a half years to prepare her submission and threw herself into it, despite facing numerous personal issues during this time. One of the most challenging experiences during this time was the illness and eventual death of her cherished sister, Anna Jaclard, in the autumn of 1887. Additionally, this period saw the arrival of Maxim Kovalevskii in her life and her collaborative work on a theatrical play with Anna Carlotta Leffler, a topic I will explore further in Chapter 7. Amidst these upheavals, the memoir was not completed by the 1 June deadline. To address this, Kovalevskaya submitted a partially finished version, asking for permission to provide a revised edition later.

The unfolding events are vividly recounted in two letters from the summer of 1888. The first, dated 11 June 1888, was from Hermite who wrote to Kovalevskaya that Bertrand and Darboux 'instructed me to let you know that you may, if you believe it necessary, send a new version of the memoir you submitted to the competition, since the committee of the Academy charged with examining the entries has complete power to decide on the deadline' (IML/CH-SK/L8). Hermite added that since it was already summer and the Academicians would be on holiday anyway, it was unlikely that her memoir would be reviewed before October, 'so that your new version or supplement to the first will be on time if it arrives at the Secretariat during the month of September or even at the end of the month'. This reflects a precedent, as we have seen in Germain's case in the early nineteenth century, where submitting supplements was not uncommon at the Paris Academy of Sciences, and that privileged information circulated within Academy circles.

Hermite finally informed her that two other memoirs had been submitted along with hers. However, in her auto/biographical sketch included in her autobiography, Kovalevskaya wrote that 'some fifteen papers were present' (Kovalevskaya 1978, 227). If there were indeed fifteen entries, it suggests that Kovalevskaya's submission was not the only one for which the committee allowed an extension beyond the June deadline.

Since anonymity was still technically required, Kovalevskaya submitted her work under the motto, 'Say what you know, do what you must, come what may'.[22] Her correspondence with Mittag-Leffler reveals that this rule was not strictly observed by the Academy. On 5 July 1888, she wrote to him: 'Last Tuesday Bertrand held a large dinner in my honour, attended by Hermite, Picard, Halphen, and Darboux', adding that 'three toasts were proposed in my honour, and Hermite and Darboux said openly that they have no doubt that I shall have the prize' (IML/SK-ML/L 242).

In the French academic community, prize competitions were sometimes tailored to publicly honour and elevate the work of distinguished scientists—a practice that, as we have seen in the earlier section on Germain, underscored a communal commitment to fostering intellectual achievement. However, this form of institutional recognition did not negate the competitive rigour and personal dedication demanded from participants. For Kovalevskaya, this meant that despite being an established mathematician, she still had to prove her mettle and deliver results that could withstand the scrutiny of her peers. Thus, the summer of 1888 became a period of intense, *agonistic* engagement for her. Embracing the spirit of intellectual struggle, she threw herself into the rigorous demands of refining her work. In Arendtian terms, this was not merely about securing a prize but about entering the public sphere where her ideas would be critically examined and contested—a test of her place within the highest echelons of mathematics. In this *agonistic* context, Kovalevskaya's work was a public declaration of her expertise and dedication, showcasing her resolve to meet and exceed the standards set by the academy.

In his memoirs of Kovalevskaya, Maxim Kovalevskii described the intense and demanding manner in which she dedicated herself to her work. He highlighted not only her relentless commitment but also the profound exhaustion she felt upon completing the initial version:

> The dissertation was written in my presence. Since she needed to hurry to finish, and I took up a lot of her time with conversations, our mutual friend, Professor of Mathematics Mittag-Leffler, convinced me to temporarily leave for Uppsala, just a few hours away from Stockholm. When I returned, considering that the memoir was completed and sent to Paris, I found S.V. suddenly aged; the mental strain she had endured had taken a heavy toll on her. She only slowly recovered from her exhaustion (Kovalevskii 1951, 389–90).

At the crucial stage of refining her initial version, it was natural for Kovalevskaya to seek the guidance of her former mentor. In their correspondence this summer, Karl Weierstrass had invited her to join him in the Harz mountains where he would be spending the summer with his sisters. 'I have already written to you recently that I urgently wish to see you this summer and spend some time with you' (IML/KW-SK /L80), he wrote on 22 June 1888. Although his health was not in a good state, he was expecting to recover and also advise her on the final submission of her paper: 'In your work, you have made significant progress since your somewhat despairing letter of May 7th', he wrote (L80). While acknowledging that 'much remains to be done' (L80) he also reassured her that submitting her paper in its current form had been wise and he urged her to press on with the further developments of her work: 'proceed with fresh courage on the further developments' (L80). On the practical side, he recommended the place as 'known for its very healthy air' and the hotel as reputed to be good and spacious, 'especially for a lady as unpretentious as you' (L80).

When Kovalevskaya arrived there, she joined a group of younger mathematicians, including Mittag-Leffler, who had also travelled to visit Weierstrass, creating a vibrant atmosphere filled with intense discussions on mathematical theories and problems. While Kovalevskaya participated in these evening conversations, her days were devoted to the meticulous work of refining her paper. It wasn't until the last five days of her stay that she had the chance to spend time alone with Weierstrass.

On 11 September 1888, Kovalevskaya sent a postcard to Mittag-Leffler, informing him that her work was nearly finished and that she would be heading to Stockholm via Berlin in a few days (IML/SK-ML/ L244). She completed and sent the refined paper to Paris before setting out on her return journey. On 18 December 1888, she received a letter from the academy announcing that she had been awarded the prize. This time, there were no delays or complications, and she was formally invited to the award ceremony in Paris on 24 December 1888. Moreover, the prize money was increased from 3,000 to 5,000 francs, in recognition of her exceptional achievement, as described in the official announcement:

> This remarkable work contains the discovery of a new case in which it is possible to integrate the differential equations describing the motion of a heavy [solid] body fixed at one of its points. Therefore, the author has not only added a result of the greatest interest to the ones established by Euler and Lagrange, the author has also made a discovery, for which we are obliged by the author's profound study which involved all the resources of modern function theory.[23]

Kovalevskaya was thus celebrated not only as the recipient of 'one of the finest and most difficult to attain'[24] prizes, but also as someone who advanced and honoured the mathematical legacy of Euler and Lagrange in this field. Her contribution was far beyond the stage of *surging knowledges*. Euler, who corresponded with Du Châtelet and inspired Agnesi and Germain, and Lagrange, who had a profound influence on Germain and whose work, along with Euler's, was studied by Somerville and Lovelace, now stood as Kovalevskaya's predecessors in a genealogical line of distinguished mathematical minds.

Regardless of their place within the intellectual hierarchies or the varying degrees of mathematical brilliance they attained, all the women explored in this chapter—and throughout the entire book—emerged as stars in the realm of mathematics in their own unique ways. But did this rise to prominence occur because they were 'exceptional'? This is the central question I wish to address in the following section. In doing so, I want to challenge and deconstruct the prevailing myths and discourses of exceptionality that often surround their achievements.

Events and *lines of flight*

So far in this chapter, I have demonstrated that women mathematicians did not enter the field of mathematical sciences as isolated, 'exceptional' figures but rather through an agonistic engagement with learning, knowledge, and research.

In her analysis of the French artist Elisabeth Vigée-Lebrun, Mary Sheriff (1996) has explored the gendered politics underlying the construction of the 'exceptional woman'. Public discourse often frames exceptions as reinforcing the rule—in this case, the notion that women were inherently unsuited for science and mathematics. However, Sheriff points out that the Latin etymology of *exceptio firmat regulam* reveals a deeper nuance. While commonly interpreted as 'the exception confirms the rule', the verb *firmare* originally means 'to strengthen, support, or fortify' (1). In this linguistic framework, *exceptio firmat regulam* implies that the exception fortifies the rule—a meaning that became embedded in the *Encyclopédie* of eighteenth-century France (1–2).

What are then the implications of reinforcing the rule of exclusion for the figure of 'the exceptional woman' in science and mathematics? She cannot serve as the foundation for overturning the exclusion of women from scientific education nor can she become a relatable role model for the majority of women restricted by this rule. As Ruth Messbarger observes regarding 'exceptional women' in science: 'Their academic brethren typically called them illustri, an epithet that in fact relegated them to positions apart from their fellows in the academy and far from the common mass of women, to whom institutional access was unfailingly denied' (2005, 11). Celebrated as 'exceptional' or *illustri*, the few women mathematicians who emerged in eighteenth- and nineteenth-century Europe were isolated from other women. This isolation came at a double cost: exceptionality was closely tied to perceptions of abnormality and unruliness, as Sheriff insightfully notes (1996, 2), affecting how they were viewed by both men and women.

Despite the challenges posed by the discourse of exceptionality, it is undeniable that these women forged new paths for women's engagement in mathematics, science, and philosophy. If not through the lens of exceptionality, then how can we understand the process of becoming a woman mathematician? In responding to this question, I propose the notion of the *event* as a modality for understanding how women were historically constituted as mathematicians, scientists, and philosophers in eighteenth- and nineteenth-century Europe. But how is the *event* configured on the theoretical and methodological plane of my analysis?

As outlined in Chapter 1, genealogy delves into the archive to uncover forgotten stories and documents, shedding light on the practices and discourses that have excluded and marginalised women from the field of mathematical sciences. Foucault (1986) conceives of genealogy as an analysis of 'descent' and 'emergence' and devotes a great deal of reflection to these two terms, tracing their various uses and connotations in Nietzsche's philosophy. 'Descent' records the true objective of genealogy, by excavating the archive to trace numberless beginnings, nuances, and subtleties that are not easily captured by the historian's eye. Indeed, if we look closely at the marginalised histories of gender and science, 'there is more to explain than we thought; there are crooked contours that we haven't spotted', as Paul Veyne has lucidly commented on Foucault's idea of the 'descent' (1997, 156).

Following the move of the 'descent' into the archive, 'emergence' seeks to capture, in Foucault's words, the very 'moment of arising' (1986, 83) when phenomena first appeared as events in history. Foucault describes emergence as 'the entry of forces' (1986, 84), illuminating the processes, power dynamics, and discourses that frame the event's appearance. Rather than marking a point in a linear historical progression, the *event* in the genealogical approach represents a pivotal moment—an episode shaped by its surrounding conditions.

In this context, Foucault's take on genealogy as 'eventalization' (1986) offers a particularly relevant framework for unravelling entanglements of gender, mathematics, and science. Unlike traditional historical approaches to *events*, 'eventalization' challenges taken-for-granted narratives by questioning cultural assumptions about how things should be. It involves 'making visible a singularity at places where there is a temptation to invoke a historical constant, an immediate anthropological trait, or an obviousness which imposes itself uniformly on all' (Foucault 1991, 76).

This disruption of self-evidences necessitates a rethinking of the power relations that, at specific historical moments, decisively shaped how things were socially and historically established. As Foucault explains, this rethinking uncovers 'a sort of multiplication or pluralization of causes' (76). Rather than treating singularities—the exceptional woman in our case—as an isolated figure to be celebrated, the genealogist examines the *event* of her emergence within the network of discursive and material practices that brought her into being. Moreover, the analysis extends to the effects of the *event* on both historical trajectories and the collective historical imagination.

Building on Foucault's genealogical concept of the *event*, I have extended its epistemological underpinning by engaging with Alfred North Whitehead's process philosophy, which redefines the world as a web of *events* rather than static entities: 'happenings rather than things, verbs rather than nouns, processes rather than substances' (Shaviro 2012, 17). This perspective emphasises the dynamic, relational nature of reality, positioning *events* as spatio-temporal unities that intertwine past, present, and future.

In Whitehead's approach, an *event* reflects its 'contemporaries', integrating the dynamics of its present context as 'a display of immediate achievement' (1967, 71). Yet, the *event* is not confined to the present; it also embodies its past, drawing on 'the modes of its predecessors as memories' that are fused into its composition (71). At the same time, an *event* anticipates the future, holding within it 'such aspects as the future throws back into the present' (72–3), encapsulating a sense of 'anticipation' (73). This interplay of temporal dimensions resonates with the experiences of the women mathematicians discussed in this book. Immersing themselves in their field, they engaged with knowledge from both past and present, while envisioning a future where their contributions might reshape and advance the mathematical sciences. Their emergence in the world of science and mathematics, thus exemplifies how an *event* becomes a synthesis of historical inheritance, present engagement, and aspirational innovation, marking a transformative intersection of temporal forces and *agonistic* intellectual work.

Gilles Deleuze's (2001) engagement with the concept of the event draws on both Foucault and Whitehead, while weaving these influences into a unique philosophical plane. For Deleuze, the *event* is not simply what happens; it is 'inside what occurs, the purely expressed'—a rupture in the ordinary that disrupts established norms and opens pathways to new possibilities (2001, 170). In this light, Deleuze's take on the *event* provides a compelling lens through which to view the emergence of women mathematicians in historical contexts. Indeed, their entry into the male-dominated domains of mathematics can be seen as *events* in Deleuze's sense: figures, moments, and processes that defied prevailing social and intellectual constraints, creating spaces for differentiation and transformation. Women's emergence in the field of mathematics disrupted the image of what it meant to be a mathematician, opening possibilities for the yet to come.

In their collaborative work, Deleuze and Guattari have further theorised the *event* as a transformative moment where existing structures are destabilised and new potentials emerge. *Events*, in their view, serve as openings to the possible—projecting alternative futures, even if those futures remain incomplete, like the May '68 *event*, which Deleuze and Guattari (1984) describe as paradigmatic. Women's emergence in the world of science and mathematics marked a historical discontinuity, brought forward radical new ways of being and becoming a woman, but also inspired alternative collective imaginations, even as the full effects of their actions unfolded unevenly, even to our own days. Situated within Deleuze's and Guattari's (1988) plane of analysis, far from isolated or 'exceptional figures', these women followed *lines of flight*—paths of escape or transformation that allowed them to break free from existing structures or limitations. In this way, they were *deterritorialised* from patriarchal *segmentarities* that confined them to certain roles or excluded them from intellectual spaces.[25]

In this line of thinking, feminist philosopher Elizabeth Grosz (2004) has offered the notion of 'untimely events'—brief ruptures, spontaneous rebellions, and unforeseen encounters that disrupt the present and provide glimpses of radical futures. Women mathematicians' emergence and becoming can indeed be seen as 'untimely events' par excellence, interruptions of the current order, and catalysts for imagining alternative possibilities. What counts in the process of their becoming a mathematician is the moment of eruption, the moment when they suddenly felt what was intolerable in their life, but also saw the possibility for something else.

In light of the above, women's emergence in mathematics was not an exception but rather an *event,* surrounded by a complex *assemblage* of ideas, relationships, and socio-historical forces converging to redefine the landscape of mathematics and beyond. As we have seen in Chapter 1, in Deleuze's and Guattari's philosophy (1988), *assemblages* (or *agencements*) refer to complex, dynamic collections of heterogeneous elements—human and non-human, material and immaterial—that come together to form a functional or relational whole, the figure of the woman mathematician in our case. By tracing women's emergence and becomings as *events* erupting from *assemblages,* I therefore agree with Messbarger's critique of the dangerous image of 'the exceptional woman' in science, a discourse that dominated the way women scientists were perceived in the past but which also reaches our days in new modalities and forms (see Zernike 2023), as I will discuss by way of conclusion.

Exceptionality reconsidered

While biographical accounts of women mathematicians have become more prevalent in recent years, the manner in which their lives are narrated and interpreted remains significant, shaping their visibility and impact. Recent studies have highlighted a wide range of discursive constructions,[26] but in my conclusion to this chapter, I want to return to the myth of exceptionality and consider its effects on women mathematicians today.

My main observation in this chapter is that the myth of the 'exceptional woman' in mathematics persists as a powerful narrative. It implies that for women to succeed in this traditionally male-dominated field, they must possess extraordinary, almost superhuman abilities, distinguishing them from their male counterparts. This myth, rooted deeply in historical precedents, as we have seen in this chapter, misrepresents women's achievements as anomalies rather than the result of intellectual rigour, perseverance, and the often systemic, albeit restricted, opportunities available to them. By framing their success as the outcome of exceptionalism, the myth perpetuates the idea that women must defy societal and institutional barriers to be recognised, which obscures the broader, structural challenges they face.

This continued focus on 'the exceptional woman' reinforces a narrow and skewed historical narrative, one that not only diminishes the role of women in the broader intellectual tradition but also undermines efforts to foster conditions that enable women in mathematics to inhabit and transform the field through processes of becoming, rather than conforming to pre-established structures. By perpetuating the myth of exceptionality, we risk further isolating women from the collective progress of the field, limiting their contributions to being framed as outliers rather than integral members of the mathematical community.

Feminist scholars over the years have critiqued this myth for its role in maintaining gendered epistemologies that place men as the norm in intellectual spaces while relegating women to the position of exceptions. However, such studies have not been particularly extended and nuanced within the world of mathematics, (see Harding 1991; Schiebinger 1999).

It is this gap in the literature that this chapter has addressed by deploying the conceptual tools of *agonism* and the *event* in analysing in detail women mathematicians' struggles and work in the eighteenth- and nineteenth-century Europe. As we have seen, women entered the field of mathematics through diverse and *agonistic* engagements, each unfolding in distinct modalities and forms. These included 'noble rivalries', polemical disputes and confrontations, initiations into the mathematical world, arduous pathways to recognition and publication, and even internal struggles with mathematical understanding and visionary aspirations. Their actions, therefore, illuminate the broader landscape of intellectual *agonism*, revealing how their work embodies the enduring challenge of carving out a place within the mathematical tradition. This was done against the backdrop of societal constraints and limitations, where each mathematician's journey was not only a quest for intellectual advancement but also a confrontation with the entrenched power structures that sought to exclude them.

By examining women's emergence in science and mathematics through the lens of the *event*, rather than limiting the discussion within the narrow discourse of exceptionalism, we gain a richer understanding of what it means to become a woman mathematician, both historically and in the present. This approach moves beyond celebrating a handful of remarkable individuals who 'broke through' and instead highlights the collective and personal struggles, efforts, and networks that have enabled women to claim spaces within mathematics. In doing so, it not only activates a more differentiated and immanent historical *assemblage*, but also charts *lines of flight* toward futures in which such participation is no longer exceptional but woven into the very *plane of consistency* of the discipline.

Notes

1 See Koutsouris et al. 2021, for a comprehensive and critical review of the literature in this area.
2 The other two winners were two Cartesian scientists, Pere Lozeran de Fiesc and M. Le Comte de Créquy. Euler's was the only submission in Latin (see Du Châtelet 1739).
3 For detailed bibliographical references and annotation of the French publications and translations of the debate as well as links to the documents, see Project Vox, https://projectvox.org/du-chatelet-1706-1749/ , particularly, section 2.5 Primary sources — reviews by contemporaries. See also the page on Du Châtelet's primary sources at https://historyofwomenphilosophers.org/project/directory-of-women-philosophers/du-chatelet-emilie-1706-1749/chatelet_primary_sources / [Both sites, accessed 31 October 2024].
4 For detailed descriptions of such events, see de Brosses 1858, 116–18; Anzoletti 1900, particularly Chapters 3&4 and Mazzotti 2007, particularly Chapter 1.
5 See Messbarger 2002, particularly Chapter 1, 'The Debate', 21–48.
6 For a discussion and comparison between Germain and Poisson, see Bucciarelli and Dworsky 1980, 37–9 and Musielak 2021, 73–4.
7 Three letters from Legendre to Germain survive from this period. The first is undated, though it precedes the other two, which are dated 19 & 28 January, 1811. These letters are held in the Bibliothèque Nationale, Correspondance de Sophie GERMAIN avec les mathématiciens et savants Cauchy, Delambre, Fourier, Gauss, Le Gendre, d'Ansse de Villoison, etc. MS. Fr., 9118 (henceforth cited as [BN 9118]), available in Gallica; also published in Germain 1896, 287–298 and also translated in Bucciarelli and Dworski 1980, Chapter 5, 42–51. For a thorough analysis of the mathematics discussed in this correspondence, refer to ibid. See also Musielak 2020, particularly Chapter 6, 72–4. All translations are mine.
8 The epigram is a translation of Bacon's *Novum Organum*, CIX (Spedding, Ellis, and Heath 1875, Vol. 8, 140–1).
9 *Georgics*, Book II, line 490.
10 *Journal des Débats,* Mardi 9 Janvier 1816, p. 2.
11 See Tamboukou 2025 for more discussion of this delayed award letter.
12 Oxford Bodleian Libraries, Somerville Collection, Dep. c. 372, Folder MSW–1, thereafter & in the text, Department number, followed by folder number. Somerville's correspondence with William Wallace has been transcribed in Stenhouse 2021, 175–8.
13 Partially transcribed in Stenhouse 2020, 5, 2021, 29. The medal is now at Somerville College, Oxford (see image in Stenhouse 2021, 32).
14 OBL/ANBL/LB/170/ff., 1–166.
15 OBL/ANBL/LB/170/ff., 167–357.
16 https://www.claymath.org/online-resources/ada-lovelaces-mathematical-papers/ [Accessed 13 November2024]

17 Unless otherwise indicated, all emphases in italics are underlined words in the epistolary text.
18 See https://www.claymath.org/online-resources/ada-lovelaces-mathematical-papers/
19 See https://www.claymath.org/online-resources/ada-lovelaces-mathematical-papers/
20 Sofia Kovalevskaya to Mittag-Leffler, Library of the Mittag-Leffler Institute [IML/SK-ML/L115] Thereafter in the text, initials of correspondents, letter number.
21 Paris Academy of Sciences, *Comptes Rendus,* 1886, vol. 103, N. 26, p. 1395, available at: https://www.biodiversitylibrary.org/item/23483#page/1401/mode/1up [Accessed 11 November2024]
22 *Dis ce que tu sais, fais ce que dois, advienne que pourra* : This motto captures a spirit of integrity, courage, and acceptance in the face of uncertainty; it reflects a stoic approach to life and a belief in acting with authenticity regardless of the outcomes. It was included in the official announcement, published in *Comptes Rendus* in December 1888, vol.107, p. 1042, available at https://www.biodiversitylibrary.org/page/4809711. [Accessed 11 November 2024]
23 *Comptes Rendus* in December 1888, vol.107, p. 1042.
24 *Comptes Rendus* in December 1888, vol.107, p. 1035, https://www.biodiversitylibrary.org/page/4809704 .
25 In Deleuze and Guattari's collective work (1988) *segmentarities* refer to the rigid, binary divisions and norms that structure society. *Deterritorialisations* and *lines of flight* express pathways of escape from fixed boundaries and conventions. See Tamboukou 2010 for a detailed discussion of these concepts in women's life narratives.
26 See, amongst others, Tamboukou 2022; Chatirichvili 2025; Sumpter 2025.

References

Agnesi, Maria Gaetana. 1801 [1748]. *Analytical Institutions for the Use of the Italian Youth.* Translated by John Colson. London: Taylor and Wilks.

Agnesi, Maria Gaetana. 2005 [1727]. 'The Studies of the Liberal Arts by the Female Sex Are by No Means Inappropriate'. In *The Contest for Knowledge*, edited and translated by Rebecca Messbarger and Paula Findlen, 117–40. Chicago and London: The University of Chicago Press.

Anzoletti, Maria Luisa. 1900. *Maria Gaetana Agnesi.* Milan: Cogliati.

Arendt, Hannah. 1998 [1958]. *The Human Condition.* Chicago: University of Chicago Press.

Bachelard, Gaston. 1964 [1938]. *The Psychoanalysis of Fire.* Translated by C. M. Ross Alan. London: Routledge and Kegan Paul.

Bölling, Reinhard. 1992. '… Deine Sonia: A Reading from a Burned Letter'. Translated by David E. Rowe. *The Mathematical Intelligencer* 14 (3): 24–30.

Brunet, Pierre. 1929. *Maupertuis, l'œuvre et sa place dans la pensée scientifique et philosophique du XVIIIe siècle.* Paris: Blanchard.

Bucciarelli, Luis L., and Nancy Dworsky. 1980. *Sophie Germain: An Essay in the History of the Theory of Elasticity.* Dordrecht and Boston, MA.

Cavarero, Adriana. 2000 [1997]. *Relating Narratives: Storytelling and Selfhood.* Translated by Paul A. Kottman. London: Routledge.

Cavarero, Adriana. 2021. *Surging Democracy: Notes on Hannah Arendt's Political Thought.* Translated by Matthew Gervase. Stanford: Stanford University Press.

Chandler, David. 1980. *Waterloo: The Hundred Days.* Oxford: Osprey Publishing.

Chatirichvili, Odile. 2025. '"To Talk of Many Things" in the feminine: A Study of Social and Narrative representations of/by Women Mathematicians, from Autobiographies to Wikipedia'. Paper presented at the Research Symposium 'Numbers and Narratives: A Feminist Genealogy of Automathographie', London, 12 May 2025. Accessed 4 June 2025. https://sites.google.com/view/numbersandnarratives/dissemination/research-symposium/proceedings/to-talk-of-many-things-in-the-feminine.

Craik, Alex D. D. 1999. 'Calculus and Analysis in Early 19th-Century Britain: The Work of William Wallace'. *Historia Mathematica* 26: 239–67.
De Brosses, Charles. 1858. *Le président de Brosses en Italie: Lettre familières écrites d'Italie en 1739 et 1740.* Paris: Didier.
Deleuze, Gilles. 2001 [1969]. *The Logic of Sense.* Translated by Mark Lester with Charles Stivale. London: Continuum.
Deleuze, Gilles, and Félix Guattari. 1984. 'Mai 68 n'a pas eu lieu. Gilles Deleuze et Félix Guattari reprennent la parole ensemble pour analyser 1984 à la lumière de 1968'. *Les Nouvelles Littéraires*, reprinted in *Chimères* 2007/2 (N°64): 23–24.
Deleuze, Gilles, and Félix Guattari. 1988 [1980]. *A Thousand Plateaus: Capitalism and Schizophrenia.* Translated by Brian Massumi. London: The Athlone Press.
Detlefsen, Karen. 2022. 'The Rise of a Public Science? Women and Natural Philosophy in the Early Modern Period'. In *The Cambridge History of Philosophy of the Scientific Revolution*, edited by David Marshall and Dana Jalobeanu, 128–45. Cambridge: Cambridge University Press.
Despeaux, Sloan Evans. 2014. 'Mathematical Questions: A Convergence of Mathematical Practices in British Journals of the Eighteenth and Nineteenth Centuries'. *Revue d'histoire des mathématiques* 20: 5–71.
Du Châtelet, Emilie. 1739. 'Dissertation sur la Nature et la Propagation du Feu' [Dissertation on the Nature and Propagation of Fire]. In *Recueil des Pièces qui ont Remporté le Prix de l'Académie Royale des Sciences en 1738* [Collection of the Pieces that Won the Prize of the Royal Academy of Sciences in 1738], edited by Académie Royale des Sciences, 85–168. Paris: Imprimerie Royale.
Du Châtelet, Emilie. 1740. *Instutitions de Physique* [Foundations of Physics]. Paris: Prault.
Du Châtelet, Emilie. 1741. *RÈPONSE DE MADAME *** A la Lettre que M. De Mairan, Secrétaire perpétuel de l'Académie Royale des Sciences, &c. lui a écrit le 18. Février 1741 sur la question des forces vives.* Bruxelles: Foppens.
Du Châtelet, Emilie. 1744. *Dissertation sur la Nature et la Propagation du Feu.* Paris: Prault Fils.
Du Châtelet, Émilie. 2009. *Selected Philosophical and Scientific Writings.* Edited and with an Introduction by Judith Zinsser, translated by Isabelle Bour and J. Zinsser. Chicago and London: The University of Chicago Press.
Findlen, Paula. 2005. 'Translator's Introduction'. In *The Contest for Knowledge: Debates Over Women's Learning in Eighteenth-Century Italy*, edited by Rebecca Messbarger and Paula Findlen, 117–27. Chicago and London: The University of Chicago Press.
Foucault, Michel. 1986. 'Nietzsche, Genealogy, History'. Translated by Donald. F. Bouchard and Sherry Simon. In *The Foucault Reader*, edited by Paul Rabinow, 76–100. Harmondsworth: Peregrine.
Foucault, Michel. 1991. 'The Ethic of Care for the Self as a Practice for Freedom'. In *The Final Foucault*, edited by James Bernauer and David Rasmussen, 1–20. Cambridge, MA: MIT Press.
Frisi, Antonio Francesco. 1799. *Elogio storico di D^A^ Maria Gaetana Agnesi, Milanese Dell'Academia Dell' Instituto Delle Scienze, e Lettrice Onoraria di Matematiche nella Università di Bologna.* Milan: Presso Giuseppe Galeazzi. Reprinted with an introduction and additional notes by Arnaldo and Giuseppina Masotti. Milan: Scuola Tipografica del Pio Istituto pei Figli della Provvidenza, 1965. Translated by Antonella Cupillari in *A Biography of Maria Gaetana Agnesi, an Eighteenth-Century Woman Mathematician: With Translations of Some of her Work from Italian into English*, 33–106. New York: Edwin Mellen Press, 2007.
Germain, Sophie. 1821. *Recherches sur la théorie des surfaces élastiques.* Paris: Huzard-Courcier.
Grosz, Elizabeth. 2004. *The Nick of Time: Politics, Evolution and the Untimely.* Durham and London: Duke University Press.

Hagengruber, Ruth. 1999. 'Émilie du Châtelet and Maupertuis: Eine Metaphysik in Briefen' [Metaphysics in Letters]. In *Pierre Louis Moreau de Maupertuis: Eine Bilanz nach 300 Jahren* [A Review after 300 years], edited by Hartmut Hecht, 187–206. Berlin: Spitz.

Harding, Sandra. 1991. *The Science Question in Feminism*. Ithaka: Cornell University Press.

Hollings, Christopher, Ursula Martin, and Adrian Rice. 2017a. 'The Lovelace–De Morgan Mathematical Correspondence: A Critical Re-appraisal'. *Historia Mathematica* 44: 202–31.

Hollings, Christopher, Ursula Martin, and Adrian Rice. 2017b. 'The Early Mathematical Education of Ada Lovelace'. *BSHM Bulletin: Journal of the British Society for the History of Mathematics* 32 (3): 221–34.

Honig, Bonnie. 1993. 'The Politics of Agonism: A critical response to "beyond good and evil: Arendt, Nietzsche, and the aestheticization of political action"' by Dana R. Villa. *Political Theory* 21 (3): 528–33.

Kochina, Pelageya. 1985. *Love and Mathematics: Sofia Kovalevskaya*. Translated by Michael Burov. Moscow: Mir Publishers.

Koutsouris, George, Lawren Stentiford, Simon Benham-Clarke, and David Hall. 2021. 'Agonism in Education: A Systematic Scoping Review and Discussion of its Educational Potential'. *Educational Review* 74 (5): 1029–54.

Kovalevskaya, Sofia. 1885. 'Uber die Brechung des Lichtes in Crystallinischen Mitteln' [On the Refraction of Light in Crystalline Media]. *Acta Mathematica* 6: 249–304.

Kovalevskaya, Sofia. 1951. *Воспоминания и письма* [Memories and Letters]. Edited and annotated by Solomon Yakovlevich Shtraikh. Moscow: USSR Academy of Sciences. [2nd ed. 1961]

Kovalevskaya, Sofia. 1978. *A Russian Childhood*. Translated by Beatrice Stillman. New York: Springer-Verlag.

Kovalevskii, Maxim. 1951. 'М. М. Ковалевский о С. В. Ковалевской' [M. M. Kovalevskii about S. V. Kovalevskaya]'. In *Воспоминания и письма* [Memories and Letters], edited and annotated by Solomon Yakovlevich Shtraikh, 388–407. Moscow: USSR Academy of Sciences.

Maindron, Ernest. 1881. *Les fondations du prix de L'Académie des Sciences*. Paris: Gauthier-Villars.

Mairan, Jean-Jacques Dortous de. 1741. *Lettre de M. de Mairan, Secrétaire perpétuel de l'Académie Royale des Sciences, &c. à Madame La Marquise du Chastellet sur la question des forces vives, en réponse aux Objections qu'elle lui a fait dans les Institutions de Physique*. Paris: Académie Royale des Sciences.

Masotti, Arnaldo. 1940. '"Maria Gaetana Agnesi", Rendiconti del Seminario Matematico e Fisico di Milano'. *Reports of the Mathematical and Physics Seminar in Milan* 14: 89–127.

Mauzi, Robert. 1961. 'Introduction'. In *Discours sur le bonheur*, édition critique et commente par Robert Mauzi, i-ccxxvii. Paris: Les Belles Lettres.

Mazzotti, Massimo. 2007. *The World of Maria Gaetana Agnesi, Mathematician of God*. Baltimore: Johns Hopkins University Press, 2007.

Messbarger, Rebecca. 2002. *The Century of Women: Representations of Women in Eighteenth-Century Italian Public Discourse*. Toronto: University of Toronto Press.

Messbarger, Rebecca. 2005. 'The Italian Enlightenment Reform of the Querelle des Femmes'. In *The Contest for Knowledge*, edited and translated by Rebecca Messbarger and Paula Findlen, 1–22. Chicago and London: The University of Chicago Press.

Mouffe, Chantal. 2005. *On the Political*. London: Routledge.

Musielak, Dora. 2020. *Sophie Germain: Revolutionary Mathematician*. 2nd ed. Cham: Springer.

Petrovich, Vesna C. 1999. 'Women and the Paris Academy of Sciences'. *Eighteenth-Century Studies* 32 (3): 383–90.

Rancière, Jaques. 2010. *Dissensus: On Politics and Aesthetics*. Translated by Steven Corcoran. London: Continuum.

Reichenberger, Andrea. 2012. 'Leibniz's Quantity of Force: A "Heresy"? Émilie du Châtelet's *Institutions* in the context of the *Vis Viva* controversy'. In *Émilie Du Châtelet: Between Leibniz and Newton*, edited by Ruth Hagengruber, 157–71. Dordrecht: Springer.

Schiebinger, Londa. 1999. *Has Feminism Changed Science*? Cambridge, MA: Harvard University Press.

Shaviro, Steven. 2012. *Without Criteria: Kant, Whitehead, Deleuze and Aesthetics*. Cambridge, MA: The MIT Press.

Sheriff, Mary. 1996. *The Exceptional Woman: Elisabeth Vigée-Lebrun and the Cultural Politics of Art*. Chicago: University of Chicago Press.

Somerville, Mary. 2001 [1873]. *Queen of Science: Personal Recollections of Mary Somerville*. Edited by Dorothy McMillan. Edinburgh: Canongate.

Souriau Étienne. 2009. [1956]. "Du mode d'existence de l'œuvre à faire". In *Les différents modes d'existence,* suivi de *De l'œuvre à faire,* presented and introduced by Isabelle Stengers and Bruno Latour, 195–217. Paris: Presses Universitaires de France.

Spedding, James, Robert L. Ellis, and Douglas D. Heath, eds. 1875. *The Works of Francis Bacon*. London: Longman & Co.

Stein, Dorothy. 1985. *Ada, A Life and A Legacy*. Cambridge, MA: The MIT Press.

Stengers, Isabelle. 2018. *Another Science Is Possible: A Manifesto for Slow Science*. Translated by Stephen Muecke. Cambridge: Polity Press.

Stenhouse, Brigitte. 2020. 'Mary Somerville's Early Contributions to the Circulation of Differential Calculus'. *Historia Mathematica* 51: 1–25.

Stenhouse, Brigitte. 2021. 'Mary Somerville: Being and Becoming a Mathematician'. PhD thesis The Open University, 2021. https://doi.org/10.21954/ou.ro.0001346f

Sumpter, Lovisa. 2025. 'The Women Who didn't Count'. Paper presented at the Research Symposium 'Numbers and Narratives: A Feminist Genealogy of Automathographie', London, 12 May 2025. Accessed 4 June 2025. https://sites.google.com/view/numbersandnarratives/dissemination/research-symposium/proceedings/the-women-who-didnt-count.

Tamboukou, Maria. 2010. 'Charting Cartographies of Resistance: Line of Flight in Women Artists' Narratives'. *Gender and Education* 22 (6): 679–96.

Tamboukou, Maria. 2022. 'Traces in the Archive. Re-imagining Sofia Kovalevskaya'. *Life Writing*, 19 (3): 341–56.

Tamboukou, Maria. 2024a. 'Decolonizing feminist theories and mapping surging feminist knowledges'. In *The Routledge International Handbook of Transdisciplinary Feminist Research and Methodological Praxis*, edited by Jasmine B. Ulmer, Christina Hughes, Michelle Salazar Pérez, and Carol A. Taylor, 24–35. London: Routledge.

Tamboukou, Maria. 2024b. 'Hidden in the Archive of Gender and Science: The Agonistics of Knowledge and Learning'. *History of Education* 53 (2): 380–402. doi/full/10.1080/0046760X.2024.2304339.

Tamboukou, Maria. 2025. 'An Award Letter too Late to Receive: Considering Sophie Germain's Mathematical Correspondence'. *Journal of Epistolary Studies* 4 (1): 33–50.

Toole, Betty Alexandra. 1992. *Ada, the Enchantress of Numbers*. Moreton-in-Marsh: Strawberry Press.

Veldman, Michael. 2023. 'Seeing Du Chatelet's Methodology by the Light of the Dissertation on Fire'. Paper presented at 'Dynamics and Reason: A Workshop on Du Châtelet and Kant, Paderborn University, 9–10 February.

Veyne, Paul. 1997. 'Foucault Revolutionizes History'. In *Foucault and His Interlocutors*, edited by Arnold I. Davidson, 146–82. Chicago: The University of Chicago Press.

Volpi, Giovanni, Antonio, ed. 1729. *Discorsi accedemici di vari autoriviventi intorno agli studidelle donne; la maggior parte recitati nell' Academia de Ricovrati di Padova* [Academic Discourses of various living authors regarding the studies of women; the vast

majority of which was read at the Academy of the Ricovrati of Padua]. Padua: Giovanni Manfrè.
Whitehead, Alfred North. 1967 [1925]. *Science and the Modern World.* New York: Free Press.
Zernike, Kate. 2023. *The Exceptions: Nancy Hopkins, MIT and the Fight for Women in Science.* New York: Scribner.
Zinsser, Judith, P. 2006. *Daring Genius of the Enlightenment, Émilie du Châtelet.* New York: Penguin.

5 Love and Mathematics

Existential Choices and Intellectual Freedom

> You wrote me once that you could come to Paris. Do you still have this project, do you imagine this possibility? It would be great to be there; we would spend our lives together, face to face, but it might be that it would be unbearable for you, especially during the first trip. If I go to Cirey this fall, or any time soon, I will have to come back here this winter to see M. de Richelieu and for my Newton, which is a very serious business and very essential for me. I can't work on it now, I'm so dazzled; I only vegetate, and I only feel that I am capable of thinking and feeling because I love you (MA.2287.59, f.9).[1]

On 5 June 1748, Émilie Du Châtelet penned a letter to her lover, Jean François de Saint-Lambert,[2] reflecting on the challenges their future together might hold. At forty-two, Du Châtelet was not only an established scientist and author but also deeply engrossed in her translation and commentary on Newton's *Principia*. Torn between her longing to spend time with Saint-Lambert at the Château de Cirey and her commitment to her scientific work in Paris, Du Châtelet's letter vividly illustrates the competing demands of personal and intellectual love. In a poignant excerpt from this lengthy twelve-page handwritten letter, Du Châtelet confesses that her love for Saint-Lambert is a vital force sustaining her life; yet, she also acknowledges that 'her Newton' was equally 'essential' to her existence. As earlier chapters have shown, her work on the *Principia* was not merely a personal endeavour but a landmark contribution to the history of science and mathematics.

This chapter explores the intricate intersections where love, as an existential force, intertwines with the pursuit of scientific, philosophical, and mathematical knowledge and emerges as a dynamic component in the process of becoming a woman mathematician. Du Châtelet's writings provide a compelling lens to examine these entanglements, illuminating the profound and often complex interplay of affects, emotions, intellect, and the labour of scholarship. Her reflections on love resonate deeply with the auto/biographical, literary, and philosophical writings of all women mathematicians examined throughout this book, each of whom navigated the emotional and intellectual landscapes of their work in unique and unrepeatable ways. In this context, the chapter unfolds in five sections.

Following this introduction, I trace threads of love in women mathematicians' auto/biographical narratives, particularly highlighting the connection between love

DOI: 10.4324/9781003533993-6

and mathematics. In the third section, I look at Du Châtelet's philosophical treatise, *Discourse on Happiness* (1961), excavating philosophical underpinnings in her conception of love. Then, I turn to her amorous correspondence with Saint-Lambert, which reveals her deeply personal engagement with both mathematics and broader existential questions. By way of conclusion, I engage with Thomas Söderqvist's concept of 'existential choices' in mathematics (1996) to reflect on the intricate interplay of intellect, emotions, and affects in the creation of knowledge. These forces, deeply entangled, shape not only mathematical inquiry but also the lived experience of becoming a woman mathematician. This perspective underscores how intellectual pursuits and affective life intersect, illuminating the transformative potential of mathematics as both a scientific and existential quest.

Auto/biographical, philosophical, and literary traces of love

Wang Zhenyi, reflecting on the challenges of mathematical study, once wrote: 'There were times when I had to put down my pen and sighed. But I *love* the subject, I do not give up' (Peterson 2000, 344, emphasis added). This brief yet profound statement captures not only the challenges intrinsic to mathematical study but also the powerful force of love that drove her persistence and intellectual pursuit. Love, for her, was not a passive feeling but an active force, fuelling her determination to overcome obstacles and continue her work. This passion rendered even the moments of frustration meaningful, transforming the act of returning to her studies into a celebration of perseverance. Zhenyi's unwavering commitment to her mathematical work echoes the experiences of Maria Gaetana Agnesi, who dedicated herself to her love for God and mathematics after liberating herself from the torment of being a public intellectual spectacle.

Indeed, Agnesi's deep engagement with mathematics becomes even more compelling when viewed through the lens of love—a love for certainty, for truth, and for intellectual fulfilment. She believed that mathematics uniquely offered the clarity and reliability necessary to uncover truth. In the sixth thesis of the *Prolegomena* to her *Philosophical Propositions*, she exalts mathematics as the supreme science in the quest for truth: 'Mathematics must absolutely be counted as a discipline of Philosophy in that it has more right than the others to be called a science, being able to lead us with well-founded certainty to the attainment and contemplation of truth, of which nothing is more delightful' (1738, 4).

Mazzotti aptly observes that for Agnesi, the contemplation of truth through mathematics provided 'the greatest earthly joy available to mankind' (2007, 66). This joy was not merely intellectual but deeply connected to the pursuit of happiness—a theme Agnesi explored in a short ethical essay found among her early notebooks in the Ambrosiana manuscripts, titled *De ultimo fine, summo bono, et felicitate naturali* [On the Ultimate Goal, Highest Good, and Natural Happiness] (VBA. 0197. SUP).[3] Although handwritten by Agnesi, the essay reflects teachings she received from her philosophy tutor, Michele Casati (see Mazzotti 2007, 46). It spans twenty pages, structured into four chapters, with the fifth—on 'Supernatural Beatitude'—announced but unwritten. Despite its brevity and unfinished state, the essay centres on the concept of natural happiness, emphasising that it derives from knowledge born of an intellectual love for God. As she writes, 'true *love* is the *love* for knowledge' (VBA. 0.197. SUP, 9r, emphasis added).

Agnesi's philosophical education in ethics and metaphysics situates her firmly within the Augustinian tradition, which interweaves love, knowledge, memory, and happiness in a dynamic relationship and has reached modernity through Hannah Arendt's work, as I have discussed elsewhere in my work (see Tamboukou 2023). For Agnesi, love—both divine and intellectual—was the animating force behind her pursuit of knowledge, her engagement with mathematics, and her vision of happiness.

In the spirit of unwavering love, Sophie Germain devoted herself wholly to mathematics, courageously defying the constraints imposed by her family. As her biographer vividly described, 'they tried in vain to curb a passion deemed extraordinary for her age and unprecedented for her gender' (Libri 1833, 11). Armed only with a copy of Bézout's *Cours de Mathématiques* she had discovered in her father's library, Germain overcame every barrier placed in her way. Her family's efforts to stifle her intellectual curiosity—whether through discouragement or the physical removal of resources—only served to strengthen her resolve.

As her biographer further recounts, Germain rose through the coldest, darkest nights, her hands numb from the biting chill that often froze the ink in her inkwell. Wrapped in layers of blankets to stave off the cold, she continued her studies by the faint light of a solitary lamp, 'her passion for mathematics' burning brightly in the shadows (12). Even when her family went so far as to remove the fire, her clothing, and the candles from her room in a desperate attempt to force her to rest, Germain refused to yield. According to Libri, 'this incident marked her first demonstration of an unrelenting *love* for the subject—a *love* so profound and pure that her family, witnessing her devotion, wisely chose not to oppose it any further' (12, emphases added). As we have already seen in previous chapters, Germain's love for mathematics was not merely a fleeting interest; it was a deep, consuming force that drove her every action, transcending the several oppositions she faced and leaving an indelible mark on her life's work. In an intriguing parallel, the humble materiality of night—candles and blankets—also marked Mary Somerville's love and passion for mathematics, as we will see next.

At the end of her autobiography, Somerville reflected on her life with poignant regret, lamenting that she had not devoted herself entirely to her true and enduring love: mathematics. She confessed, 'Mathematics are the natural bent of my mind. If I had devoted myself exclusively to that study, I might probably have written something useful, as a new era had begun in that science' (Somerville 2001, 286). For Somerville, mathematics was not merely a discipline but her true and abiding love that began with intensity, and throughout her autobiography, she vividly recounted the obstacles she overcame to nurture it.

Amidst her household responsibilities—managing chores and mending clothes—she carved out time for mathematics with fierce dedication. She wrote, 'I had to take part in the household affairs, and to make and mend my own clothes. I rose early, played on the piano, and painted during the time I could spare in the daylight hours, but I sat up very late reading Euclid' (87). However, her nocturnal pursuits were discovered when the servants reported to her mother that the stock of candles was dwindling rapidly because 'Miss Mary sat up reading till a very late hour' (87). In response, her mother ordered that her candle be taken away as soon as she was in bed. Her father, alarmed by her obsessive interest, expressed concern that she might end up in 'a strait jacket' (87), instructing her mother to put an end to her studies entirely. Yet, Somerville remained undeterred, even amid the social whirl of Edinburgh's dances and soirées. She recalls, 'I rose at daybreak, and after dressing, I wrapped myself in a blanket from my bed on account of the excessive cold—having no fire at that hour—and read algebra or the classics till breakfast time' (100).

This enduring love for mathematics was not only a private passion but also a social and intellectual journey. During her first trip to France in 1817, as discussed in Chapter 3, Somerville encountered a thriving community of French mathematicians. Her vivid recollections of le Marquis and la Marquise de Laplace, visits to their home in Arcueil, and long discussions with Laplace himself about his *Mécanique Céleste* reveal how these experiences deepened her commitment. The autographed copy of *Système du Monde* gifted to her by Laplace, alongside convivial dinner parties with French mathematicians and their spouses, became tangible manifestations of her lifelong love for mathematics—woven into the overall process of becoming a mathematician.

For Somerville, mathematics was a profound source of joy, a lens through which she viewed the world, and a passion that shaped her life despite every effort by societal norms and familial expectations to suppress it. Her story is a testament to how love—persistent, unyielding, and transformative—can sustain intellectual ambition and illuminate the path to the advancement of science and knowledge. It also seems that the energy of her love for mathematics was transmitted to her young protégée Ada Lovelace in her own configuration of the discipline as poetic science.

In the final verses of the third canto of her father's renowned poem *Childe Harold's Pilgrimage*, Lovelace is celebrated as 'the child of love, though born in bitterness/And nurtured in convulsion' (CXVIII, l.1094). Yet, despite this tempestuous beginning, love emerged as a vital force in her life, deeply intertwined with her enduring passion for mathematics. While direct references to love are scarce in Lovelace's extant correspondence and writings—whether in her early letters with her husband or in her essays and poems archived in the Bodleian Libraries—signs of love as a transformative and creative force can be traced in her archival documents.

A playful glimpse of the connection between love and mathematics appears in a congratulatory letter from Reverend Samuel Gamlen, the minister who officiated her wedding in 1835. Gamlen humorously enquired whether there might be a 'Royal Road to Love', suggesting that no such easy path existed for mathematics (LB 72, f. 3).[4] Unfortunately, Lovelace's response to this question, if given, has not survived.

On 14 August 1843, during the demanding final stages of preparing the Menabrea paper, Lovelace shared deeply reflective thoughts on love in a letter to Charles Babbage (MS 37192, f. 426)[5]. Her words reveal a profound devotion to truth, a spiritual connection to God, and an abiding belief in the inherent goodness of humanity. Lovelace contrasts her values with what she perceives as Babbage's priorities, articulating her 'uncompromising principle' of placing love for truth and God above worldly recognition, such as fame or glory. Gently and thoughtfully, she observes that while Babbage, too, loves truth and God, his passion for acclaim and honours seems to surpass these ideals in practice. Far from reproachful, her tone reflects both honesty and respect, allowing for the possibility that his ambitions might be 'noble & beautiful'. This nuanced critique not only highlights Lovelace's own values but also underscores her capacity for introspection and her ability to engage with differing perspectives. Her reflections reveal that her perception of love is rooted in authenticity, integrity, and a profound sense of connection to both the intellectual and the spiritual dimensions of life.

> My own uncompromising principle is to endeavour to love *truth & God before fame & glory* or *even just appreciation*; & to believe generously & unwaveringly in the good of human nature~ (however dormant & latent it may often seem). *Yours* is to love truth & God (yes, deeply & constantly); but to love *fame, glory, honours, yet more*. You will deny this; but in all your intercourse with every human being (as far as I know & see of it), it is a *practically paramount sentiment*. Mind, I am not *blaming* it. I simply state my belief in the *fact*. The fact may be a very *noble & beautiful* fact. *That* is another question (MS 37192, f. 426, emphases in the text).

Lovelace's idea of love extended beyond the abstract, finding vibrant expression in her profound passion for music—a lifelong devotion deeply intertwined with her scientific imagination. As discussed in Chapter 3, she envisioned how the Analytical Engine could process and generate music, showcasing her extraordinary foresight. This visionary idea not only connected her love for the arts with her work in mathematics but also anticipated a field that would flourish more than a century later.

In the July 1991 special issue of *Computer*, dedicated to computer-generated music, Denis Baggi aptly highlighted Lovelace's pioneering insight, writing: 'Computer music has existed as a formal application of computer science for at least 35 years. The idea has existed much longer. Ada Lovelace suggested the use of machines in composition a century and a half ago' (1991, 6). This acknowledgement underscores her ability to merge passion and intellect, imagining possibilities that bridged the realms of art and technology with a love both profound and enduring.

Though the archival record provides limited insight into Lovelace's expressions of maternal love, an undated letter addressed to her mother offers a poignant glimpse into her affections. This short note reveals the intertwining complexities of her roles as mother, daughter, and intellectual thinker. While mathematics and music often commanded the spotlight in her life, Lovelace's understanding of love extended into the metaphysical, as reflected in her evocative musings: 'You cannot think how charmed I am with my *metaphysical* child, & how I have thought of her. If she will only be kind enough to be a metaphysician & a mathematician instead of a silly minikin dangling *Miss* in leading strings, I shall love her mind too much to care whether her *body* is male, female or neuter' (LB 42, f. 138, emphases in the text).

In this letter, Lovelace envisions love as transcending conventional roles and relationships, grounding it instead in the realms of intellect and imagination. Her reflections reveal entanglements of personal affections with philosophical and creative pursuits, suggesting that for Lovelace, love was a metaphysical force—abstract yet profoundly forceful. While her 'metaphysical child' bore little resemblance to her actual children, it symbolised her love for intellectual exploration, especially in mathematics, which she famously regarded as 'a poetical science'. This perspective is further illuminated in an undated epistolary fragment to her mother, where Lovelace muses on the union of disciplines:

> Are we d'accord now, do you think, about poetry & music & philosophy? I don't consider that as yet I have made anything like full *use* of music; and often much foolish *abuse* indeed.
>
> You will not concede me *philosophical poetry*. Invert the order! Will you give me *poetical philosophy, poetical science*?
>
> (LB 44, f. 210, emphases in the text)

Here, Lovelace frames her passion for mathematics as an artistic and emotional endeavour, rooted in her capacity to see the beauty of interconnected ideas. Her poetic vision of science, coupled with her philosophical inclinations, reflects a form of love that defies boundaries, merging the ethereal with the analytical, and the personal with the universal.

While Lovelace's image of love hovers between the realms of reality and imagination, for Sofia Kovalevskaya, the profound connection between love and mathematics became a defining theme of her legacy. This intersection was expressively encapsulated in the title of *Love and Mathematics* (1985), the influential biography of Kovalevskaya written by Pelageya Kochina, a renowned mathematician in her own right. Kochina's work not only documented and analysed Kovalevskaya's life through the lens of a fellow mathematician but also celebrated the profound interplay between passion and intellect that defined her remarkable journey.

In her obituary of Sofia Kovalevskaya, written shortly after her death in 1891, Swedish feminist and writer Ellen Key fondly recalled how her friend described her intellectual and spiritual heritage: 'My thirst for knowledge comes from my Hungarian ancestor, King Matthias Corvinus; mathematics, musicality, and lyricism from my German great-grandfather, the astronomer Schubert; my love of freedom from Poland; from a gypsy ancestress, the wanderlust and resistance to conventional norms—the rest is from Russia' (Key1891, 70). This vibrant self-portrait crafted by Kovalevskaya herself—at least in Key's memory—illustrates a profound interplay between heritage, intellect, and the self, all suffused with affects—love for knowledge, for freedom, for mathematics, and for the arts.

Kovalevskaya's acknowledgement of diverse ancestral influences reflects an almost poetic gratitude. Each facet of her intrinsic nature is imbued with passion: a thirst for knowledge linked to her Hungarian lineage, a love for mathematics and music traced to her German heritage, and a fierce freedom attributed to her Polish ancestry. Interestingly enough, her love for mathematics is intertwined with musicality and lyricism, as I will further discuss in Chapter 7. Perhaps most striking, however, is her embrace of her 'gypsy' [sic] ancestress's wanderlust and defiance of norms, which resonates with her life as a cosmopolitan mathematician and unconventional thinker. This connection to a nomadic spirit can be seen as a metaphor for her own intellectual and emotional restlessness, a form of love that refuses boundaries and thrives in exploration.

As I have examined extensively in previous writings (Tamboukou 2022), the mythographies surrounding Sofia Kovalevskaya often present her in complex and contradictory roles. In her autobiography, she recalled feeling unloved as a child: 'I have often heard Nanny say that Anyuta and Fedya are Mama's favourites, while I am not loved' (Kovalevskaya 1978, 53). She also wrote about her youthful infatuation with Dostoevsky, an unrequited love that caused her great distress when she learnt that he had proposed to her sister Anyuta and was utterly stunned when her sister rejected him (192).

Kovalevskaya's deep love and admiration for her sister Anyuta have left traces in their correspondence. On 29 September 1868, writing from St. Petersburg, where she had recently moved with her fictitious husband, she expressed her affection tenderly: 'I wish you knew, Anyuta, how I love you' (in Kochina 1985, 51). Later that same year, as they both prepared to flee to Europe, her admiration poured forth in another heartfelt letter: 'Darling Anyuta, you are so nice, so fine, so great that I am ready to bow before you to the earth. Anyuta, my best, my purest, most sensible ideal' (36).

In the discursive context of her auto/biographical narratives, Kovalevskaya emerges as both eager and vulnerable in her pursuit of love, often finding herself drawn to connections that were not fully reciprocated. This recurring sense of emotional neglect eventually created the discursive figure of an unfulfilled woman, haunted by a poignant longing for love and deeper connection. Iulia Lermontova evocatively encapsulated this image in her memoir of Kovalevskaya: 'When in aftertimes Sonia spoke to me about her past life, her bitterest complaint was, "Nobody has really loved me!" And when I objected, "But your husband did love you most fondly and truly," she always said, "He only loved me when I was with him, but he could do quite well without me"' (Lermontova 1951, 384).

At the same time, Kovalevskaya has often been celebrated as a prodigious girl-genius, wholly absorbed by her passion for mathematics. Yet, this portrayal frequently comes at the expense of her personal relationships, painting her as seemingly indifferent to other aspects of life. While her marriage is often framed by narratives of detachment and coldness, her romantic relationship with Maxim Kovalevskii offers a striking contrast, revealing, at least in her friend Ellen Key's imagination, a glimpse of her 'essence'.

> I saw her every week in the course of several years, but I met Sonya Kovalevskaya essentially only once. It happened to be at a concert where Beethoven's Ninth Symphony was performed. Contrary to her habit, Sonya was elegantly attired: she wore a black silk dress with lace [...] The object of her love, her countryman, was sitting beside her. The divine sounds of Beethoven's music flew around. The commonly nervous features of Sonya Kovalevskaya were lucid and calm. She seemed transformed. She was in love, and the music carried her away into a world of bright dreams [...] I only saw this expression on her face twice, the other time Sonya Kovalevskaya was dead (in Kochina 1985, 277).

The concert setting, marked by the grandeur of Beethoven's *Ninth Symphony*, provides a vivid backdrop to Kovalevskaya's rare moment of emotional fulfilment. Music, often considered an abstract language akin to mathematics, becomes the medium that amplifies her love, transcending the mundane and carrying her into 'a world of bright dreams'. This transformative moment suggests that love—both romantic and artistic—offers Kovalevskaya an ephemeral escape from the struggles and isolation that defined much of her life. The striking juxtaposition of this radiant moment with the only other instance of similar calm—Kovalevskaya's death—draws an unsettling connection between love and mortality. Death, in this account, becomes the ultimate stillness, a cessation of the woman mathematician's restless energy. Just as her love momentarily brought tranquillity and transcendence, so too does death, albeit in a permanent and immutable form.

While mathematics is not explicitly mentioned, its implicit presence looms large in Key's narrative. Her rare encounter with Kovalevskaya's 'true self' emphasises how seldom she allowed herself to escape the intellectual rigour and societal challenges tied to her public image as a mathematician. Mathematics, much like the music of Beethoven, represents a higher order of beauty and harmony. However, Kovalevskaya's transformation at the concert reveals that love and artistic beauty provide a more personal and visceral transcendence than mathematics alone. This haunting connection between love and death also resonates in Fritz Laffler's poem, 'Death of Sonya Kovalevskaya', copies of which were circulated at her funeral on 16 February 1891:

> Has Your thought gone up, designing,
> When the stars were saying 'love you,
> When great Saturn's ring was shining
> In the dark blue sky above you?
> Will the numbers' supreme spheres
> Analytic functions, dears,
> Open to You in condescendence
> Immortality's transcendence?
>
> (in Kochina 1985, 284)

Overall, these contrasting narratives, oscillating between emotional vulnerability and intellectual singularity, underscore the tensions in how Kovalevskaya's life has been remembered and mythologised through the trope of love. As I will explore further in Chapter 7, Kovalevskaya's plays and poetry offer rare but insightful glimpses into her image of both romantic and intellectual love. Her literary adventures were deeply influenced by her mentor, Karl Weierstrass, as their correspondence reveals.

Indeed, Weierstrass' letters to Kovalevskaya frequently reflect a philosophical and poetic sensibility, blending his mathematical rigour with a profound appreciation for beauty and love. In a letter dated 14 December 1885, Weierstrass provided a detailed account of his jubilee celebration, an event neither Mittag-Leffler nor Kovalevskaya, despite being on the organising committee, were able to attend. Among the speeches and events he recounted, Weierstrass included a poem by August von Platen, interwoven with his own verses to create what he described as a 'poetic toast' (153). This artistic melding of a poet's work with the contributions of a 'poet-mathematician' reflects an intricate tapestry of love and mathematics, illustrating their shared role in elevating the human spirit.

The poet's verses highlight beauty as an essential force that imbues life with meaning and harmony. Whether manifested through art, love, or the elegant truths of mathematics, beauty emerges as the glue that binds existence together. Without it, the poet suggests, the world would descend into chaos and silence—a sentiment that resonates deeply with both Kovalevskaya's intellectual pursuits and her emotional struggles. This blending of poetry and mathematics reveals not only the interconnectedness of these domains but also their power to inspire and sustain the human soul (153)

> Beauty is world's sacred secret,
> lure in image, lure in word.
> Drive it out of life forever,
> love will never then be heard.

Weierstrass, as a poet-mathematician, then turns to the 'Men of Science', whose joy and purpose arise from their deep understanding of the world's harmony—a harmony revealed through mathematics and reason. The poet-mathematician's verses highlight love's transformative role, comparing it to the pursuit of knowledge, suggesting that inspiration arises from the interplay of affection and thought (154).

> So said the poet. Men of Science
> who, by Gods will, feel the glee
> With their inner understanding
> of the world, its harmony
> [...]
> Magic words of love they tell us -
> train of best ideas starts.

Ultimately, the poetic interfaces of verses between the poet and the poet-mathematician celebrate the synergy of love and mathematics, portraying them as forces that bring meaning, beauty, and progress to life. Together, they elevate human existence, showing that neither emotion nor intellect alone is sufficient for true fulfilment. Instead, it is their elegant fusion that sparks creativity, wisdom, and the unending quest for understanding.

Overall, women mathematicians' narratives collectively underscore the central argument of this chapter: that love—in its many forms—is not merely a backdrop to intellectual pursuits but a driving force, shaping and sustaining the labour of scholarship. Their writings reveal that passion for knowledge is not only compatible with emotional intensity but is often fuelled by it, forging a profound and enduring connection between the heart and the mind in the history of science and mathematics.

While love is a forceful motif weaving together women mathematicians' auto/biographical narratives in this study, the focus of this chapter shifts to the analytics of love, particularly as it converges with *automathographies*. Such insights have been mostly traced in Émilie Du Châtelet's philosophical and epistolary writings, as I will discuss in the next two sections.

Love, knowledge, and happiness

'Learned women, even the happiest, have ignored true happiness', wrote Émile Fage, a member of the Société des lettres, sciences et arts de la Corrèze, in his 1894 review of Rebière's *Les femmes et la science*, published earlier that year (1894, 255). This work included profiles of women such as Du Châtelet and Germain, who were also featured in the expanded second edition of the book, published in 1897. As Jenny Boucard has observed, Fage was among the many writers who engaged with Rebière's portraits, often appropriating them to articulate their own views on women's involvement in science (2020, 206). Boucard further notes that women scientists, mathematicians, and philosophers became subjects of significant interest in essays, books, and articles at the turn of the eighteenth century. More intriguingly, their biographies were often weaponised in broader debates—both to defend and to challenge women's rights, particularly regarding education. This surge of interest is unsurprising: though their numbers were few, women scientists disrupted conventional gender norms, emerging as anomalies that demanded scrutiny and explanation, as explored in Chapter 4.

These discussions, comparisons, and juxtapositions unfolded within a complex web of discourses and themes. Central to these were debates about the appropriate level and nature of women's education, the compatibility of scientific or mathematical pursuits with domestic life and happiness, and the perennial question of whether mathematics aligned with 'female nature' (208). Among the multifarious narratives surrounding the French *femmes philosophes*, this section focusses on the question of happiness and its intricate connection with love—not through the perspectives of biographers, historians, or commentators, but through Du Châtelet's own philosophical treatise, *Discours sur le bonheur* [*Discourse on Happiness*]

(1961). Likely written over a period spanning the late 1730s and early 1740s, during and after her romantic relationship with Voltaire, the essay was completed in the spring of 1748[6] By that time, Du Châtelet had fallen in love again, this time with Saint-Lambert, to whom the work was presented as a gift.

Although several manuscripts of *Discours sur le bonheur* circulated among Du Châtelet's friends during her lifetime, the essay was first published posthumously in 1779, thirty years after her death. It is perhaps her most renowned work, appearing in numerous editions and translations. Despite its later prominence, the essay was largely ignored upon its initial publication and faced predominantly negative critiques from the early nineteenth century into the twentieth century.[7]

In writing her *Discours sur le bonheur*, Du Châtelet contributed to what Robert Mauzi, editor of the essay's critical edition, described as the French Enlightenment's 'glorious discovery' of happiness (1960, 255). As Ritchie Robertson (2022) observes, Enlightenment intellectuals viewed happiness as an earthly pursuit and a central purpose in life, reflecting the broader context of Western secularisation. Yet, in his survey of seventeenth- and eighteenth-century French approaches to happiness, Mauzi found most writings on the subject insignificant, pedantic, and of poor quality—'pitiful', in his words—with one notable exception: Du Châtelet's treatise. He regarded it as both original and the only work on happiness from that period worth reading in the twentieth century (1960, 9). It is unsurprising, then, that Mauzi chose to publish a critical edition of her work, even though he made only limited reference to her ideas in his own study of happiness (see Whitehead 2006, 255).

Barbara Whitehead, in her exploration of the intellectual milieu surrounding Du Châtelet's engagement with the question of happiness, highlights three men in her immediate circle who likely influenced her thinking. Each had written on happiness in poems or philosophical treatises: Voltaire; Claude Adrien Helvétius, a literary theorist and philosopher who was Voltaire's protégé during the Cirey years; and the physician-philosopher Julien Offray de La Mettrie. Helvétius dedicated his *L'Épitre sur l'amour de l'étude* (1730) to Du Châtelet, while La Mettrie opened his 1747 edition of *Traité de l'âme* with a letter addressed to her (Whitehead 2006, 257).

Du Châtelet's intellectual exchanges with Voltaire, particularly during their time at the Château de Cirey, have been extensively examined in the literature (see Wade 1941, 1947). Despite their mutual influences, their perspectives on happiness reveal striking differences. As Mauzi succinctly observed, 'Madame du Châtelet is undoubtedly Voltaire's friend, but her book is in no way a "Voltairean" book: it does not imply any provocation towards established ideas or powers' (1961, lxxxvi).

Both Mauzi (1961) and Whitehead (2006) argue that La Mettrie's ideas had little to no influence on Du Châtelet's approach, despite some similarities in their views on passions. Tracing Du Châtelet's intellectual connection to Helvétius is equally challenging, as they were contemporaneously writing on happiness (Whitehead 2006, 257). Additionally, Helvétius's poems on love were met with negative reviews, even in the eighteenth century. What emerges from these comparisons, however, is what Mauzi highlights as 'Madame Du Châtelet's entire originality' (1961, civ).

Despite the originality of her ideas, Du Châtelet's *Discours sur le bonheur* clearly aligns with the broader Enlightenment discourse on happiness, though with two significant interventions. First, she wrote as a woman, thereby challenging and gendering the presumed abstractness and universality of philosophical discussions. Second, she grounded her treatise in her lived experiences, weaving personal insights into her philosophical arguments.

As Mauzi observed in his introduction, a parallel study of the *Discours* and Du Châtelet's correspondence reveals it as 'a lived book' (1961, lxxxiv), deeply rooted in its author's experiences. This personal dimension, while enriching the text, also limits its universality. Mauzi writes, 'Not only does Madame Du Châtelet compose it with maxims suggested by her own experience, but she expresses in it an ideal of life that is hardly valid except for those whose major inclinations coincide with hers: the need for love and a passion for study' (lxxxiv).

Although deeply rooted in the author's lived experiences, Du Châtelet's *Discours sur le bonheur* firmly aligns with the tradition of moral philosophy, particularly a strand that Mauzi identifies as 'modern Epicureanism'—notably devoid of Christian moral undertones, lxxxv). This is especially evident in her striking rejection of repentance. She writes, 'This feeling of repentance is one of the most useless and most disagreeable that our soul can experience' (1961, 16–17). She further asserts that 'there is no point in looking back, and we should always brush from our mind the memory of our errors' (17).

According to Mauzi, her morality is natural, embracing pleasure, comfort, and the joys of high living. However, it lacks a combative edge: 'Émilie hardly cared to fight, in the name of natural morality, for her personal liberation and for that of man in general', (lxxxvi). Viewed in this light, *Discours sur le bonheur* positions itself as a practical 'manual of good living' rather than a theoretical treatise on human nature (lxxxvi).

Mauzi's observations underscore how Du Châtelet's situated approach to happiness marked a radical departure from the universalistic perspectives of her contemporaries, including Voltaire and La Mettrie. However, as Whitehead critically notes, the *philosophes'* notion of the universality of happiness was riddled with paradoxes and contradictions. She observes that 'In effect, they move from the belief that all can be happy to the rigid position that in order to be happy one must follow a very precise path that only a subset of humanity can walk, and therefore only the few can attain happiness' (2006, 259–60). Du Châtelet's refusal to advocate for universal forms of happiness is precisely where the originality and enduring value of her *Discours sur le bonheur* lie—a point I will explore in greater depth later in this discussion.

Let us return to the *Discours sur le bonheur* and examine how Du Châtelet both aligns with and diverges from the philosophical traditions of her time. Her essay begins by addressing a central question in Enlightenment thought: whether it is possible to achieve happiness, a theme that, as Robertson notes, was foundational to the era's discourse on the pursuit of happiness (2022, 4). Interestingly, Robertson refers to Du Châtelet's treatise, but he frames it within the discourse of 'the unhappy woman mathematician' that we discussed at the beginning of the section: 'Her essay on happiness makes sad reading for it was written at some time in the late 1740s during the breakdown of her long relationship with Voltaire, and is really an analysis of her own unhappiness', (5). This perspective reduces her philosophical contributions to mere reflections of her personal struggles, thereby overshadowing the intellectual rigour and originality of her work with an overemphasis on her emotional and biographical context.

Du Châtelet's 'discourse on happiness' unfolds quite differently, however. While she acknowledges the challenges of attaining happiness, she emphasises the role of reason in guiding one's actions towards it: 'We commonly believe that it is difficult to be happy, and we have much reason to believe this; but it would be much easier for men to become happy if they could reflect on and planned their conduct before acting' (1961, 3).

This opening firmly situates her within the rationalist tradition, or as Mauzi describes it, a blend of rationalism and 'modern Epicureanism' (1961, lxxxviii). However, Du Châtelet's approach introduces a paradox. While advocating for deliberate, reasoned action in the pursuit of happiness, she simultaneously embraces the value of passions and illusions—elements that seemingly defy the calculated rationality she promotes. This tension reveals the nuanced complexity of her thought, a theme that will be explored further.

> To be happy, it is necessary to have freed oneself of prejudices, to be virtuous, and in good health, to have tastes and passions, and to be susceptible to illusions; for we owe most of our pleasures to illusion, and unhappy are those who lose it. Far then, from seeking to dispel illusion by the lantern of reason, let us try to thicken the glaze that it lays on most objects, more necessary for them than are the cares and adornments of our bodies (1961, 4).

Mauzi finds a paradox at the heart of Du Châtelet's 'modern Epicureanism', particularly in how she links happiness to the realm of sensations and passions. According to Du Châtelet, 'the only point of living is to experience sensations and agreeable feelings; and the livelier the agreeable feelings are, the happier one is' (6). The core of her argument in the *Discours* is strikingly simple: 'one must have passions to be happy', (19). She urges her readers not to fear passions but to embrace them, advising that they should be 'susceptible to passions', (6).

Du Châtelet's own intellectual and cultural background—shaped by her education and active participation in the *salon* culture—fostered her passion for the arts, particularly theatre and opera. Throughout her life, she championed the pleasures of immersing oneself in the illusions of the performing arts as a pathway to happiness. As she writes: 'But what pleasure would we have at another spectacle where all is illusion, if we were not able to abandon ourselves in it?' (15). Here, Du Châtelet uses the theatre as a metaphor for her broader argument: illusions, far from deceiving us, open up new ways of seeing and experiencing the world, offering alternative routes to happiness. In her view, the beauty of a spectacle lies in our collective willingness to suspend disbelief, allowing ourselves to become absorbed in its illusory world.

As Whitehead notes, Du Châtelet's focus on passions and illusions sets her apart from typical Enlightenment thinking on happiness (2006, 270). The *Discours* raises important questions about the nature of passions, particularly their role in achieving happiness. Du Châtelet distinguishes between 'passions' and 'tastes' (*goûts*), positing a hierarchy of importance. 'One is only happy because of satisfied tastes and passions; I say tastes because one is not always happy enough to have passions, and lacking passions, one must be content with tastes' (1961, 4–5). For Du Châtelet, passions do not equate to suffering; rather, they represent intense feelings and affects that are essential for experiencing great pleasure (6).

In her reading of Du Châtelet's *Discours*, Véronique Le Ru identifies Cartesian influences in her treatment of passions. Like Descartes, Du Châtelet does not reject passions but emphasises their regulation, acknowledging them as an integral part of human nature (2019, 63). Whitehead also observes that Du Châtelet's reference to 'the great machines of happiness' (1961, 4) aligns her with the Epicurean tradition, as noted by Mauzi. In this regard, Du Châtelet rejects what she sees as the futile exercise of contemplating death with equanimity, a stance that sets her apart from both Voltaire and Helvétius, but brings her closer to La Mettrie's views on happiness (Whitehead 2006, 266–7).

However, I suggest that Du Châtelet's exploration of the complex relationship between passions and happiness also reflects Spinozist influences, particularly in her emphasis on joy as central to the ethics of a fulfilling life. This connection between Descartes and Spinoza is unsurprising, given that Spinoza, in his early drafts of the *Ethics*, adapted Descartes' taxonomy of emotions, although he distilled the primary affects into joy, sadness, and desire, omitting the Cartesian categories of wonder, love, and hate (Shapiro 2020).

For Spinoza, as Mathew Kisner (2012) explains, passions are unavoidable elements of human existence; they can be beneficial and, ultimately, serve as indicators of our power to live virtuously, depending on whether they enhance or diminish our capacity for action. Du Châtelet echoes this view, advising her readers that passions must be harnessed to serve their happiness (1961, 19). She warns against destructive passions such as ambition, which should be avoided, and vices such as hatred, vengeance, and rage, which diminish our ability to achieve happiness. As she puts it: 'one must never be vicious if one does not want to be unhappy' (13).

Passions notwithstanding, Du Châtelet was also a rationalist, and for her, autonomy, clarity of mind, and the ability to make decisions were essential conditions for happiness. 'Be decided on who you want to be and on what you want to do, this is what most people lack; it is however the condition without which there is no happiness', she wrote, describing what she called 'the great machines of happiness' in her *Discours sur le bonheur* (16). The metaphor of 'the machine' resonates with a Deleuzian mode of thought, where happiness emerges not as a fixed state or inner essence but as a process of becoming, a continuous *assemblage* of flows, connections, and intensities, as we have already seen in Chapter 1.

For Du Châtelet, happiness is *machinic*—it is an active, productive process that involves constructing *assemblages* between desires, actions, and external forces. It is not a static goal but an open-ended experimentation with life, requiring the individual to navigate and reconfigure the conditions of their existence. In this sense, happiness operates as a field of immanence, where subjectivity emerges through the interplay of forces, decisions, and the transformative potentials of thought and action.

While framing happiness as relational, Du Châtelet also asserts that 'our happiness will always depend on others' (20). However, she continues, 'the less our happiness depends on others, the easier it is for us to be happy' (20). It is important to note that, although happiness is relational in the *Discours*, men and women experience it differently. Among the few sources of happiness available to women, the love for study is the most independent and constant. This theme runs throughout the entire *Discours*. 'The *love* for study holds within it a passion from which an elevated soul is never entirely exempt, that of glory' (20, emphasis added), Du Châtelet wrote. Men, she observes, can find 'infinite sources' (20) for glory and, by extension, happiness, but 'women are excluded, by their state, from any kind of glory, and when, by chance, there is someone who was born with a rather elevated soul, only study remains to console her for all the exclusions and for all the dependencies to which she is condemned' (21).

In walking the fine line between dependence and independence, Du Châtelet suggests that 'for this reason of independence, the *love* of study is, of all the passions, the one that contributes the most to our happiness' (20, emphasis added). Her own serious engagement with the study of science and mathematics serves as a living example of how the pursuit of happiness materialised in her life. As we have seen in previous chapters, Du Châtelet's scientific publications, the products of her love for study, became her independent sources of happiness.

Beyond her personal experiences, Du Châtelet's emphasis on the love of study as a pathway to happiness also places her squarely within the Enlightenment project. Gottfried Wilhelm Leibniz, the mathematician and philosopher who significantly influenced her own approach to 'living forces' [vis viva], had similarly linked wisdom, science, love, and happiness: 'Wisdom is the science of happiness [scientia felicitatis] … A good man is one who *loves* all human beings. To *love* is to find delight in the happiness of another. To find delight is to feel harmony' (cited in Antognazza 2009, 115, emphases added). By positioning herself as an Enlightenment intellectual, Du Châtelet also gestures to the classical tradition of her education: 'Cicero is right to say: The pleasures of the senses and those of the heart are, without doubt, above those of study; it is not necessary to study to be happy: but there is perhaps a need to feel that we have this resource and this support within us' (1961, 23). Du Châtelet had indeed engaged with Cicero's philosophical works from an early age, and according to Voltaire's memoirs, it was her father who had taught her Latin (see Hamel 1910, 22–3).

Apart from her love for study, Du Châtelet also delves deeply into the notion of love itself. Her famous statement, 'we know more of *love* through the unhappiness it causes than by the often-obscure happiness it spreads over the lives of men' (1961, 6, emphasis added), often interpreted as an expression of disillusionment with her affair with Voltaire, as we have seen above with Robertson (2022), is far more than a personal reflection. In her discourse, happiness through love is experienced rather than proclaimed: 'happy people do not seek anything and will not notify others of their happiness' (1961, 5). Here, love becomes 'a sixth sense', described as 'the finest, the most delicate, the most precious of all', a feeling that intertwines happiness and pleasure into an inseparable whole. It follows, then, that life loses meaning without love, and 'you have to give up life when you lose that happiness' (1 (23). However, since it is rare for two souls to meet in mutual love, happiness can also be found in 'the pleasure we feel in giving ourselves up to our feelings of tenderness' (30).

In this section of the *Discours*, Du Châtelet reveals herself not only as a philosopher but as a woman who loved a man for ten years without realising she was loving for two: 'I spent all my time with him, and my heart, unsuspecting, enjoyed the pleasure of *loving* and the illusion of believing myself *loved*' (32, emphases added). The author's vulnerability is evident as she confesses the depth of her suffering: 'the wound in my heart has bled for a long time', she writes, yet despite her pain, her advice to readers is to embrace love, for 'it would be ridiculous to refuse oneself this pleasure for fear of an unhappiness that perhaps will only be experienced after having known great happiness' (35). It is no wonder that Mauzi found Du Châtelet's essay so moving, as it not only revealed personal truths but also subtly depicted the process of suffering, the mastery of passions, and eventual transformation (9). While Du Châtelet's sincerity and boldness about love are clear, she also suggests that we may eventually need to 'leave *love* behind' when it ceases to bring happiness. When this day inevitably arrives, a cultivated 'taste for study' will ensure that 'our happiness depends only on ourselves' (1(39, emphasis added).

By gendering love and, consequently, happiness, Du Châtelet disrupts the universal concept of happiness that is central to the Enlightenment tradition. It is not only gender but also social class and age that emerge as key markers of difference in the experience of happiness within the *Discours*. Early in the essay, she clarifies her position by stating, 'I do not aim to write for all kinds of social conditions and for all people; not all ranks are susceptible to the same kind of happiness' (7). Later, in her discussion of the passion for gambling, she differentiates between the pleasures of youth and those of old age: 'Every age has its own pleasures; those of old age are the most difficult to obtain: gambling, studying, if one is still capable of it, the enjoyment of fine foods, consideration, those are the mainsprings of old age' (38). While some critics have dismissed her approach as elitist and disconnected from a universal notion of human happiness, Whitehead has argued that Du Châtelet's originality lies in her engagement with the gender, class, and age disparities in the experience of happiness: 'Those who held to a universal view would never have to confront the issue of the differences in material conditions of men and women, the well off and the poor. Mme du Châtelet's personalized approach allows for consideration of these differences' (2006, 263).

Time also plays a crucial role in Du Châtelet's conception of happiness, with past, present, and future coexisting in its subjective experience. As she writes, 'we are happy in the present moment, not only by our actual pleasures but by our hopes, by our reminiscences' (1961, 22). This view aligns with the broader Enlightenment project that champions the possibility of happiness within one's earthly life. John Locke had expressed a similar sentiment in 1677: 'The business of man [is] to be happy in this world' (in Aaron and Gibb 1936, 88). For Du Châtelet, time was not only central to happiness but also to preserving her own legacy. Given that her scientific work was both her source of glory and happiness, she ensured that it would be preserved for future generations, as we have seen in Chapter 1. Her concern about the future of her work was deeply intertwined with her anxieties over her love affair with Saint-Lambert, a relationship that shaped the final stages of her emotional and intellectual life. As I will explore next, these personal uncertainties seemed to parallel her apprehension about the legacy of her scholarly contributions.

Epistolary love

'Yesterday you said things so tender and so touching that you have penetrated my heart, so love me always in this way. Believe that, when you love me, I adore you. I spent the most agreeable night that I could spend without you, you never left my thoughts' (MA 2287.27, f. 1). This tender letter fragment, likely written in the summer of 1748, reveals the intensity and depth of Émilie Du Châtelet's emotional connection to Saint-Lambert. It is one of the ninety-nine surviving letters she wrote to him between 1748 and 1749, capturing the force of a passionate love that defined the final chapter of her life. In this brief note, Du Châtelet expresses her longing and devotion with disarming sincerity: 'You wish me to tell you about what I will do today. What I want to do every day of my life: I will see you, I will love you, I will say it to you. But I want to read that in the charming eyes that I adore' (f. 1).

Du Châtelet likely encountered Saint-Lambert during the winter of 1748 while staying at the Château de Lunéville in Lorraine.[8] She and Voltaire had been invited to join the court of Stanislas Leszczyński, the former king of Poland. By this point, her romantic relationship with Voltaire, which had profoundly shaped her intellectual and personal life, had already transitioned into a deep and enduring friendship. Du Châtelet's love relationship with Voltaire was, of course, a remarkable *event* that radically changed the route of her life and grandly initiated the process of becoming a woman mathematician and scientist. The mythographies around it are immense and often contradictory.[9] The absence of preserved correspondence between Du Châtelet and Voltaire, however, means that their relationship can only be analysed indirectly, including through her *Discourse on Happiness* (1961).

This philosophical treatise, as discussed earlier, offers glimpses into Du Châtelet's evolving views on love, autonomy, and fulfilment. Her love for Saint-Lambert, however, was of a different tenor—more immediate, tender, but also visceral, transversed by intense forces of passion and desire. It is in these letters that we witness a woman not only grappling with the uncertainties of her final love but also reflecting on the fleeting nature of happiness and the enduring power of affection. Indeed, Du Châtelet's encounter with Saint-Lambert marked a profound moment of convergence in her life, where the temporal dimensions of past, present, and future intertwined in what could be described as an exemplary Bergsonian *durée*.[10] This encounter, coupled with her enduring love for study and the happiness derived from the glory surrounding the publication of her *Institutions de physique*, created a rich tapestry where her intellectual and emotional worlds seamlessly merged.

In this dynamic *durée*, Voltaire belonged to her past, a chapter that had profoundly shaped her intellectual identity and personal growth. Their relationship, though no longer romantic, had laid the foundation for her achievements as a scholar and scientist. In contrast, Saint-Lambert represented the promise of the future, a new and intense passion that reawakened her capacity for romantic love. Her present, however, was a fusion of these temporal strands, filled with the dual pleasures of loving and writing. On one hand, she was deeply engrossed in her correspondence with Saint-Lambert, crafting tender and heartfelt letters that revealed her innermost thoughts and emotions. On the other hand, she was immersed in her scientific pursuits, most notably working on *her Newton*, a project that would secure her place in the annals of scientific history, as we have already seen in previous chapters.

In this context, Du Châtelet's life became a vivid embodiment of Bergson's *durée* (1970), where the past informed the present, the present shaped the future, and all moments coexisted in a fluid and living continuity. Her ability to weave the threads of love, intellect, and ambition into a coherent and fulfilling existence underscores the complexity and richness of her character—a woman who lived fully in every dimension of time.

Read in the context of Liz Stanley's *epistolarium* (2004), Du Châtelet's letters to Saint-Lambert embody the quintessential characteristics of the *lettre d'amour*.[11] Their uniqueness lies in how the passion of love infuses the writer's individual style, making each letter a vivid expression of emotion and intellect. These letters charm and seduce, their words carefully crafted to convince the recipient of her ardour. They alter the fabric of reality, transcending the immediate moment with the intoxicating promise of expectation. Through their poetic rhythm, they spark the imagination, intensify desire, and amplify emotional resonance. The letters sustain and rekindle the dynamic interplay of love, while also bearing the unmistakable imprint of temporal and spatial separation, the unavoidable constraints of their earthly existence.

In this light, the epistolary exchange between Du Châtelet and Saint-Lambert operates as a shared experience of Bergsonian *durée* (1970). As readers and writers, in turn, they exist in a fluid temporality, where the present moment of reading and writing envelops them in pleasure. Simultaneously, the letter acts as a vessel for memory, reviving the sweetness of past encounters, even as it projects their union into an imagined future filled with the promise of renewed affection and desire. In this way, Du Châtelet's amorous correspondence becomes a bridge across time, a medium through which love endures and thrives, transcending the physical distance that separates the correspondents.

What sets Du Châtelet's *lettres d'amour* apart, however, is the way they intertwine the intensity of romantic devotion with the intellectual fervour of her scientific pursuits. This unusual blend is poignantly evident in her letter to Saint-Lambert dated 16 June 1748, where her love-stricken voice shares space with the anxieties of a scholar immersed in the completion of her magnum opus. Her dual passions—love and study—coalesce, revealing a complex portrait of a woman whose heart and mind were equally inflamed, each feeding the other in a singular expression of human creativity and longing.

> I am leaving all my business affairs, and my book, which you must have the fairness to regard as a business, because it is still very essential for me. This book is awaited, promised, it begun two years ago, my reputation depends on it. It was certainly nothing less than necessary to undertake it, but it is essential to finish it, and to do it well. And it is a work of which what remains for me to do demands the greatest contemplation and the greatest concentration (MA 2287.61, f. 1).

The letter above was sent from Paris, where Du Châtelet had travelled to dedicate herself to the completion of her book, alongside attending to other pressing matters. Yet, her separation from Saint-Lambert proved agonising, leaving her tormented by the emotional distance between them. His letters had plunged her into despair, leading her to confront the heartbreaking realisation 'that you no longer love me, that you no longer want to love me, that you repent of having loved me' (f. 1) His words, she lamented, were 'the harshest and most distressing things' (f. 1) , piercing her heart to the point of making her feel as though she would 'die of grief' (f. 1).

Roland Barthes famously observed that the lover's discourse 'exists only in outbursts of language' (2002, 3–4), and Du Châtelet's letters to Saint-Lambert are no exception. They overflow with emotional outbursts, each one a testament to the rawness of her passion and the depth of her love and suffering. Her words become a conduit for her desire for emotions that transcend the ordinary—a longing for feelings so intense they consume and transform. In these letters, Du Châtelet's eloquence reaches its zenith, as she pours her soul onto the page with a fervour that mirrors the scientific precision she brought to her scholarly work. Her discourse moves fluidly between the intellectual and the visceral, combining the rigour of a philosopher with the fragility of a lover. The result is a body of correspondence that not only captures the tumult of her romantic entanglement but also elevates it to a universal meditation on the nature of love, longing, and the human capacity for enduring both ecstasy and heartbreak. Through her words, Du Châtelet offers a poignant reminder of the excessiveness and power of love as well as its ability to inspire and devastate in equal measure. Although I am perhaps more a geometer than you, I am not so stiff. I will not tell you that I will always love you in proportion to what I will be loved; but I will tell you that I cannot be happy loving you if you do not love me excessively. Remember that when it comes to love, enough is never enough (MA 2287.66, f. 2).

What stands out as rare and unique in these amorous epistolary exchanges is that Saint-Lambert's true rival was neither Voltaire, the lifelong companion, nor a new lover—it was 'her Newton', the remarkable intellectual work that occupied her mind and kept her away from her beloved. In the above letter dated 16 June, Du Châtelet's emotional conflict is palpable: she declares her decision to abandon her scientific pursuits momentarily and rush to Commercy, where Saint-Lambert was staying—a dramatic gesture of love that symbolised the ultimate sacrifice of her intellectual devotion. Casting aside 'her Newton' was, in her view, the clearest proof of her passion and commitment.

Life at Commercy, however, offered little time for study or contemplation. Du Châtelet and Voltaire were consumed with staging theatrical productions to entertain King Stanislas and his court. In a letter to her friend the Comte d'Arsenal, dated 20 July 1748, she humorously excused her lack of correspondence, explaining, 'it is true that twenty-four hours are not enough to rehearse two or three operas as well as so many comedies' (in Kölving and Brown 2018, vol. 2, 339).

This frenetic pace of courtly life made her decision to join Saint-Lambert a temporary escape from her rigorous intellectual routine. However, her dedication to science and philosophy was unyielding; once the summer distractions faded, Du Châtelet returned to her work with renewed intensity in early 1749, determined to complete her magnum opus. This brief interlude of love and theatre, while seemingly at odds with her scholarly pursuits, highlights the profound tensions she navigated between her roles as a lover, a courtier, and a trailblazing intellectual.

On 15 February 1749, writing to the mathematician François Jacquier from Paris—where she had returned to dedicate herself fully to completing 'her Newton'—Du Châtelet confessed that the project had consumed her entirely. She admitted that much work remained, as she had lost an entire year in Lorraine, where the distractions and constrained lifestyle had made it impossible to work 'in the middle of the dissipation and the cut-off life that one leads there' (Kölving and Brown 2018, vol. 2, 395). Her letter outlined the structure of the forthcoming book, particularly her efforts on the preface to the second volume, where she aimed to present and explain Newton's system 'without figures and without algebra' (395). As already discussed in Chapter 3, she would also include sections on the figure of the Earth, as well as the theory of the moon and comets, incorporating the latest discoveries, particularly those of AlexisClairaut. Du Châtelet assured Jacquier that as soon as these sections were ready, she would send him her manuscript for his review.

That same day, she penned another letter to the renowned Swiss mathematician, Johann II Bernoulli, reaffirming her commitment to the project: 'I have come here to finish my Newton, and I won't leave until it is finished. You will certainly receive one of the first copies. I am very sorry not to have been able to bring your insights into the work, but I would like you to find it worthy of you' (392). These parallel letters to two eminent scholars served as more than just updates; they functioned as a binding epistolary pact, solidifying her determination to complete her monumental work.

Yet, just three days later, on 18 February 1749, the pull of her emotional life interrupted her intellectual resolve. Du Châtelet wrote a deeply impassioned letter to Saint-Lambert, apologising for previous 'cruel' letters she had sent and expressing her fear that he was on the verge of breaking up with her. This juxtaposition of letters—one set addressed to celebrated mathematicians, showcasing her intellectual rigour, and the other to her lover, revealing her passion and vulnerability—captures the profound affective entanglements of her life at this time. Torn between the pursuit of scholarly glory and the consuming intensity of her love affair, Du Châtelet traversed these competing forces with an intensity that unfolded the becoming of her fierce and passionate self, a self in perpetual motion, shaped by the flows of desire and the forces of thought.

> I don't know what to make of your last two letters. Are you detached from me? I will not believe it until you tell me again, I will never console myself [...] If you send me back my portrait, which I had the imprudence to ask for again in my last letter, you will give me a death blow. However, I hardly doubt it anymore, by the coldness of your last letter, and the way you seem to be heading for a breakup (MA 2287.82, f. 4).

The letter above, along with others penned by Du Châtelet during this period, reverberates with what Mikhail Bakhtin theorises as the *discourse of pathos* (2002). This mode, rooted in the baroque novel, unfolds through a blend of apologia and polemic, creating a discursive terrain where the writer simultaneously justifies and indicts. For Bakhtin, pathos is inherently prosaic—a dialogic struggle with other voices, perspectives, and truths. It is marked by resistance and tension, entwined with acts of self-justification and accusations, persistently negotiating with 'alien discourses' that contest its dominance (394).

Du Châtelet's love letters embody this pathos, dissolving the boundaries between lived reality and fiction. While addressing the immediate exigencies of absence and longing—bridging spatial and temporal gaps—they also construct a fictionalised world for the amorous relationship. This world exists exclusively within the textual space of the letter, where love and its tribulations are enacted, dramatised, and lived. Viewed through Bakhtin's lens, this *epistolary pathos* reveals its novelistic core: dialogic, shaped by suffering, despair, and passion, woven into rhetorical performances of a profound desire to be seen, heard, and reciprocated.

In these letters, Du Châtelet emerges not merely as a lover but as the heroine of an epistolary novel of her own making. Within this textual drama, she defends herself, accuses her beloved, contests his logic, and dismantles his resistance. As Linda Kauffman notes of the heroines in amorous epistolography (1986, 45), Du Châtelet occupies a textual stage where love's intensity is both weapon and vulnerability. These letters are zones of emotional density where the force of love and exposure of self fuse. They transcend simple assertions of identity to become dynamic *assemblages*, capturing the oscillations of affective existence.

Du Châtelet's epistolary acts, in this sense, are akin to what Karen Barad (2007) describes as 'agential cuts'—emergent, processual forces that resist fixity, evolving through negotiation and adaptation. Her passionate expressions of love, far from signifying vulnerability or passivity, reflect a radical openness to the unpredictable, a capacity to affect and be affected. Her letters chart a trajectory of becoming—becoming-lover, becoming-writer, becoming-visible—each movement generating new connections and intensities.[12] On a Deleuzian plane of analysis, her correspondence resists binaries, unfolding instead as a rhizomatic network where desire and expression intermingle in continuous flows. Here, *becoming* replaces static categories, emphasising processes over fixed states. The amorous and intellectual dimensions of her writings traverse and transform one another, creating multiplicities rather than oppositions. In this unfolding, the correspondence acts as an *assemblage* of movements, intensities, and affects, illustrating the dynamic and non-linear nature of thought and feeling. Through this lens, her letters do not represent a unified subject but instead embody an ongoing process of relational and creative becoming.

But Du Châtelet's letters also align with Jessica Benjamin's conception of love as the yearning 'to be known as oneself' by the beloved (1983, 290). Pulsating with this intimate demand for recognition, her letters navigate the interstices of desire and performativity. They become vivid sites of relational and rhetorical struggle, where love is simultaneously an act of writing and a process of being written into existence—a self that writes itself (*s'écrit*), as Philippe Lacoue-Labarthe (1989) frames it, a theme I will expand upon in Chapter 7.

Further read through a narratological lens, Du Châtelet's letters construct a layered narrative of emotional intensity, where *epistolary pathos* transforms ineffable affects into textual expressions. Repetition—of themes, arguments, and stylistic devices—is a key strategy through which her desires, fears, and doubts become not only legible but viscerally perceptible. These recurring motifs are not redundancies but deliberate rhythms externalising inner turmoil, creating a pulsating affective texture throughout her correspondence. The interplay between the constancy of her amorous story and the evolution of her discourse imbues the letters with dynamic depth. Repetition becomes a choreography of passion, striving to articulate the immediacy of her emotional state.

Occasionally, Du Châtelet reflects on the insufficiency of her language, as when she writes on 25 April 1748: 'This letter is full of inconsistencies; it only feels too much for the turmoil you have put in my soul' (MA 2287.52, f. 3). Such meta-narrative gestures underscore her acute self-awareness as a writer and the limitations of language in capturing raw affect. Her letters, thus, emerge as polyphonic texts—confessions, rhetorical pleas, aesthetic creations—interwoven with the spiraling temporality of her emotional world.

Her concern for her lover's health, expressed on 18 February 1749, reflects the dialogic essence of Bakhtin's framework: 'You are ill, and I am eighty leagues away from you! You are sick, and I am not at your bedside! All my anger is over, and I only feel my worry and my despair' (MA 2287.82, f. 1). This anguished declaration encapsulates the tension and multiplicity of pathos, oscillating between apology and polemic. It foregrounds the interplay of presence and absence, love and distance, while simultaneously resisting the limitations of language and physical separation. Thus, in the broader dialogic landscape of her correspondence, Du Châtelet's concern for her lover's health becomes a motif that animates the ongoing conversation between her and her beloved. It is not merely an expression of worry but a performative act that seeks to reaffirm the relational bond, to collapse the distance between them through the sheer force of language, even as it acknowledges its limits.

> No, I don't have anger anymore, no I never will. No, all I have now is emotion, pain at having displeased you. My last letter will have afflicted you,

> perhaps revolted you, but it came in the wake of so many others so tender that it is impossible that we have not unraveled, in the fury that reigned there, all the love that had dictated it. No, I believe your heart is as sincere and tender as it has always seemed to me (f. 1).

In deploying a diffractive reading of Du Châtelet's amorous correspondence through different angles and perspectives, my argument is that her letters resist reductive interpretations of her as a desperate marquise, hopelessly devoted to her beloved, as certain strands of scholarship on her romantic entanglements have suggested.[13] Instead, as I have discussed elsewhere, epistolary narratives carry the marks of passionate intensity and unspeakable affects, but they do not represent reality—they respond to it (Tamboukou 2010, 26). They are shaped by the dialogic interplay of voice and silence, presence and absence, opening a space for the creative exploration of subjectivity beyond conventional frameworks.

In her letters, Du Châtelet simultaneously embodies and transcends the trope of the inconsolable lover. While she writes with raw emotionality about her anguish, she also frames her intellectual pursuits as an existential anchor. For instance, in a letter to Saint-Lambert on 16 June 1748, she declares: 'I always try to keep my soul in such a situation that I find resources in my courage, in my philosophy, and especially in my love for study, if you abandon me' (MA 2287.61, f. 3). Here, her subject position is not confined to that of a forlorn lover; it expands into the realm of a thinker and a creator who finds resilience and meaning in the life of the mind. Her letter to Saint-Lambert sent from Paris on 21 May 1749 gives an idea of her fervent work schedule:

> My departure does not depend absolutely on me, but on Clairaut and the difficulty of what I do. I sacrifice everything to that, even my looks. I beg you to remember that if you find me changed. Have you any idea of the life I have led since the departure of the king? I rise at 9 o'clock, sometimes at 8, I work until 3, I have my coffee at 3 o'clock; I take up work again at 4, I leave it at 10 in order to have a little to eat alone, I chat until midnight with M. de V., who attends my supper, and I take up work again from midnight to 5 o'clock. Sometimes I wait for M. Clairaut, and I attend to my affairs and read through my proofs. Mme Du Deffand, Mme de B, everybody without exception is denied for supper and I have made a rule for myself not to go out to supper, in order to be able to finish my work (MA 2287.89, ff, 2–3).

This interplay of love and intellectual passion resists the portrayal of love as merely a disciplinary technology that subjugates women within the heteropatriarchal regimes of modernity.[14] While feminist critiques have long interrogated the ways romantic relationships have historically been used to overshadow women's intellectual contributions, Du Châtelet's letters and philosophical writings suggest a more complex entanglement. Love, for her, is not solely a site of subjection but also a source of becoming, a transformative flow of desire that continually reshapes her subjectivity.

By framing Du Châtelet's letters within Bakhtinian, narratological, and feminist paradigms, what I argue is that they epitomise a dialogic tension between pathos, self-realisation, and existential confirmation. These letters are not static records of a romantic relationship but dynamic negotiations of affect, intellect, and becoming, revealing a subjectivity in constant flux, propelled by the creative intensities of love and study, the central theme of her *Discourse on Happiness*, as we have seen in the previous section.

Mathematics as an existential choice

Historians, sociologists, and philosophers of science have often marginalised or outright denied the role of emotions, affects, and feelings in the creation of scientific knowledge and the enactment of scientific practices, as Paola Govoni has observed (2014, 15). Challenging this perspective, Thomas Söderqvist's proposal of existential scientific biographies reframes the passions of scientists as central to understanding their lives and work. He argues that 'the passions of the scientist—both negative emotions such as anguish, anxiety, despair, and dread, and positive emotions such as joy, hope, and *love*' (1996, 65, emphasis added)—reveal the existential uniqueness and embodied singularity of their lives. From intense feelings of frustration in grappling with unsolved problems to the euphoria of discovery, and the accompanying fears and anxieties along the way, Söderqvist underscores that emotions are not peripheral but integral to the making of scientific knowledge (67). In this view, scientific biographies transcend mere factual accounts; they become narratives that explore existential questions, choices, and struggles, both for their subjects and for those who write about them:

> the aim of science biography is to provide us with stories through which we can identify ourselves with other human beings who have chosen to spend their lives in scholarly or scientific work. Such stories can make us understand and change ourselves - scientists, historians of science and laymen alike. In that sense, biographies of scientists are useful as exemplars for what Kierkegaard considered to be the core of freedom in an ethical life: the continuous renewal of oneself (76).

Drawing on Richard Rorty's (1980) notion of 'edifying literature', Söderqvist suggests that science biographies go beyond historical documentation to engage readers in reflection about the purpose and value of scientific work: 'existential biographies of scientists may be edifying; they may provide us with opportunities for reorienting our familiar ways of thinking about our lives in unfamiliar terms' (1996, 76). These narratives inspire, provoke, and encourage a rethinking of the roles of science and scientists in society, 'they take us out of our old selves by the power of strangeness, to aid us in becoming new beings' (Rorty 1980, cited in Söderqvist 1996, 77). Söderqvist's invocation of Rorty's notion of 'edifying literature' enriches our understanding of the narratives of women mathematicians. These auto/biographical texts do more than chronicle historical achievements; they provoke reflection on the broader significance of their lives.

Extending Söderqvist's argument to women mathematicians' auto/biographical, literary, and philosophical writings reveals how these narratives serve as edifying texts, offering deep insights into the human condition, the nature of mathematical and scientific inquiry, and the ethical and philosophical dimensions of lives dedicated to mathematics and science. Söderqvist particularly highlights Evelyn Fox Keller's *A Feeling for the Organism* (1983), on the life and work of the Nobel laureate Barbara McClintock, 'as a model for women scientists trying to cope with their own lives in science' (Söderqvist 1996, 77) .

Writing about her experience, Keller remembered that when she submitted her manuscript, the publisher returned it with the suggestion for a change of title among several corrections, 'marking with a red pencil to cut out just about everything that had to do with the emotional side of her scientific life' (Keller 2014, 36). Having borrowed the phrase 'a feeling for the organism' from McClintock herself, Keller insisted that feeling should not be taken as 'a feminine trait', arguing that 'feeling' was an important component of McClintock's scientific work, 'a tradition that was particularly associated with the "great minds" of science' (36). Keller also remembers that when her book was eventually published, the Chair of the Department of Mathematics at Northeastern University, where she was working at the time, 'came and thanked me for giving him permission to do mathematics the way he always wanted to do it' (36–7). Indeed, if we patiently follow traces of women mathematicians' becomings, we discern the design left behind the life of their passions, alongside the life of their mathematical mind.

Women mathematicians' auto/biographical, philosophical, and literary writings exemplify what Söderqvist (1996) describes as 'existential projects'. Their choice to pursue mathematics often represented a radical act of defiance, challenging gender norms and carving out intellectual spaces where they could assert themselves as subjects of mathematical knowledge. Their letters and personal writings reveal mathematics as more than an intellectual pursuit: it became a sanctuary, a source of joy, and a way to navigate turbulent emotional and societal landscapes. Söderqvist's emphasis on dialogue allows us to understand love and mathematics as entangled, existential discourses. Love, with its emotional intensity and relational intricacies, contrasts yet complements mathematics, which demands abstraction, precision, and logical rigour.

Throughout history, women mathematicians have had to balance complex roles—motherhood, partnership, intellectual pursuits—inhabiting liminal spaces where familial and scientific expectations collided. This dynamic illuminates what Söderqvist (1996) might describe as the existential negotiations that define a life. Their writings reflect not only their mathematical struggles but also their broader existential uniqueness, what Hannah Arendt would describe as their 'whoness' within the web of human relations (1998).

Viewed through an existential lens, women mathematicians' narratives become edifying texts that challenge the traditional compartmentalisation of 'the personal and the scientific', as we have seen in Chapter 2. They illustrate how love and mathematics are deeply entwined in their lived experiences, serving as mediums for making sense of their world and asserting their presence in the male-dominated sphere of mathematics. They further underscore the transformative power of their 'existential choices' (Söderqvist 1996), decisions that inspire us to reimagine intellectual freedom and the integration of emotions, subjectivity, and knowledge creation. By foregrounding the dialogic tensions in their lives, we see their struggles as not merely personal but profoundly universal, illuminating the entanglements of gender, love, and intellectual pursuit. These narratives continue to provoke and inspire, redefining the possibilities of a life dedicated to both intellect and humanity.

Notes

1 Émilie Du Châtelet to Jean François de Saint-Lambert, Morgan Library and Museum, Department of Literary and Historical Manuscripts, in the text and thereafter (MA 2287).
2 For biographical details on Saint-Lambert, see Poirier (2001).
3 Veneranda Biblioteca Ambrosiana, Manoscritti, Agnesi, Maria Gaetana <1718-1799> Compendium Ethicae sive Studi di etica seguiti da Maria Gaetana Agnesi nella prima sua età e scritti di sua mano, thereafter and in the text VBA, followed by codice number.
4 Oxford Bodleian Libraries, Archive of the Noel Byron and Lovelace Families, Dep. Lovelace Byron. In the text and thereafter, LB followed by the folder and folio number.
5 British Library, Manuscripts Collection, Correspondence of Charles Babbage, Augusta Ada King, daughter of Lord Byron; wife of William, Lord King, 1st Earl of Lovelace : Letters to C. Babbage: 1836–51, Add MS 37189-37192.
6 In his introduction to the 1961 critical edition of the *Discours,* Mauzi discusses at length the possible periods of its composition (1961, lxxiv–lxxxiv), but reaches the conclusion that the most probable date would be perhaps sometime in 1747. See also Whitehead 2006, 256 and Zinsser 2009, 345.
7 See Mauzi 1961, Whitehead 2006, and Zinsser 2006, 2009 for a discussion of the essay's reception.
8 For more details of their meeting and relationship, see Zinsser 2006, particularly Chapter 6: 260–79.
9 See amongst others, Wade 1941, Mitford 1957 , and Bodanis 2006.
10 In his philosophy of time, Henri Bergson (1970) has made the distinction between 'objective', calendar time and la durée [duration], which is the time of our lived, subjective experiences.
11 For a rich discussion of the 'love letter', see, amongst others, Kauffman 1986, Grassi 1998, and Tamboukou 2010, 2023.
12 Gilles Deleuze extensively theorised the concept of *becoming* in several of his works as well as in his collaborative work with Felix Guattari. See, amongst others, Deleuze 2001, 2004 and Deleuze and Guattari 1988.
13 See amongst others, Edwards 1970; Mitford 1957.
14 See Tamboukou 2023 for a discussion of this literature.

References

Aaron, Richard, I., and Jocelyn Gib, eds. 1936. *An Early Draft of Locke's Essay.* Oxford: The Clarendon Press.

Agnesi, Maria Gaetana. 1738. 'Proposizioni Filosofiche'. Accessed 29 November 2024. https://www.centrostudimariopancrazi.it/images/pubblicazioni/proposizioni_filosofiche/proposzioni_filosofiche.pdf.

Antognazza, Maria Rosa. 2009. *Leibniz: An Intellectual Biography.* New York: Cambridge University Press.

Arendt, Hannh. 1998 [1958]. *The Human Condition.* Chicago: Chicago University Press.

Baggi, Denis, L. 1991. 'Computer Generated Music - Guest Editor's Introduction'. *Computer*, Special issue on Computer Generated Music 24 (7): 6–9.
Bakhtin, Michail. 2002. *The Dialogic Imagination: Four essays.* Edited by Michael Holquist and translated by Caryl Emerson and Michael Holquist. Austin: The University of Texas Press.
Barad, Karen. 2007. *Meeting the Universe Halfway: Quantum Physics and the Entanglement of Matter and Meaning*. Durham: Duke University Press.
Barthes, Roland. 2002 [1979]. *A Lover's Discourse: Fragments*. Translated by Richard Howard. London: Vintage Books.
Benjamin, Jessica. 1983. 'Master and Slave: The Fantasy of Erotic Domination'. In *Powers of Desire: The Politics of Sexuality*, edited by Ann Snitow, Christine Stansell and Sharon Thompson, 280–99. New York: Monthly Review Press.
Bergson, Henri. 1970 [1899]. *Time and Free Will: As Essay on the Immediate Data of Consciousness*. Translated by Frank L. Pogson. London: George Allen & Unwin.
Bodanis, David. 2006. *Passionate Minds: Émilie du Châtelet, Voltaire, and the Great Love Affair of the Enlightenment*. New York: Three Rivers Press.
Boucard, Jenny. 2020. 'Arithmetic and Memorial Practices by and Around Sophie Germain in the 19th Century'. In *Against All Odds. Women's Ways to Mathematical Research Since 1800*, edited by Eva Kaufholz-Soldat and Nicola M. R. Oswald, 185–230. Cham: Springer, 2020.
Deleuze, Gilles. 2001 [1969]. *The Logic of Sense*. Translated by Mark Lester. London: Continuum.
Deleuze, Gilles. 2004 [1968]. *Difference and Repetition*. Translated by Paul Patton. London: Continuum.
Deleuze, Gilles, and Felix Guattari. 1988 [1980]. *A Thousand Plateaus: Capitalism and Schizophrenia*. Translated by Brian Massumi. London: The Athlone Press.
Du Châtelet, Émilie. 1961. *Discours sur le bonheur, édition critique et commente par Robert Mauzi* [Discourse on Happiness, Critical Edition and Commentary by Robert Mauzi]. Paris: Les Belles Lettres.
Edwards, Samuel. 1970. *The Divine Mistress: A Biography of Émilie du Châtelet, The Beloved of Voltaire*. New York: David Mckay.
Fage, Émile. 1894. 'Chronique des livres et des revues. 1. Les Femmes dans la science [...], par A. Rebière'. *Bulletin de la Société des lettres, sciences et arts* 16: 245–55.
Govoni, Paola. 2014. 'Crafting Scientific (Auto)Biographies'. In *Writing about Lives in Science: (Auto)Biography, Gender and Genre,* edited by Paola Govoni and Zelda Alice Franceschi, 7–30. Göttingen: V&R Unipress.
Grassi, Marie-Claire. 1998. *Lire l' épistolaire.* Paris: Dunod.
Hamel, Frank. 1910. *An Eighteenth-Century Marquise: A Study of Émilie du Châtelet and her Times*. New York: James Pot & Co.
Kauffman, Linda, S. 1986. *Discourses of Desire: Gender, Genre and Epistolary Fictions*. Ithaca and London: Cornell University Press.
Keller, Evelyn Fox. 1983. *A Feeling for the Organism: The Life and Work of Barbara McClintock*. San Francisco: W. H. Freeman and Company.
Keller, Evelyn Fox. 2014. Pot-holes Everywhere: How (not) to Read My Biography of Barbara McClintock'. In *Writing about Lives in Science: (Auto)Biography, Gender and Genre*, edited by Paola Govoni and Zelda Alice Franceschi, 34–42. Gottingen: V&R Unipress.
Key, Ellen. 1891. 'Sonja Kovalewsky'. *Tidskrift* Dagny 3: 1–17. Accessed May 15, 2024.
Kisner, M. J. 2012. 'Spinoza: Defender of the Passions?'. Accessed 15 January 2023. https://emotionsblog.history.qmul.ac.uk/2012/09/spinoza-defender-of-the-passions.
Kochina, Pelageya. 1985. *Love and Mathematics: Sofia Kovalevskaya*. Translated by Michael Burov. Moscow: Mir Publishers.

Kölving, Ulla, and Andrew Brown, eds. 2018. *La Correspondance d' Émilie Du Châtelet.* 2 vols. Ferney-Voltaire: Centre international d'étude du xviiie siècle (vol. 1, pp. 1–633; vol. 2, pp. 1–593).

Kovaleskaya, Sofia. 1951. *Воспоминания и письма* [Memories and Letters]. Edited and annotated by Solomon Yakovlevich Shtraikh. Moscow: USSR Academy of Sciences. [Second edition 1961].

Kovalevskaya, Sofia. 1978[1895]. *A Russian Childhood.* Translated by Beatrice Stillman. New York: Springer-Verlag.

Lermontova, Iulia, V. 1951. 'Memories of Sofia Kovalevskaya'. In *Воспоминания и письма* [Memories and Letters]. Edited and annotated by Solomon Yakovlevich Shtraikh, 375–87. Moscow: USSR Academy of Sciences.

Lacoue-Labarthe, Philippe. 1989. *Typography: Mimesis, Philosophy, Politics.* Edited by Christpher Fynsk. Cambridge MA: Harvard University Press.

Le Ru, Veronique. 2019. *Émilie du Châtelet, phiolosophe.* Paris: Classiques Garnier.

Libri, Guglielmo. 1833. 'Notice sur Mlle Sophie Germain [Notice on Mademoiselle Sophie Germain]'. In *Considérations Generales Sur l'Etat Des Sciences Et Des Lettres Aux Différentes Époques de leur Culture* [General Considerations on the State of the Sciences and Letters at Different Periods of Their Development], edited by Sophie Germain, 11–16. Paris: Lachevardiere.

Mauzi, Robert. 1960. *L' Idée du bonheur dans la littérature et la pensée françaises au XVIII' siècle* [The Idea of Happiness in French Literature and Thought in the 18th Century]. Paris: Librairie Armand Colin.

Mauzi, Robert. 1961. 'Introduction'. In *Discours sur le bonheur, édition critique et commente par Robert Mauzi.* [Discourse on Happiness, Critical Edition and Commentary by Robert Mauzi], ix–cxxvii. Paris: Les Belles Lettres.

Mazzotti, Massimo. 2007. The World of Maria Gaetana Agnesi, Mathematician of God. Baltimore: Johns Hopkins University Press.

Mitford, Nancy. 1957. *Voltaire in Love.* London: Hamish Hamilton.

Peterson, Bennet, Barbara. 2000. 'Wang Wang Zhenyi'. In *Notable Women Of China: Shang Dynasty to the Early Twentieth Century*, edited by Barbara Bennett Petrson with He Hong Fei, Han Tie, Wang Ziyu, and Zhang Guangyu, 341–46. New York: M. E. Sharp.

Poirier, Roger. 2001. *Jean François de Saint-Lambert, 1716–1803: Sa vie, son oeuvre.* Sarrerquemines: Pierron.

Robertson, Ritchie. 2022. *The Enlightenment, The Pursuit of Happiness, 1680–1790.* London: Penguin.

Rorty, Richard. 1980. *Philosophy and the Mirror of Nature.* Princeton: Princeton University Press.

Shapiro, Lisa. 2020. 'Descartes and Spinoza on the Primitive Passions: Why So Different?' In *Freedom, Action and Motivation in Spinoza's Ethics*, edited by Noa Naaman-Zauderer, 62–81. New York and London: Routledge.

Söderqvist, Thomas. 1996. 'Existential Projects and Existential Choice in Science: Science Biography as an Edifying Genre'. In *Telling Lives: Studies in Scientific Biography*, edited by Michael Shortland and Richard Yeo, 45–84. Cambridge: Cambridge University Press.

Somerville, Mary. 2001 [1873]. *Queen of Science: Personal Recollections of Mary Somerville.* Edited by Dorothy McMillan. Edinburgh: Canongate.

Stanley, Liz. 2004. 'The Epistolarium: On Theorizing Letters and Correspondences'. *Auto/Biography* 12 (3): 201–35.

Tamboukou, Maria. 2010. *Nomadic Narratives, Visual Forces: Gwen John's Letters and Paintings.* New York: Peter lang.

Tamboukou, Maria. 2022. 'Traces in the Archive: Re-imagining Sofia Kovalevskaya'. *Life Writing* 19 (3): 341–56.

Tamboukou, Maria. 2023. *Epistolary Narratives of Love, Gender and Agonistic Politics: An Arendtian Approach.* London: Routledge.

Wade, Ira. 1941. *Voltaire and Madame du Châtelet: An Essay on the Intellectual Activity at Cirey*. Princeton: Princeton University Press.
Wade, Ira. 1947. *Voltaire and Madame du Châtelet* and *Studies on Voltaire with Some Unpublished Papers of Mme Du Châtelet*. Princeton: Princeton University Press.
Whitehead, Barbara. 2006. 'The Singularity of Mme du Châtelet: An Analysis of the *Discours sur le bonheur*'. In *Emilie Du Châtelet Rewriting Enlightenment Philosophy and Science*, edited by Judith P. Zinsser and Julie Candler Hayes, 255–76. Oxford: Voltaire Foundation.
Zinsser, Judith P. 2006. *Daring Genius of the Enlightenment, Émilie du Châtelet.* New York: Penguin.
Zinsser, Judith P. 2009. 'Volume Editor's Introduction'. In *Selected Philosophical and Scientific Writings*, edited by Judith, P. Zinsser and translated by Isabelle Bour and Judith P. Zinnser, 345–6. Chicago and London: The University of Chicago Press.

6 Philosophical Ruminations and Process Epistemologies

'Time preserves only the works which defend themselves against it', Sophie Germain insightfully observed in her philosophical writings, a reflection that her editors deemed significant enough to position first in the *Diverse Thoughts* (*Pensées Diverses*) section of her posthumously published works. While Germain's contributions to mathematics, particularly in number theory and elasticity, are widely acknowledged, her philosophical musings remain overlooked.[1] These writings, which delve into the nature of knowledge, creativity, and intellectual perseverance, have yet to be fully explored or even translated into English, underscoring the broader neglect of women's intellectual legacies.

This neglect is emblematic of a systemic failure to recognise the contributions of women mathematicians throughout history. Germain's dual accomplishments in mathematics and philosophy highlight a broader issue: the undervaluation of interdisciplinary work by women and the lack of preservation and dissemination of their lesser-known endeavours. Germain's situation mirrors that of other women mathematicians, such as Émilie Du Châtelet's *Institutions de Physique*, which has only been revisited in the last two decades, or Maria Gaetana Agnesi, whose *Instituzioni Analitiche* was celebrated during her lifetime but later faded from the academic canon, as we have already examined in Chapter 3.

Throughout this book, we have also seen that women mathematicians often faced structural barriers to institutional acknowledgement, and their works were frequently underpublished, mistranslated, or relegated to obscurity. As already discussed in Chapter 3, the lack of accessible translations in English of Germain's *Œuvres philosophiques*, or Agnesi's *Philosophical Propositions,* illustrates how language and editorial practices can compound this neglect. When compared to the prolific availability of translations and studies of male philosophers and scientists, the disparity becomes starkly evident.

The omission of women's broader intellectual contributions—philosophical, mathematical, or otherwise—stems not only from the biases of their time but also from ongoing patterns in historiography and academia. Recovering and foregrounding these voices, as well as their works, require intentional efforts, including the translation, publication, and contextual analysis of their philosophical writings. Only by addressing these gaps can the contributions of women mathematicians receive the comprehensive acknowledgement they deserve.

DOI: 10.4324/9781003533993-7

This chapter contributes to the growing body of scholarship that seeks to bridge this gap by examining the philosophical reflections of women mathematicians, positioning their work within the broader landscape of epistemological theories. As discussed in Chapter 5, Émilie Du Châtelet's philosophical essay on love and happiness and Maria Gaetana Agnesi's *Philosophical Propositions* manifest how these thinkers integrated reflections on love, happiness, truth, and mathematics into their broader inquiries. These women worked within a context where the boundaries between science and philosophy were fluid—particularly in the eighteenth century—despite the increasingly rigid disciplinary divisions that emerged in the nineteenth century. This chapter, however, focuses specifically on women mathematicians' engagement with process epistemologies. What I argue is that their writings invite a re-evaluation of the intersections between science, philosophy, and creativity, offering insights that remain resonant for contemporary discussions in the history of ideas. By emphasising the philosophical dimensions of their work alongside their mathematical achievements, the analysis of the chapter contributes to a more holistic understanding of their legacy.

This chapter unfolds in six sections, beginning with this concise introduction. The second section examines Émilie Du Châtelet's *Institutions de Physique*, as a pathbreaking work in the history of philosophy. Next, Maria Gaetana Agnesi's *Philosophical Propositions* is analysed, with attention to her nuanced metaphysical arguments and their connection to her mathematical work. Sophie Germain's *Œuvres philosophiques* (1896) follows, revealing the philosophical depth and intellectual courage behind her reflections on time, creativity, and knowledge. The fifth section considers the epistemological underpinnings of Mary Somerville's *On the Connexion of the Physical Sciences* and reads Ada Lovelace's short essay on imagination, emphasising her visionary take on the interplay between analytical reasoning and creative thought. Although Somerville and Lovelace were not recognised as philosophers nor authored philosophical treatises, their epistemological insights contribute to the assemblage of process epistemologies that anchor this chapter. The concluding section reflects on the transdisciplinary endeavours of these women mathematicians, linking their intellectual pursuits to their broader cultural and historical contexts.

By crossing the traditional boundaries of mathematics, philosophy, and science, this chapter illuminates the deeply interconnected nature of these fields as navigated by these trailblazing women scholars. Their works transcend disciplinary silos, revealing profound insights into the shared principles that underpin diverse domains of inquiry. These women not only contributed groundbreaking ideas within their respective disciplines but also engaged with broader metaphysical and epistemological questions, reshaping our understanding of the intellectual landscape of their time.

Each section examines the philosophical reflections of these mathematicians in both the context of their historical era and their relevance to broader philosophical trajectories, many of which resonate with contemporary discussions. By situating their work within evolving philosophical frameworks, the chapter underscores the interplay between mathematical reasoning and philosophical inquiry, demonstrating how these disciplines mutually inform and enrich each other.

Together, these explorations aim to showcase the depth and breadth of women mathematicians' contributions to the history of ideas, challenging conventional narratives that have often marginalised their intellectual achievements. The chapter finally highlights the transdisciplinary nature of their endeavours, revealing how their integrative approaches continue to inspire and shape the modern understanding of the intertwined evolution of mathematics, philosophy, and science.

Émilie Du Châtelet: an *œuvre à faire*

'We rise to the knowledge of the truth, like those giants who climbed up to the skies by standing on the shoulders of one another' (2009, 118), Du Châtelet wrote in the Preface of her *Institutions de Physique*, paraphrasing Newton's often quoted statement from his 1675 letter to Robert Hooke.[2] With this remark, she laid the groundwork for her book, which encompasses a wide range of topics, including epistemology, metaphysics, methodology, mechanics, and physics. As Katherine Brading has pithily noted, the *Institutions* addresses several critical questions of its era, such as the existence of atoms, the respective roles of God and hypotheses in scientific inquiry, the mechanisms (if any) through which bodies exert influence on one another, and the nature of gravity—specifically, whether it operates as an action-at-a-distance force (2019, 1). Brading particularly highlights the fact that 'Du Châtelet sought to resolve these issues within a single philosophical framework that builds on her critique and appraisal of all the leading alternatives (Cartesian, Newtonian, Leibnizian, and so forth) of the period' (1).

Brading further comments that the *Institutions* stands out as the first scientific project to undertake such an ambitious and comprehensive synthesis, further distinguished by its exceptional accessibility and clarity (1). Its lucid and approachable style makes complex scientific and philosophical ideas intelligible to a broader audience, reflecting a groundbreaking effort to bridge rigorous scholarship with effective communication, rather than simply popularise science and mathematics, as her friend Francesco Algarotti had done with his *Newtonianism for Ladies (1737)*, a book that both Du Châtelet and Agnesi had criticised, as we have seen in Chapter 2.

While the significance of Du Châtelet's synthesis is undeniable, Brading has strongly contended that the *Institutions de Physique* transcends its characterisation as a mere synthesis. In addressing what she identified as the most pressing philosophical problems, Du Châtelet not only employed the intellectual frameworks of her time but reimagined and transformed them. Through this process, she developed original positions and reshaped the philosophical discourse surrounding the theory of matter, leaving a lasting impact on the intellectual landscape. Her multifaceted approach—a synthesis of explication, critique, and creation—demonstrates her ability to navigate and advance the philosophical and scientific debates of her time and highlights her philosophy of knowledge: 'Each philosopher has seen something, and none has seen all; no book is so bad that nothing can be learned from it, and no book is so good that one might not improve it' (Du Châtelet 2009, 122). Scientific knowledge in this light could only advance through hypotheses, experiments, and gradual understanding, a long process in the development of scientific methods, but also an adventure in speculative philosophy, where physics and metaphysics inevitably meet: 'there are probably some truths not made to be perceived by the eyes of our mind, just as there are objects that those of our body will never perceive' (120), she wrote.

Judith Zinsser has identified three primary objectives behind Du Châtelet's decision to write the *Institutions de Physique* (2009, 105–6). The first, and most immediate, was addressing the glaring absence of comprehensive and accessible educational texts in natural philosophy, a deficiency that became especially apparent to her while teaching her son mathematics and science, as discussed in Chapter 3. The second objective was to establish the metaphysical foundations that she believed Newtonian mechanics lacked. While Newton's theories revolutionised physics, Du Châtelet saw them as incomplete, particularly in their explanation of phenomena such as gravity. Finally, she aimed to refine and challenge established theories, critically assessing and, where necessary, correcting the ideas of both Newton and Descartes throughout the *Institutions*.

However, Zinsser's interpretation of Du Châtelet's second objective has been contested by Brading. Brading cautions that such a reading risks portraying Du Châtelet as a passive compiler of ideas from Leibniz, Wolff, and Newton, combining their theories without contributing significant originality (Brading 2019, 9). This interpretation, according to Brading, underestimates the philosophical depth and creativity of Du Châtelet's work. Instead, Brading argues that the *Institutions* represents a profound intellectual endeavour, wherein Du Châtelet drew on the most advanced philosophical resources of her time—spanning Cartesian, Leibnizian, and Newtonian traditions—not to simply reconcile these frameworks but to leverage their strengths in addressing pressing philosophical and scientific questions. Through this integrative and critical approach, Du Châtelet crafted original solutions to enduring problems in natural philosophy and metaphysics, underscoring her active role in shaping Enlightenment thought.

> A careful reading of the philosophical content of the *Foundations* shows that she is active in her engagement with the resources on which she draws. She makes deliberate choices of which resources to use (and, equally importantly, which to omit) in light of her philosophical goals. She transforms those resources in the service of those goals. And she thereby arrives at new philosophical positions on a wide variety of issues and at a new overall philosophical system (Brading 2019, 16).

In light of the above, Brading contends that the question of scientific methodology forms both the structural framework and the unifying theme of Du Châtelet's *Institutions*, (17). She proposes a three-step approach to interpreting the text: it begins with an exploration of method, applies this methodological lens to the problem of understanding bodies, and extends it further to address the most contentious issues in contemporary physics. These include Newtonian action-at-a-distance gravitation and the *vis viva* controversy, topics discussed in earlier chapters. Viewed through this perspective, Brading argues, the *Institutions* unfolds as 'a sustained argument concerning bodily action' (8).

In thus challenging Zinsser's interpretation that Du Châtelet aimed to provide a metaphysical foundation for physics, Brading emphasises that Du Châtelet's main objective was to develop a methodological framework capable of tackling the pressing scientific challenges of her era. In doing so, however, Du Châtelet inevitably adopted metaphysical positions that diverged significantly from those of her Newtonian contemporaries. Exploring how she arrived at these positions offers a richer appreciation of her philosophical contributions, Brading asserts (17).

The *Institutions* integrates these objectives seamlessly within its text. In the *Preface*, Du Châtelet underscores the critical importance of education, writing: 'I have always thought that the most sacred duty of men was to give their children an education that prevented them at a more advanced age from regretting their youth, the only time when one can truly gain instruction' (2009, 116). She took on this responsibility for her son, who was 'in this happy age when the mind begins to think' (116), expressing frustration at the lack of suitable educational resources. Lamenting this void, she noted her surprise that in a nation as intellectually rich as France, no one had yet undertaken such a project: 'I am often surprised that so many clever people as France possesses have not preceded me in this work that I embark upon for you today' (117).

Du Châtelet's pedagogical vision was clear—to make science accessible and relatable. She emphasised the need to disentangle scientific knowledge from the abstract language of algebra, which, she argued, 'separates things from images, eludes the senses, and speaks only to the understanding' (116). Her aim was to bridge this gap, offering a text that could serve as both an educational foundation and a methodological guide to the philosophical implications of natural phenomena. By blending her methodological rigour with a commitment to education, Du Châtelet positioned her *Institutions* as both a practical resource and a sophisticated philosophical endeavour, bridging gaps in scientific inquiry while addressing the urgent need for accessible knowledge.

In this regard, she challenged prevailing ideas by advancing a rigorous epistemology of hypotheses. This defence of hypotheses is particularly prominent in Chapter IV of the *Institutions*, where she argues against the Newtonian rejection of hypotheses in scientific inquiry. Newton, in his *Principia*, famously stated, 'Hypotheses non fingo' [I frame no hypotheses], expressing scepticism about speculative reasoning and emphasising reliance on empirical evidence.[3] Du Châtelet, however, argued that hypotheses are essential as a preliminary scaffolding in the process of building scientific theories: 'One of the mistakes of some philosophers of our time is to want to banish hypotheses from physics; they are as necessary as the scaffolding in a house being built; it is true that, when the building is completed, the scaffolding becomes useless, but it could not have been erected without it' (121). Her perspective reflects a progressive view of scientific methodology, emphasising that speculation and provisional ideas are crucial for advancing understanding and discovering new truths: It is true that hypotheses become the poison of philosophy when they are made to pass for the truth [...] But an ingenious and bold hypothesis, which has some initial probability, leads human pride to believe it, the mind applauds itself for having found these subtle principles, and next uses all its sagacity to defend them (121).

Du Châtelet's insistence on the importance of hypotheses has been largely ignored in the history of science, but as Gianni Paganini (2022) has argued, her epistemology of hypotheses is an autonomous contribution and a major innovation in epistemology, methodology as well as the philosophy of science. Indeed, her argument anticipated modern views in the philosophy of science, where hypotheses are understood as necessary tools for the formulation and testing of theories. By advocating for this approach, Du Châtelet not only defended a Leibnizian perspective but also contributed to broader methodological debates in the Enlightenment era (see Reichenberger 2012, 161). This position was particularly bold for a woman scientist in her time, highlighting her intellectual independence and innovative thinking.

Du Châtelet's *Institutions* further progressed into a sophisticated exploration of metaphysics, addressing the nature of space, time, and substance. She critiqued traditional views, offering a deeper, more nuanced understanding of these concepts that was in harmony with the empirical discoveries of her era. Acknowledging that Leibniz's philosophy was not widely understood in France, she nonetheless championed its clarity and utility. She admitted certain 'obscurities' in metaphysical discussions but asserted that Leibniz's thought offered a 'compass capable of guiding us in the moving sands of this science' (Du Châtelet 2009, 123).

Additionally, Du Châtelet delved into the philosophical debate surrounding the nature of force, introducing her own conceptualisation of *vis viva* (living force) as a critical foundation of natural law, as we have already seen in previous chapters. Building on Leibniz's arguments, she developed a theory of matter that diverges from both Cartesian and Newtonian models, offering an alternative interpretation that is consistent with her broader metaphysical views (see Reichenberger 2012).

In the final sections, Du Châtelet applied her mathematical and philosophical insights to natural phenomena, such as the motion of pendulums, free fall, and planetary orbits, demonstrating how mathematical principles can be used to understand these physical behaviours. Through this work, she not only contributed to the scientific discourse of her time but also laid the groundwork for future developments in physics.

Du Châtelet's *Institutions* presents a sophisticated critique of Cartesian and Newtonian methodologies while emphasising the superiority of mathematical reasoning in resolving empirical inconsistencies. However, she acknowledged the limitations of applying mathematical precision to metaphysical inquiry. Reflecting on this, she observed: 'We lack a system of calculation for metaphysics similar to that which has been found for mathematics, by means of which, with the aid of certain givens, one arrives at knowledge of unknowns. Perhaps some genius will one day find this system' (Du Châtelet 2009, 123–4). This admission underscores her forward-thinking approach, recognising both the power and the boundaries of existing methodologies. While she championed mathematical rigour in science, she was keenly aware of the complexities and uncertainties inherent in metaphysical exploration, leaving room for future innovations to bridge this gap.

Her dedication to advancing science within a metaphysical framework reflects her belief in the unity of scientific and philosophical inquiry—a perspective that resonates with later transdisciplinary approaches. She notably grappled with the *principle of sufficient reason* to ground her theories, articulating a vision of physics as governed by logical imperatives rather than arbitrary forces. Central to Leibniz's philosophy, the *principle of sufficient reason* governs not only physical phenomena but also metaphysical and logical reasoning. Leibniz formulated it as part of his broader philosophical system, particularly emphasising that the world is ordered in such a way that every event or truth has a rational basis.[4] Du Châtelet embraced and elaborated upon Leibniz's *principle of sufficient reason*, weaving it into her critical synthesis of Descartes, Newton, and Wolff.[5] She regarded this principle as a foundational tool for navigating both scientific and philosophical inquiry:

> When asking someone to account for his actions, we persist with our own question until we obtain a reason that satisfies us, and in all cases we feel that we cannot force our mind to accept something without a sufficient reason, that is to say, without a reason that makes us understand why this thing is what it is, rather than something completely different (2009, 128–9).

Du Châtelet applied the principle across natural philosophy, particularly in discussions on the nature of matter and motion, and extended its relevance to human behaviour, arguing that rational decision-making must align with this law. For her, the principle was not merely a theoretical concept but a practical guide for hypothesis-driven reasoning. She wrote emphatically that 'this principle is the only thread that could guide us in these labyrinths of error the human mind has built for itself in order to have the pleasure of going astray' (131). This conviction underscores her belief in the interconnectedness of reason, science, and metaphysical inquiry as essential to progress in knowledge.

Du Châtelet's *Institutions* was more than a groundbreaking scientific treatise; it also served as a deliberate challenge to the systemic intellectual marginalisation of women. Through incisive correspondence and robust public defences—such as her well-known rebuttal to Dortous de Mairan, discussed in Chapter 4—she claimed her space within the male-dominated Republic of Letters. In addition to her dispute with Mairan, the publication of the *Institutions* provoked her former tutor Samuel König to allege that she had merely taken dictation from him. As examined in Chapter 2, Du Châtelet countered this accusation in a letter to Johann Bernoulli II, dated 1 February 1740, where she eloquently defended her authorship, showcasing her intellectual independence and repudiating König's claims (Kölving and Brown 2018, vol. 1, 569).

This controversy, however, underscores the significant challenges women faced in establishing intellectual authority. Despite these hurdles, Du Châtelet's *Institutions* not only demonstrated her mastery of complex philosophical and scientific ideas but also solidified her status as an innovative thinker capable of reshaping contemporary debates on natural philosophy and gender. By challenging prevailing stereotypes about women's intellectual capabilities, Du Châtelet asserted her authority and garnered acknowledgement from leading thinkers across Europe. Du Châtelet's *Institutions*, thus, serve as a remarkable example of how women in the Enlightenment carved intellectual spaces despite structural and cultural barriers.

Karen Detlefsen has aptly highlighted Du Châtelet's philosophical rigour, observing that she 'exhibits a combination of precision and creativity that is one mark of some of the best philosophy' (2017, 131). This fusion of analytical clarity and innovative thought not only positioned her as a significant contributor to Enlightenment science and philosophy but also as a figure who reshaped perceptions of women's intellectual contributions in her era. Overall, Du Châtelet's interdisciplinary approach prefigures modern debates about the integration of science and philosophy, emphasising the necessity of metaphysical grounding in scientific inquiry. Her methodology—balancing synthesis, critique, and innovation—remains an inspiring model for addressing epistemological challenges today. Moreover, her philosophical engagement with figures like Newton, Leibniz, and Wolff enriches our understanding of eighteenth-century thought and its contribution to the development of critical philosophical currents in later centuries (see Hagengruber 2012).

More significantly, the so-called 'Paris manuscripts' of Du Châtelet's *Institutions de Physique* offer an extraordinary and invaluable glimpse into the evolution of her intellectual journey, capturing pivotal moments in her becoming a scientist and philosopher. It is particularly noteworthy that Du Châtelet initially planned to publish the *Institutions* in the autumn of 1738. Although the manuscript had received royal approval and was sent to press, she halted its publication and postponed it for two years. What transpired between September 1738 and December 1740, when the first edition was finally published? A revealing clue comes from a brief preface by her printer and librarian, Prault, which sheds light on the circumstances surrounding this delay:

> This first volume of the *Institutions de Physique* was ready to be printed as early as September 18, 1738, as indicated by the Approval, and the printing had even begun at that time. However, the author, wishing to make some changes, had me suspend it; these changes concerned the Metaphysics of Mr. Leibniz, a brief exposition of which can be found at the beginning of this volume (Du Châtelet 1740, f. 13).

From its inception, the *Institutions de Physique* underwent a rigorous and intricate process of revision, driven largely by Du Châtelet's deep engagement with Leibnizian metaphysics. This engagement not only reshaped her approach to physics but also necessitated substantial alterations to the work's content and structure. These evolving iterations of the manuscript have now been meticulously integrated into the first online historical-critical edition of the *Institutions*, officially launched in December 2024, while this book was in progress (Du Châtelet 2024). This landmark edition encompasses the analysis of multiple key sources: the original manuscript, the 1740 partial page proofs, the Paris edition published later that year, and the second edition printed in Amsterdam in 1742—collectively referred to as 'The Paris manuscripts'[6], as already noted above.

A close study of these manuscripts reveals a dynamic visual archive of Du Châtelet's working methods. Layers of deletions, strikethroughs, marginal notes, restructured sections, and multiple versions of specific chapters reflect her iterative intellectual process. These textual interventions illuminate not only the complexity of her revisions but also her persistent efforts to refine and expand her ideas. The analysis of the *Paris manuscripts* underscores the remarkable scope and depth of Du Châtelet's theoretical framework, which seamlessly interweaves metaphysics, epistemology, physics, philosophy of science, and pedagogy. It allows readers to trace her intellectual dialogues with influential predecessors such as Descartes, Leibniz, Newton, and Wolff, showcasing her ability to critically evaluate and synthesise their ideas while contributing original insights to Enlightenment thought.

Furthermore, the critical-historical edition highlights Du Châtelet's methodological rigour and her integrative approach to knowledge creation. By documenting her iterative revisions and contextualising her work within the broader intellectual currents of her time, the edition positions her as an innovative and influential figure in the scientific and philosophical debates of the era. Far from merely adapting the ideas of others, Du Châtelet emerges as a pioneering thinker whose contributions rival those of the intellectual giants she engaged with, solidifying her legacy as a central figure in the Enlightenment project.

In light of the above, what I therefore argue is that the *Institutions de Physique* was an *œuvre à faire* [work to be done] in Étienne Souriau's philosophical vocabulary (2009). This concept encapsulates the idea that a creation exists as a potentiality before it is fully realised. It foregrounds the entanglement between the creator and the work itself, suggesting that the work has an inherent 'call' that guides its own unfolding. Such a perspective disrupts the conventional boundaries between creator and creation, reimagining artistic or scientific production, as in Du Châtelet's case, as a dynamic and iterative interaction rather than a straightforward or finite act of completion. This sense of anticipation, of the *œuvre à faire*, is what ignites the flight of imagination. As Isabelle Stengers (2011, 462), drawing on Souriau's notion, has remarked, it is an 'adventure of human experience'. Souriau himself captures this sentiment beautifully:

> If the poet did not already love the poem a bit before writing it, if all those who think of a future world that is to be brought into life did not find, in their dreams on this subject, some amazed premonitions of the presence called for, if, in a word, the waiting for the work was amorphous, there would no doubt be no creation (Souriau 2009, 206).

For Souriau, the *œuvre à faire* is neither about finality nor mere futurity, for these notions exclude the richness of the lived experience in the making. As he writes, 'If you consider the *œuvre à faire* as a project, you miss the delights of discovery, of exploration, in short, the experiential input in the historical route of the advancement of the work' (207). The trajectory of the work encompasses every encounter along the way—the fidelity to an idea, the painful acceptances, the onerous refusals, and the creative tensions that fuel the process. What matters is not the fixed endpoint but the *instauration*, the act of bringing something entirely new into being—a dramatic exploration and a spontaneous adventure (209).

It was precisely this waiting, this process of making, that defined not only Du Châtelet's relationship with the *Institutions* but also her entire journey of becoming—a woman philosopher, mathematician, and scientist at a time when such roles were fraught with societal constraints. The *Institutions* was not merely a static product of intellectual labour; it was a living entity that mirrored her transformation. It was animated not by its final form but by the ideas that shaped it, the forces that propelled it forward, and the ongoing intellectual dialogue it demanded. This unfinished and ever-evolving engagement embodied the essence of the *œuvre à faire*—a work constantly in flux, guided by its own unfolding potential and sustained by the profound imaginative love of its creator. In this light, the *œuvre à faire* becomes a metaphor for becoming—a process of constant negotiation, reinvention, and discovery. For Du Châtelet, as for all women mathematicians that this book revolves around, their works were not only contributions to knowledge but also profound acts of self-realisation, where the boundary between creator and creation dissolved into a shared journey of potentiality and transformation. In this light, Du Châtelet's *Institutions* as an *œuvre à faire* resonates deeply with Agnesi's philosophical adventures, as I will discuss next.

Maria Gaetana Agnesi: becoming a mathematical philosopher

In the previous section, we traced Du Châtelet's intellectual journey from Newton to Leibniz, highlighting the profound influence of these two towering figures on her work. This intellectual trajectory, however, was not unique to Du Châtelet. Maria Gaetana Agnesi's philosophical and mathematical contributions were similarly shaped by the intellectual currents defined by Newtonian and Leibnizian thought. This dual alignment is evident not only in Agnesi's own writings but also in contemporary scholarship that engages with her contributions (see Minonzio 2006; Zen 2017).

As Stefano Zen notes, the influence of the Riccati family was instrumental in shaping Maria Gaetana Agnesi's decision to adopt Leibnizian differential notation over Newtonian fluxions in her *Instituzioni Analitiche* (2017, 16). This choice was not merely technical but also deeply philosophical, reflecting a conscious alignment with the Leibnizian tradition. It situates Agnesi's work within a broader European discourse on mathematical methodologies and their underlying philosophical implications.

As discussed in Chapters 2 and 3, Agnesi's engagement with these competing traditions demonstrates her capacity to navigate, synthesise, and contribute to complex intellectual currents, underscoring her pivotal role in the history of ideas. This section delves further into that theme by focusing on the creation of her *Philosophical Propositions*, a treatise in which Agnesi not only articulated her philosophical ideas, but also passionately defended the intellectual potential of women in philosophy and the sciences, making a strong case against their exclusion from scholarly pursuits:

> No one will deny that even the weaker sex has achieved great merits in all branches of philosophy. Indeed, we know that, besides the approximately seventy highly learned women mentioned by Ménage, many others have distinguished themselves throughout history, receiving the highest praise for their intellect in philosophical disciplines. Therefore, Nature has endowed female minds as well with the capacity for every kind of science and knowledge. Consequently, it is rather unjust for some to entirely forbid them from the study of the liberal arts, especially since such studies are not only not harmful to private and public life but are, in fact, highly beneficial (Agnesi 1738, III, 2–3).

This passage from the *Prolegomena*—the Preface—of the *Philosophical Propositions* (1738) presents a forward-thinking perspective on women's role in philosophy, particularly striking given the constraints on women's education during her time. Agnesi advocates for women's intellectual capacities, critiquing societal norms that excluded them from engaging with liberal arts and philosophy. She argues that such pursuits benefit both individuals and society, challenging prejudices that relegated women to inferior roles.

Her reference to Gilles Ménage's enumeration of learned women underscores the historical precedent for female intellectual achievement.[7] However, as Emanuele Costa notes, while Agnesi praises Ménage, she also critiques his list for being incomplete (2024, 8). She contends that it fails to fully acknowledge women's intellectual contributions, calling instead for recognition of their equality in education and scholarship. By juxtaposing the relative equality of women in ancient times with the barriers they faced in her own era, Agnesi highlights the regressive attitudes of her time, making a compelling case for the advancement of women in intellectual life.

The assertion that 'Nature endows female minds' encapsulates key Enlightenment ideals of natural equality, placing Agnesi's argument firmly within broader debates on gender, education, and rationality, themes we have explored in earlier chapters. As Costa (2024, 8) notes, Agnesi's advocacy for women's equal participation in intellectual pursuits aligns with a broader Enlightenment discourse, exemplified by figures like Marie le Jars de Gournay. In *Equality of Men and Women* (2002), de Gournay similarly employed historical arguments to defend women's right to education and intellectual development. Within this framework, Agnesi's stance reflects the Enlightenment's commitment to reason and the universal capacity for intellectual growth, while also foreshadowing later feminist demands for women's education and equality. Agnesi goes further by criticising those who categorically prohibit women from engaging in the liberal arts, denouncing such restrictions as unjust. She asserts that allowing both men and women to contribute to the advancement of philosophy and science benefits society as a whole.

Overall then, the passage above serves as a robust defence of women's intellectual potential, advocating for their rightful inclusion in scholarly and philosophical discourse—an extension of the themes she first articulated in her *Oratio*, as discussed in Chapter 4. To fully grasp the significance of this work, however, it is crucial to consider the historical and cultural context in which it was created.

Agnesi's *Propositiones Philosophicae*, published in 1738, represented the culmination of her philosophical education and demonstrated her mastery of the intellectual traditions of her time. As noted in Chapter 4, Agnesi's education was shaped by some of the most eminent scholars in the Italian states. Her curriculum, while closely aligned with that of contemporary Jesuit schools, included notable modifications, as Massimo Mazzotti has pointed out (2007, 55). The *Propositiones* reflect this pedagogical influence, drawing on the Jesuit practice of having students compile and defend a series of philosophical theses as part of their intellectual training.

In this context, Paula Findlen describes the *Propositiones* as Agnesi's 'virtual degree' (2011, 252), serving as a symbolic substitute for a university qualification. For her family, tutors, and Milanese society, the publication represented a remarkable academic accomplishment in a city without its own university and at a time when women had no pathway to graduate from the satellite university in Pavia (253). This achievement not only marked Agnesi's intellectual maturity but also challenged the educational norms of her era.

Written entirely in Latin, the treatise exemplifies the rigour and depth of Agnesi's scholastic training. Elena Rossi has commented that 'Agnesi's prose is of a purely ciceronian cast, with fairly lengthy sentences and a lexicon typical of classic literature' (2015, 20). Agnesi's stylistic choices in the *Propositiones* are not merely aesthetic but also deeply methodological. Her use of Latin—a language of universal intellectual exchange in early modern Europe—positions her work within a broader scholarly tradition while asserting her place in a male-dominated intellectual landscape. The text's lucidity reflects her ability to distil complex philosophical arguments into accessible formulations, an achievement that underscores her mastery of both content and form. Franco Minonzio has particularly highlighted the clarity and precision of her Latin prose:

> I believe that the Latin of the Propositions, that orderly and clear Latin, whose simplicity presupposes mastery of the conceptual matter, and whose concise precision bears the mark of a balanced selection, seems to prefigure the core of formal values that will constitute the most remarkable merit of the *Analytical Institutions* (2006, 62–3).

The publication represented the peak of years devoted to refining Agnesi's education, but it was also a strategic move by her family to bolster their social standing. As Mazzotti has noted (2007, 12), the *Philosophical Propositions* was released the same year her father, Pietro, acquired an imperial fief, granting him the title 'Don' and enabling him to construct a noble lineage for the family. The publication, therefore, was not only an intellectual milestone but also a calculated step in the family's pursuit of upwards mobility.

According to Michele Maylender, whom Findlen describes as 'the most comprehensive historian of the Italian academies' (2011, 251), the publication of Agnesi's *Propositiones* was associated with a prestigious event at the Palazzo

Agnesi. This gathering, attended by Milan's leading scholars and nobility, reportedly featured Agnesi presenting an oral synthesis of her philosophical views, which were recorded in her *Propositiones* (Maylender 1926–1930, vol. I, 109). While this narrative has been influential, it has also been the subject of scholarly debate. Minonzio, for example, has questioned whether such a definitive summation and formal oral defence ever occurred or whether the *Propositiones* should instead be understood as a more nuanced and layered record of Agnesi's intellectual trajectory over time (2006, 40). Mazzotti however, has definitively shown that as part of the Jesuit tradition of compiling a list of philosophical theses at the end of their course, 'these theses would be published and partially defended in a public ceremony' (2007, 64). This contextual framing supports the idea that the *Propositiones* were both a scholarly exercise and a public demonstration of Agnesi's erudition, aligning with contemporary academic customs while showcasing her intellectual accomplishments.

The diverse and often conflicting interpretations surrounding Agnesi's *Propositiones* not only affirm its historical significance but also reveal a persistent tendency to mythologise women's intellectual achievements within constrained and oversimplified narratives. By critically revisiting and interrogating these accounts, scholars illuminate the broader intellectual currents and cultural dynamics that influenced Agnesi's contributions to philosophy and mathematics, offering a more nuanced understanding of her legacy. But let us return to the text itself.

The *Propositiones* occasionally reveals Agnesi's knowledge of classical Greek, underscoring her deep engagement with classical languages and humanistic traditions and showcasing her wide-ranging mastery of both humanistic and scientific disciplines, despite her young age. Rossi has further noted that the structure of *Propositiones* appears inspired by Edmond Pourchot's 1695 university textbook, the *Institutio philosophica*, although Agnesi diverges from him on several key topics) (2015, 251). The text comprises 191 theses organised into 29 chapters, each adhering to a systematic format: a proposition is formulated, followed by corollaries and a discussion of potential objections. The treatise is notable for its interdisciplinary breadth, engaging with a variety of topics, including logic, ontology, pneumatology—akin to what we now refer to as theories of mind—as well as general and particular physics.

As Findlen observes, 'there was virtually no domain of philosophical knowledge excluded from this fantastic exercise in erudition' (2011, 252). Agnesi actively engaged with key intellectual debates of her time, such as Vallisneri's theories on freshwater sources, Leibniz's concepts of force, Newton's principles of gravitation and optics, and the innovative geodetic expeditions by members of the Paris Academy of Sciences aimed at determining the Earth's shape (252). Overall Agnesi's approach integrates classical influences with contemporary scientific advancements, demonstrating her ability to navigate and synthesise diverse intellectual currents. This precision and interdisciplinary scope not only solidify her contributions to the history of ideas but also provide a window into the intellectual vibrancy of early modern Europe.

Building on these broad inquiries, Agnesi's prologue to her *Propositiones* accentuates her conviction that mathematics is the supreme philosophical discipline. As discussed in Chapter 5, Agnesi argued for reserving the term *scientia* exclusively for mathematics, particularly algebra and geometry. Her perspective on logic, meanwhile, emphasises its profound integration with mathematical thought, illustrating her alignment with both classical and emerging mathematical traditions:

> It is evident both why it is the duty of Logic to enlighten the mind and direct its reflections towards the search for truth, for which reason it is also called the art of thinking, or natural doctrine, and the fact that it is especially necessary to successfully and completely acquire other faculties, provided you embrace those of Mathematics; for indeed, Mathematics does not depend on Logic, but rather Logic originates from Mathematics, as is well known to those who, through excellent work, have derived the laws of Logic from the very sources of Geometry (Agnesi 1738, VII, 5).

Logic is thus framed as a guiding discipline that illuminates the mind, enabling it to pursue truth effectively. This idea aligns with its designation as the 'art of thinking', underlining its centrality in philosophical and intellectual pursuits. According to Mazzotti, Agnesi 'labelled as "logic" what was really a series of basic epistemological considerations' (2007, 65). In this context, the claim that logic originates from mathematics rather than the other way around is significant. It suggests that mathematical reasoning provides the axiomatic and structural foundation upon which logical systems are built. This perspective echoes the influence of figures like Euclid and later formalisations of logical systems in mathematical contexts. In formulating her argument, Agnesi incorporates epistemological considerations, adhering to the Cartesian principles of clarity and distinctness as criteria for truth.

> Whoever seeks truth must use an analytical method, which is very rarely achieved by those who, intending to investigate the principles and causes of things, accept certain general arguments as valid without combining them with the effects and phenomena of Nature. However, in order to present the discovered truth to others, one must use a synthetic method, which naturally proceeds step by step from simpler to more complex things. Additionally, care must be taken not to propose doubtful matters in place of those known with certainty and to ensure that later propositions are correctly derived from preceding ones—a task that can be accomplished accurately only in mathematical disciplines (Agnesi 1738, XVI, 10).

In Section XVI of the Logic, Agnesi outlines a dual methodology for discovering and communicating truth, presenting a sophisticated approach to epistemology. She advocates an analytical method for the initial pursuit of truth, emphasising the necessity of precision and caution against generalisations that lack empirical evidence. Her critique of relying solely on 'general arguments' divorced from observable phenomena highlights her insistence on integrating empirical data into philosophical and scientific investigations. This method reflects a commitment to rigour, ensuring that inquiry remains grounded in the natural world.

Once truth is discovered, Agnesi advocates for the use of the synthetic method to communicate it effectively. This method involves presenting knowledge in a systematic and logical order, progressing from fundamental principles to increasingly complex concepts. Such an approach mirrors pedagogical practices rooted in deductive reasoning, prominently seen in mathematical proofs and geometric constructions, where the logical flow moves from established axioms to derived theorems.

Agnesi's analytical and synthetic methods, as described in Section XVI of her *Logic*, align closely with the epistemological concepts of deduction and induction. The analytical method, akin to deduction, begins with general principles or hypotheses and logically derives conclusions from them. In contrast, the synthetic method, akin to induction, builds towards more complex truths by starting with specific observations or facts, progressing step by step. This dual methodology bridges the empirical and deductive dimensions of knowledge, reflecting Agnesi's intellectual rigour and her innovative contributions to Enlightenment thought. By integrating abstract reasoning with tangible observation, she offers a holistic framework for understanding and sharing truths.

Her method exemplifies a balanced approach to truth-seeking and its dissemination, rooted in both rationalist philosophy—particularly Descartes' emphasis on clarity and distinction—and the empirical spirit of early modern science. Agnesi critiques unfounded generalisations and advocates for evidence-based reasoning, positioning her thought within scientific realism and countering speculative philosophies disconnected from observable phenomena. In this way, her work integrates analytical rigour with synthetic clarity, highlighting the foundational role of mathematics in achieving and conveying reliable knowledge.

As Minonzio observes (2006, 47), Agnesi's formative education with renowned scholars, as discussed in Chapter 4, cultivated her deep engagement with physical-mathematical disciplines. This education instilled a steadfast commitment to clarity and intellectual rigour, which made the publication of the *Propositiones* a natural culmination of her scholarly journey. It further solidified her dedication to mathematical pursuits, as explored in Chapter 2.

Indeed, Agnesi's assertion that mathematical disciplines uniquely enable rigorous derivation highlights her belief in the unmatched clarity and reliability of mathematics. Mazzotti emphasises that Agnesi distinguished between empirical knowledge, which she viewed as inherently fallible and subject to ongoing debate, and mathematics, where she believed truths could be discovered and contemplated with absolute certainty (2007, 66). This contrast underscores her view of mathematics as a domain where abstract reasoning leads to conclusive, universally valid results, in stark contrast to the natural sciences, grounded in observation and subject to revision. Her stance also reflects a profound confidence in the objectivity and precision of mathematics, setting it apart as a discipline where irrefutable truths are accessible, a notion not applicable to the more fluid nature of empirical science.

Moving from Logic to the section on Ontology, Agnesi delves into foundational questions, grappling with the nature of being, universality, and the metaphysical status of essences. Her argument reflects a nuanced understanding of how entities, despite being singular, can be represented in multiple ways depending on their relationships and connections: Every being is unique and particular; however, since the same thing can have some relationship or connection with entirely different things, it can also be presented through different concepts. Hence arises the distinction of reason, which should not at all be separated from the virtual distinction (Agnesi 1738, XVII, 11).

Agnesi's recognition of the 'distinction of reason' reflects her engagement with Aristotelian and scholastic traditions, while her incorporation of 'virtual distinction' demonstrates a nuanced metaphysical inquiry, notably influenced by Cartesian thought. She critiques the concept of a singular, universal nature equally applicable to multiple entities, asserting that what we perceive as 'universal' originates more from human abstraction and perceived similarity than from an authentic metaphysical unity. 'There is no universal nature presented, not even by the mind' (XVIII, 11), she asserts. By describing universality as 'improper and metaphorical' (XVIII, 11), Agnesi challenges Platonic ideals, aligning instead with an empirical and logical perspective.

Agnesi's exploration of the metaphysical and physical interpretations of essences bridges classical metaphysics and the emerging scientific paradigms of her era. She contends, 'there is no reason to separate nature from subsistence, as though this entity were added to nature' (XIX, 12), challenging a rigid dualism and advocating for a view of substance as inherently self-sufficient. This perspective resonates with Nicolas Malebranche's axiom of *occasionalism*, encapsulated in the assertion that 'God alone has any efficacy', as Alain Badiou has aptly put it in his seminar on Malebranche, which situates occasionalism within a broader historical and philosophical context (2019, 21). In this framework, the existence and interactions of entities are understood as entirely dependent on divine causation, emphasising God as the sole true agent in all occurrences.

The influence of Malebranche's *occasionalism*, which denies any direct causal interaction between mind and body, is particularly evident in Agnesi's section on pneumatology. Here, she delves into key philosophical issues concerning the nature and immortality of the human soul, engaging deeply with both metaphysical and epistemological debates. The theses in this section reflect Agnesi's nuanced engagement with both Cartesian and Augustinian traditions, offering a synthesis that positions her thought within the broader Enlightenment discourse on metaphysics, theology, and human nature. While Agnesi adopts Cartesian principles of clarity and distinction, as discussed earlier, she does not shy away from critiquing the *cogito*: The Cartesians achieve nothing when they strive to prove that the human mind thinks perpetually and that its essence is situated in actual thought. While the mind indeed becomes known to itself only through its thoughts, actual thinking does not appear to pertain to its essence any more than motion pertains to the essence of the body (Agnesi 1738, XXII, 13).

Agnesi's critique here challenges Descartes' identification of the mind's essence with actual thought (*cogito*). By distinguishing the essence of the mind from continuous actual thinking, Agnesi maintains a more flexible understanding of mental faculties. This view avoids reducing the mind's nature to perpetual activity, allowing for potentiality rather than constant actuality. In thus searching for the nature of the mind, Agnesi turns to Augustine, defining the soul's essence in terms of its rational capacity and its role as the governing principle of the body: 'Thus, to one who asks what the mind is, it can be answered with St. Augustine's words: *It seems to me to be a certain substance, participating in reason, and adapted to governing the body*' (XXIII, 14, emphasis in the text).

Malebranche's assertion that there is no direct relationship between body and soul, coupled with his proposition that God acts as the intermediary orchestrating all perceived interactions, profoundly influences Agnesi's portrayal of knowledge acquisition as fundamentally passive. As Mazzotti aptly notes (2007, 65), Agnesi emphasises divine facilitation over human agency in her epistemological framework. She conceptualises cognition within a voluntaristic model, asserting: 'Our mind does not act while it perceives something, nor does it produce ideas, but receives them from the outside. Although attention, which often accompanies thought, is indeed an action, it derives from the will, not from the intellect' (1738, XXVII, 15).

This perspective underscores Agnesi's effort to harmonise metaphysical traditions with theological and empirical frameworks, showcasing a distinctive intellectual synthesis characteristic of her philosophical approach. Her *Propositiones* then seamlessly transitions from epistemological discussions to broader explorations of general and particular physics, reflecting her comprehensive engagement with the scientific and philosophical inquiries of her time (XXXVII–CXCI, 20–119).

In these sections—which I do not consider in more detail, as they lie beyond the scope of this chapter focussed on women mathematicians' epistemologies—Agnesi critiques the Cartesian vortex theory and aligns herself with Newton's gravitational theory, which she describes as 'a most beautiful and simple theory' (1738, CXVII, 81). Her discussions encompass topics such as mechanics, gravity, motion, hydrostatics, and the collision of bodies, reflecting her strong adherence to Newtonian principles. She also explores optics, favouring Newton's corpuscular theory of light over Descartes' wave theory, and engages with experimental debates like those between Newton and Giovanni Rizzetti.

Agnesi's intellectual pursuits extend well beyond physics, encompassing fields such as astronomy, meteorology, geology, botany, and anatomy. As Rossi has noted (2015, 24), she challenges the mechanistic view of animals as automatons devoid of consciousness, instead attributing to them an 'animal soul' endowed with will and intellect. Her engagement with the works of prominent scientists like Marcello Malpighi and Antonio Vallisneri further enhances her sophisticated understanding of reproduction and anatomy. This wide-ranging inquiry highlights Agnesi's interdisciplinary approach, positioning her at the confluence of natural philosophy, experimental science, and theological reflection. Despite the impressive breadth and depth of her *Propositiones*, its reception has been mixed.

While some commentators have celebrated its encyclopaedic scope and intellectual ambition, others have critiqued its lack of theoretical originality. It is this nuanced and, at times, ambivalent legacy that I now turn to explore in greater detail.

Maria Luisa Anzoletti has praised the *Propositiones* for their good Latin diction, characterised by conciseness, simplicity, order, and clarity—qualities reflective of Agnesi's intellect (1900, 160). However, she has significantly pointed out that these printed theses do not offer any original contributions to philosophy. Overall, she considers the *Propositiones* as an outline of Agnesi's philosophical studies that do not go into depth, but are merely descriptive 'programmes' of the lively disputes she had held at the family Palazzo over the years, as we have seen in previous chapters:

> As they are, these philosophical theses only show the mechanism of the dispute; they are merely paragraphs of a few lines, in which the topic to be discussed is stated. We do not even know if these paragraphs were composed by Agnesi or merely compiled. In any case, we cannot say that we have before us an original imprint of her thought, nor an evident exercise of her own reflection (159).

Arnaldo Masotti similarly acknowledges the *Propositiones Philosophicae* as an encyclopaedic and comprehensive work, yet he aligns with Anzoletti in asserting that it lacks the theoretical originality necessary for greater scholarly distinction (1940, 96). This evaluation is both fair and balanced. As discussed earlier, however, the text's primary aim was not to present groundbreaking philosophical innovation but to demonstrate Agnesi's extensive erudition and mastery of academic traditions. Within the intellectual context of her time, the *Propositiones* served as a capstone to her rigorous education, reflecting her broad engagement with contemporary philosophical and scientific discourse.

However, as Anzoletti insightfully suggested, glimpses of Agnesi's original philosophical contributions might emerge if we had access to the actual debates underpinning the *Propositiones*. Drawing on Agnesi's own words in the Preface, Anzoletti (1900, 160) emphasises the author's candid reflection on her performance during these disputes. In her dedication to Count Belloni, Agnesi openly expresses concern about whether she had fully met the expectations of her audience and her mentor. She questions whether she had effectively articulated various philosophical perspectives, compared them with sufficient rigour, convincingly supported her own views, or adequately refuted opposing arguments:

> Whether I, in fulfilling my role, have rightly explained the various opinions of the philosophers on the proposed questions, compared them among themselves, sufficiently established my own views with valid arguments, or fully refuted objections, I do not entirely know, and I fear that I may have disappointed the expectations of the audience and failed to fulfil your trust, since you were accustomed to make great promises about me to your friends (Agnesi 1738, iv).

This introspective passage reveals Agnesi's methodological precision and philosophical engagement, offering a rare glimpse into her self-reflective awareness as a scholar and public intellectual. While primarily demonstrating her intellectual breadth, it also underscores her humility and deep sense of responsibility to her audience. Agnesi acknowledges the high expectations placed upon her, likely intensified by her groundbreaking role as a woman in the intellectual sphere of eighteenth-century Italy. Her concern about living up to the trust of her mentor highlights the social and cultural pressures she faced, balancing her intellectual obligations with the societal expectations of a woman excelling in philosophy and mathematics.

Moreover, her awareness of the dynamics between performance, scholarship, and expectation emerges clearly in her words. The act of disputation was not just an academic exercise but a scrutinised and performative endeavour, demanding not only intellectual rigour but also rhetorical clarity and persuasive engagement with an audience. Her reference to her mentor's promises further emphasises how her intellectual reputation was intricately tied to her social network, adding additional layers of pressure to her scholarly pursuits. Ultimately, this self-reflection captures Agnesi's dual role as both a pioneering woman navigating uncharted academic territory and a diligent scholar devoted to the ideals of philosophical inquiry. It invites us to consider her legacy not only in terms of the history of women in science and philosophy but also in relation to the evolving expectations of intellectual discourse in her era.

Beyond the window that Agnesi's dedication to Belloni opens into her psychosocial constitution as a woman mathematician and scientist, the philosophical framework underpinning the *Propositiones* operates within a distinctly encyclopaedic paradigm. It embodies a comprehensive conception of philosophy that was widely embraced across diverse intellectual traditions of the sixteenth and seventeenth centuries. Unlike certain strands of the Newtonian tradition, where philosophy often converges with and is subsumed by physics, the *Propositiones* maintains a broader philosophical scope. As Minonzio (2006, 47) aptly observes, the text emphasises the pivotal influence of physical-mathematical knowledge, which occupies a central and commanding role in its intellectual structure.

In this light, the *Propositiones* reveal several facets of Agnesi's intellectual importance. Her emphasis on mathematical precision, her early advocacy of Newtonian theories—at a time when they were still not widely accepted in the Italian states—and her integration of Newtonian and Leibnizian principles demonstrate her progressive and forward-thinking approach. Minonzio underscores this intellectual transformation, remarking: 'The years when discussing universal gravitation, the fall of heavy objects, or tides in quantitative terms was groundbreaking science were as astonishing in the Newtonian era as it had been for a man in the sixteenth century to claim that the Earth revolves around the Sun' (2006, 62). This paradigm shift, Minonzio argues, was one that Agnesi actively embraced, in contrast to the dominant scholarly preoccupations of her time, which largely revolved around debates over the validity of Aristotelian syllogisms or, at best, supported the Cartesian conception of matter as composed of particles. Within this broader context, the originality and importance of Agnesi's *Propositiones* emerge as a reflection of her engagement with the most advanced scientific and philosophical ideas of her era.

Moreover, her seamless integration of scientific inquiry, mathematical diligence, and theological reflection underscores her unique contributions to eighteenth-century intellectual history, particularly to the trend that Mazzotti (2007) has configured as the Milanese version of the Catholic Enlightenment. As Mazzotti has commented, Agnesi's effort to reconcile reason with faith and critical analysis with religious dogma was not a mere pragmatic balancing act but 'the very essence' of the enlightened Catholic culture in which she was immersed (66). Her *Propositiones* embody this synthesis, blending rigorous mathematical reasoning and scientific inquiry with a deeply rooted theological perspective. This harmony reflects the intellectual currents of her time, where the pursuit of knowledge was often framed within a context of divine order and universal truths. Agnesi's ability to navigate these realms with precision and thoughtfulness underscores her role as a key figure in advancing both the scientific and philosophical discourse of the eighteenth century, while remaining firmly grounded in her faith. Her writings exemplify how an enlightened Catholic worldview could integrate critical inquiry with spiritual devotion, presenting a model of intellectual engagement that sought coherence rather than conflict between these domains.

Drawing on Minonzio's and Mazzotti's detailed analyses of the *Propositiones*, Findlen has compellingly argued that these early works reveal Agnesi's intellectual formation as 'a *mathematical philosopher*' (2011, 253, my emphasis). Indeed, Agnesi emerges as a staunch defender of the mathematical sciences, presenting them as the most certain and reliable form of knowledge. In her view, mathematics possesses a unique capacity to guide the mind in its pursuit of truth, offering a rigorous foundation for philosophical inquiry. Findlen thus underscores that Agnesi's advocacy for mathematics was not merely technical but deeply philosophical, reflecting her belief in its universal applicability and transformative potential. This early commitment to uniting mathematical precision with philosophical depth laid the groundwork for her later contributions, positioning Agnesi as a pivotal figure in the intersection of Enlightenment science, mathematics, and philosophy.

Findlen further emphasises the process of Agnesi's constitution as a mathematical philosopher after the publication of the *Propositiones*, which were met with widespread acclaim in Milanese intellectual circles and across Italy and Europe. This recognition led to valuable relationships with prominent Italian male natural philosophers. According to her biographer Francesco Frisi (1799, 51) Agnesi achieved a kind of celebrity status through the publication of the *Propositiones*, earning the title of *la filosofessa milanese*, a counterpart to Algarotti's (1737, 6) designation of Laura Bassi as *la bolognese filosofessa*, as Zen has noted (2017, 55).

Zen has further highlighted the intellectual exchanges following the publication of the *Propositiones*, which cemented Agnesi's reputation as a mathematical philosopher. The mathematician and physicist Giovanni Battista Bertucci, for example, sent her his manuscript *De Telluris ac Syderum Vita* [On the Life of the Earth and the Stars], praising the intellectual achievements of women in modern philosophy and mathematics. He incorporated Agnesi's suggested revisions and commended her expertise in geometry and mechanics (56).

Similarly, Agnesi's correspondence with Giovanni Bianchi, a physician-naturalist from Rimini, showcases her elevated position in the Italian scientific community, but also highlights her commitment to revising her work and refining her knowledge. In 1741, Bianchi sent her his research on tides and shell formations, seeking feedback and wider recognition for his work. When Agnesi replied with copies of her *Propositiones*, she admitted to having revised her views on key topics since its publication. Specifically, she mentioned altering her opinion on the Earth's shape after reviewing new observations from the Paris Academy, informed by Maupertuis' expedition to Lapland, which contradicted earlier conclusions by Cassini and others (see Findlen 2011, 254).

These exchanges exemplify Agnesi's active and evolving participation in contemporary scientific discussions, demonstrating her openness to adapting her views in response to new evidence. This processual approach to knowledge highlights her commitment to ongoing intellectual development, characteristic of a mindset that sees knowledge as a continuous becoming rather than a fixed set of truths. Her ability to revise and refine her ideas in dialogue with others reflects an innovative epistemology, one that emphasises fluidity and responsiveness—an approach that was ahead of its time and not yet integral to the intellectual climate of her era.

Minonzio has astutely observed that Agnesi's *Propositiones* have seldom been studied as a standalone work and are often superficially linked to her more renowned *Institutiones Analitiche* (2006, 52). Addressing this significant gap in the literature, Iulia Mihai and colleagues have undertaken a new project, 'Agnesi on Natural Philosophy and Mathematics', which aims to reposition Agnesi's *Propositiones* within the history of philosophy. According to its initial statement, this project contends that rather than aligning herself strictly with Newtonianism or Cartesianism, Agnesi turns to Stoicism to navigate and resolve key philosophical questions.[8]

Thinking with this project's initial hypothesis in mind, one could argue that several aspects of Agnesi's *Propositiones* reflect a Stoic legacy in philosophy. When Agnesi underscores the capacity of the human mind—women's minds included—for intellectual pursuits and philosophical inquiry, she aligns with the Stoic emphasis on rationality as the defining trait of humanity and a crucial pathway to virtue and fulfilment. Her argument that philosophical and liberal studies enhance both private and public life resonates with the Stoic ideal of living a life of active engagement, where individuals contribute to the community through reason and virtue. Moreover, Agnesi's assertion that 'Nature has endowed female minds as well with the capacity for every kind of science and knowledge' (Agnesi 1738, III, 2–3), as previously discussed, echoes the Stoic principle of equality. For the Stoics, all human beings, regardless of gender or social standing, possess reason and are therefore equally capable of attaining virtue. Finally, Agnesi's propositions suggest that intellectual clarity and the pursuit of knowledge are essential to achieving happiness—a view that mirrors the Stoic belief that true happiness arises from living in accordance with reason and aligning one's will with the rational order of nature. As a preliminary reflection, these elements highlight Agnesi's engagement with Stoic ideas in the development of her philosophical and educational vision.[9]

While anticipating more concrete findings from the development of this emerging project mentioned above, it is significant to note that Agnesi is gradually being recognised as an essential figure in the canon of Western philosophy. In this context, more recent research on Agnesi's *Propositiones* has shifted attention to the philosophical foundations of her views on women, gender, and science. As already noted at the beginning of the section, Costa (2024) has argued that themes of proto-feminism, gender equality, and women's right to education run consistently throughout Agnesi's philosophical and scientific writings. Costa suggests that the Preface to the *Propositiones* 'is perhaps its most interesting and original section' (8). By reading the *Propositiones* through a gendered lens (ibid., 9), he highlights how Agnesi engages with Cartesian dualism, affirming the distinctness of mind and body while also underscoring their unity as part of God's perfect design, when she writes:

> We most certainly know that it was established by God the Creator as a law that the thoughts of the mind and the movements of the body excite each other and succeed one another. Beyond this law, there is nothing that can account for the union of mind and body, as the most acute men, Descartes, Malebranche, and Leibniz, have believed (Agnesi 1738, XXIV, 14).

By asserting that the mind is capable of abstract reasoning without necessarily interacting with the body, Agnesi reinforces the idea that intellectual capacity is independent of gender. Costa (2024, 9) further suggests that Agnesi's philosophical stance resonates with the arguments of other proto-feminists, such as François Poullain de la Barre (2005), who asserted that the mind, being non-gendered, enables women to excel in any intellectual field, and Anna Maria van Schurman (2013), who famously argued that reason is a human capacity shared equally by both men and women.

I agree with Costa that Agnesi's rigorous philosophical defence of women's intellectual abilities and their right to equal opportunities in learning, teaching, and knowledge creation across scientific and humanistic fields is one of the most compelling aspects of the *Propositiones*. However, I do not believe this section alone defines the treatise's originality. As Costa has shown, Agnesi was engaging with a proto-feminist tradition that had been developing for over a century. What I find truly innovative about the *Philosophical Propositions* is the way it envisions and facilitates the process of women's *becoming* in science and mathematics—opening new possibilities for their transformation, participation, and recognition in fields traditionally closed to them.

Here, it is essential to remember that the *Propositiones* were published in 1738, just a year after Algarotti's popular book, *Il Newtonismo per le dame* (1737). Algarotti's book stemmed from his replication of Newton's optical experiments at the Institute for Sciences in Bologna in 1728 and, although officially published in Naples, was actually printed in Milan. Despite his ties to the Milanese publishing world, Algarotti neither met nor corresponded with Agnesi. While he maintained a friendly relationship with her tutor, Count Belloni, to whom the *Propositiones* were dedicated, he was never invited to the Palazzo Agnesi in Milan.

Mazzotti observes that 'there were important differences between the uses of science, and Newtonianism in particular, in Algarotti and in the Agnesi *conversazione*' (2007, 57). He further highlights the varied reception of Newton's ideas within the Italian context. While Newton's theories had many supporters in Rome, they faced greater resistance or selective adoption in institutions where the Galilean experimental tradition remained influential, such as the University of Pisa. Meanwhile, other institutions, like the University of Padua, were predominantly shaped by Leibnizian thought (60).

Agnesi was a Newtonian, though her scientific thought navigated between Newton and Leibniz, as Zen has consistently demonstrated (2017). However, her interpretation of Newton's theories differed markedly from Algarotti's. His book belonged to a genre designed to encourage women's engagement with science, but always under the male gaze and guidance. This literary tradition originated in France with the publication of Bernard le Bovier de Fontenelle's *Conversations on the Plurality of Worlds* in 1686. Fontenelle's work, a successful and influential effort to popularise scientific knowledge, introduced a new scientific persona. As Findlen has observed, Fontenelle's philosopher was no longer an academic figure rooted in the university, a scholastic teacher addressing an audience of male disciples; instead, he became a charismatic figure, charming women with his wit and making science accessible through relatable cultural analogies (2003, 59–60). In this context, the image of the woman natural philosopher was shaped within a literary discourse that oscillated 'between fiction and reality' (60). In Algarotti's book, this figure took the form of a Marchesa—a noblewoman eager to learn and understand Newton's theories by asking questions:

> And you ask me, who was truly the Marchesa of my dialogues? I respond: A grandniece of Fontenelle, or perhaps one of my own daughters, made by me just as I would have liked her, with a lively and sophisticated mind, very curious, and able to listen carefully. But if the Marchesa of my dialogues is imaginary, the one to whose judgment this work is now submitted, is not (59).

As we have seen throughout this section, Agnesi's approach to Newton's theories diverged fundamentally from the paternalistic framing often associated with works like Algarotti's. Rather than positioning Newton's ideas as something to be passively received by women, Agnesi framed his 'most beautiful and simple theory' (1738, CXVII, 81) as a model for active engagement in scientific experimentation and laboratory practice. Her work embodies what Zen aptly terms a *Newtonianism of women* rather than a *Newtonianism for women* (2017, 55), foregrounding women as active participants in the scientific process rather than mere spectators.

Through the *Propositiones*, Agnesi enacted what, following Karen Barad (2007), we might call an *agential cut*—a reconfiguration of the boundaries between women, mathematics, science, and knowledge creation. By rejecting the literary figure of the woman natural philosopher as a passive questioner, Agnesi redefined the potentialities of women's actionability, emphasising their capacity to shape the actualities of mathematical and scientific discoveries. Her intervention as a 'mathematical philosopher' (Findlen 2011, 253) was not merely about granting women access to knowledge but about redefining their role within it, transforming them into co-creators of epistemic frameworks.

In Deleuzian terms, Agnesi's work can be seen as initiating new *becomings*—dynamic processes through which women could actualise their intellectual potential. Far from being confined to predetermined roles or societal expectations, her approach created a space for women to navigate the transition from potentiality to actuality, asserting their right to pursue knowledge in all its forms. This is evident in her radical advocacy for women's freedom to follow *lines of flight*: 'a right to regain possession of their time and to pursue any kind of knowledge in opposition to the idea of studying as a privilege', as Zen aptly notes (2017, 15).

Through the *Propositiones*, Agnesi articulated a vision that transcends mere inclusion, offering a dynamic reimagining of women's participation in science and mathematics as fluid, evolving, and inherently transformative. This perspective aligns with a process-oriented epistemology, foregrounding knowledge as a continual unfolding rather than a fixed attainment. It is precisely this conception of the epistemological process that I will further explore in Germain's work in the next section.

Sophie Germain: philosophy and mathematical reasoning

> We would imperfectly appreciate the high range of Mademoiselle Sophie Germain, if we limited ourselves to consider her as a mathematician [géomètre], whatever the eminent merit she demonstrated in mathematics. Her excellent posthumous discourse, published in 1833, on the state of science and the letters in the different periods of their culture, indicates in her a very lofty philosophy, both wise and energetic, of which very few superior minds have such a clear and profound feeling today. I will always attach the highest value to the general conformity that I saw in this writing with my own way of conceiving the whole intellectual development of humanity (Comte 1835, 604n1).

Auguste Comte included this lengthy reference to Germain's philosophical work in the second volume of his major corpus *Cours de Philosophy Positive,* first published in 1835, only four years after Germain's untimely death in 1831. What made the founder of positive philosophy praise so highly a woman who was mostly known as a mathematician in the Parisian academic circles in the first half of the nineteenth century? This is what I want to discuss in this section by mapping Germain's philosophical writings in the field of process epistemologies.

'These pages, found in Mademoiselle Germain's papers, were not intended for printing. She wrote them in the moments when the severe pains to which she succumbed did not allow her to devote herself in the study of mathematical sciences that had made her famous' (Germain 1833, 5), Jacques-Amant Lherbette, the editor of the first publication of her philosophical work, wrote in introducing his aunt's

Considérations Générales, in 1833. For Lherbette then, Germain's essay was an unfinished substitute for her 'real' scientific work, which she was unable to pursue in the last months of her life, when suffering from breast cancer, and it was because of her 'lack of time' (5) that the work remained unfinished.

Guglielmo Libri, a fellow mathematician and friend, as we have seen in previous chapters, seems to follow Lherbette's evaluation of this work, and he only devotes one paragraph to it in his obituary: 'We have also found in her papers, immense works on history, on geography, in particular that of the ancients, and on the natural sciences, and also very fine philosophical reflections because she had been much occupied with metaphysics', he wrote (1832, 2). What we understand from this short reference is that there was a selection of 'the pages' that were initially published in 1833, and perhaps the *Pensées Diverses*, which were added in the first full edition of her *Œuvres philosophiques* in 1879, were amongst these 'immense works' that Libri's obituary refers to. Seen as a mathematical set of thought fragments, Germain's *Pensées* can be configured as a philosophical annex to her *Considérations*, a poetic rendition of the ideas that her philosophical work either develops or leaves out at least in its extant unfinished form, but it can also serve as an exemplar of the interwoven intellectual processes between science, literature, and the arts.

Unlike Libri, who only makes a passing reference to Germain's philosophical work, Hippolyte Stupuy, the editor of the second publication of her philosophical work, discusses Germain's posthumous discourse at length, criticising Libri for not mentioning her *Considérations* 'in his obituary (1896, 45). In disagreement with Lherbette, Stupuy argues that this work must have started much earlier: 'it is without temerity to suppose that, imperfect as it still was, as to the execution, when death tore the quill from the hands of the writer, a work of such great significance had been conceived long before, at length thought through, often revised and retouched' (44). There is no evidence for this assertion, only clues that Stupuy takes by studying in depth not only the text of the posthumous essay but also Germain's *Pensées*, particularly drawing on the following fragment:

> If the men who have advanced the sciences through their work, those to whom it has been given to enlighten the world want to retrace the path they have taken, they will see that the most beautiful, the greatest ideas are the ideas of their youth, matured by time and experience. They are enclosed in their first attempts like the fruits in the buds of spring (Germain 1896, 208–9).

We cannot be sure that a fragment of Germain's *Pensées* can sustain Stupuy's argument about the period when she wrote her philosophical work. Perhaps this essay was 'the outgrowth of her early habit of writing down pensées', as Louis Bucciarelli and Nancy Dworsky have suggested (1980, 112). Dora Musielak has considered the idea that Germain perhaps followed the lead of the seventeenth-century mathematician and philosopher Blaise Pascal's *Pensées* in writing down her own (2020, 144). Indeed, both her philosophical treatise and her *Pensées* include her reflections on the complexities of the human condition; the possibilities, but also the limitations, of human reason; as well as the existential dimensions of human experience, as I will further discuss. Jil Muller (2023) has further explored similarities between Germain's and Pascal's take on space and time, notions that are treated both mathematically and philosophically in their work, as she has noted (85).

Since Germain was first and foremost a mathematician, she drew, of course, on the field of mathematical sciences, and thus the concept of 'symmetry' is at the heart of her philosophical thesis and thought fragments. As mathematician Edward Frenkel has argued, symmetry is a key concept in mathematics, a property that allows any object 'to keep its shape and position unchanged, even when we apply changes to it' (2013, 15). In this context, symmetry is a unifying and organising principle that allows mathematicians, scientists, and, in our case, philosophers to discern patterns, simplify problems, and reveal the underlying structure of reality, including mathematical objects. Moreover, symmetry contributes to the elegance and beauty of mathematical reasoning, which Germain extends to philosophy and the reality of the cosmos.

Musielak has added analogy as a mathematical concept in Germain's philosophical toolbox: 'the key concept that unifies her text is the "analogy" that she believed allows one to sort and discover the laws of the universe', she has noted (2020, 141). Analogy involves drawing parallels or making comparisons between different mathematical structures and concepts when solving problems or proving theorems. While symmetry and analogy are distinct mathematical concepts, they are often interconnected in mathematical thinking in general and Germain's mathematical logic in particular, as I will further discuss in this section.

In the light of Germain's distinct mathematical logic then, let us consider the content, context, and form of her philosophical writings in their interrelation. Her unfinished essay comprises two chapters: in the first, she formulates the thesis that the human mind is subject to laws and the character of truth is a spontaneous feeling of order and proportion: 'a profound feeling of order and proportions becomes for us the trait of truth in all things' (1896, 78). In tackling the question of truth, Germain highlights here two important mathematical concepts: order and proportion. We all know, even from our basic mathematics education, how crucial the order of operations in arithmetic is, while in dealing with the equality of ratios, proportion provides a framework for comparing quantities in a meaningful way. What is important in Germain's discourse, however, is that these mathematical concepts are transposed to the epistemological realm in her engagement with the philosophical problem of truth. It is in this light that she argues that both the sciences and the letters are dominated by this feeling which is common to them. What is more, Germain adds simplicity in configuring the universal triangle of 'order, proportion and simplicity', which frames her own take on aesthetics and epistemology: 'The oracles of taste and the judgments of reason resemble each other; order, proportion and simplicity do not cease to be intellectual necessities. The subjects are different, but judgment is constantly based on this universal type which belongs equally to the beautiful and the true' (79).

Mathematicians and scientists often strive for simplicity as a guiding principle that can lead to greater clarity, elegance, and deeper understanding. In Germain's words, 'the mind demands clarity, [...] it requires an order easy to understand; it delights in simplicity, the source of elegance', as she noted in her *Considérations* (196), while her *Pensées* include the fragment that ideas are sublime when they are simple (196). Perhaps the DNA scientist James Watson had this principle in mind when he famously stated that 'the truth, once found, would be simple as well as pretty' (2001, xi).

Following the configuration of the tripartite schema of order, proportion, and simplicity, Germain compares the impressions we get from fictional and scientific works and concludes that there are no important differences between them as 'the human mind is guided in all its conceptions by the foresight of certain results, towards which all its efforts are directed' (1896, 81), and therefore obeys 'the laws of its own existence' (97). In this light, in all the strokes of genius, 'in eloquence, in the sciences, the fine arts, or literature, what pleases us is the discovery of a multitude of relations which we had not yet perceived' (82).

In making these comparisons, the author draws on the mathematical notions of symmetries and analogies, as already discussed above, and carefully demonstrates the identity of intellectual processes both in poetry and in science by showing that there is a continuous interchange of feelings [*sentiments*], imagination, and rational reasoning in the way they unfold. For the poet, there is 'a tumultuous struggle' of abstract images and opposing projects until a simple idea finally emerges (82). For the mathematician, there is also a simple, 'fruitful idea' that arises through [his] struggle with imagining a new problem in areas already researched and established:

> he [sic] sees results he cannot yet achieve; his imagination leaps to seize them, along the paths it has cleared for itself; he fears he has gone astray, he doubts his initial perceptions, he retraces his steps and tries to rethink the clues that had first guided him; a great number of ideas have joined those that were the first; they complicate the subject, divide attention and suspend judgment. But, through this chaos of thoughts, genius discerns a simple idea; his choice is irrevocably fixed, he knows that this idea will be fruitful (83).

Germain follows these parallel intellectual movements between poetry and science in the realisation of the work: 'while outlining his plan, the poet will never lose sight of the principal idea. It will give his work the unity of interest and action, the source of all true beauty' (83). By way of analogy, the mathematician 'pays close attention to the happy idea that directs his research', by unfolding a chain of truths, already contained in the first truth of his initial idea (84). In tracing the process of creation, Germain also points to the importance of the choice of style and makes reflections of remarkable accuracy on the perfection of language in literature and of 'the language of calculations' in mathematics: 'the man of letters will take care of the choice of words, their arrangement, the harmony of the verse or that of the sentence' (86). But the mathematician also needs to attend to the demands of style, since the language of calculations also has its own aesthetics: the choice of words in literature analogically corresponds to the choice of mathematical expressions, which can be 'more or less elegant' (87) as 'not all authors write with the same degree of perfection' (86).

In concluding the first chapter of the essay, Germain is therefore able to exclaim: 'Ah! Let's no longer doubt it, the sciences, the letters and the fine arts were born out of one and the same feeling' (90). The notion of feeling [*sentiment*] is a central concept in Germain's philosophical essay; it is deployed throughout the text in different modalities and contexts and is crucial in her processual thought, as we will further see.

In the second chapter, Germain follows a historical investigation of her principles through different periods in science and culture. In this context, she recounts how under the initial reign of imagination, poetry first recounted the most remarkable events and painted the great scenes of nature. Imagining an action would come later for the poet, she notes, but the need was soon felt to discover the rules, which would later become the precepts of art: 'unity of action, unity of interest and clarity of exposition' (92). As [he] found himself 'thrown to the earth in the midst of the immensity of things', man marvelled at himself and, seeking his own image everywhere, he personified inanimate and intellectual beings, rendering them 'children of his imagination' (92). This is how the human type became universal: 'faithful to his constant thought, man has never ceased to regard his own existence as the archetype of all other existences' (94).

Already in the first chapter of the *Considérations*, Germain had made several references to the faculty of imagination, stressing how it had served 'as a guide to reason' (78) for the human mind. As we have already seen, imagination is not only important for poetic ideas (82), but also crucial for the mathematician, whose imagination 'leaps' to seize results that 'he cannot yet achieve' (83). Germain had first-hand experience of the importance of mathematical imagination. As Bucciarelli and Dworsky have pointed out about her work, she often had brilliant ideas, but due to her lack of rigorous mathematical education, her proofs were 'awkward and clumsy when viewed against the background of available mathematics at the time' (1980, 7).

It is perhaps through her lived experiences that Germain went in depth to demonstrate the importance of imagination for the mathematical mind in her *Pensées*: 'Depth of vision, accuracy of judgment, vivid imagination, these are the qualities of the mathematician' (1896, 225) she wrote, particularly highlighting the connection between them: 'what gives this depth, what exercises this judgment, is the imagination, not the one that merely plays on the surface of things, animating them with its colours, diffusing brightness, life, and movement, but an imagination that operates just as well inside bodies as it does outside them' (226). It is only when imagination has done the difficult work of 'penetrating nature' and forming 'the anatomy of things' that the mathematician can proceed. Glesser and colleagues have particularly highlighted the importance of imagination 'as the fundamental duty of being a mathematician' (2020, 219) in Germain's thought, making connections between her *Pensées* and Hardy's *Apology* (2004 [1940]).

In further developing the second chapter of her thesis, Germain engages with Kant's 'transcendental idealism', positing that our knowledge is shaped by the structures of our own minds.[10] She thus raises the question of whether the conditions of possibility for knowledge and understanding are 'the immediate result of the laws of being, or they derive only from a relation between any other reality and that of our existence' (1896, 105). In pointing out that this is an area of debate between philosophers, she refutes Kant's thesis that 'the most conclusive arguments can be attributed either to necessary relations, or to the forms of our understanding [since] in this regard, any rational decision seems to be forbidden to us' (106).

Her Kantian critique notwithstanding, Germain considers the conditions of possibility for knowledge and understanding and engages with the problematics of *a priori reasoning*: 'one cannot deny the legitimacy of philosophical doubt, for this doubt is founded on the impossibility of comparing any other judgment with that of man' (106). She thus offers some preliminary observations, which are to be taken as inductions in her distinct method of mathematical logic and reasoning. In this background, Germain considers space and time as constant and measurable entities in their interrelation in her *Pensée's*: 'space and time, this is what man sets out to measure; one circumscribes his momentary existence; the other accompanies his successive existence' (196). Germain is still far from taking space/time as a continuum. And yet, she points to movement as 'a necessary relation' (196) linking space and time, further suggesting that unlike human beings who are continuously modified, 'as soon as it is constant and uniform, space is known by time, time is measured by space' (196). As Muller has noted, there are three main ideas linking space and time in Germain's philosophical thought: '(1) space and time are measurable (2) they are linked by motion, and (3) humans have no constancy nor uniformity' (2023, 88). Given that space and time are measurable, 'they play a role in the understanding and circumscribing of existence', Muller has further noted (89). Moreover, while humans have no constancy, their intellectual judgements function within the mathematical framework of order, proportion, and unity, and so do space and time.

In the context of *a priori reasoning*, Germain also observes that 'our logic is composed of rules dictated by universal reason' (1896, 106) and thus Kant's transcendental idealism 'tends to undermine in its foundations the absolute reality of all the certainties that we can obtain' (108). To ground her argument, Germain traces the universality of reason in the works of antiquity and the Middle Ages: from the first astronomical knowledge to the foundation of Cartesian geometry and Newton's discoveries, amidst 'the thousand deviations' of reason that the history of science has pointed to (113). Like Agnesi, she highlights the importance of mathematics in offering truth and nothing but the truth: 'From their birth, the mathematical sciences have offered the human mind the full realization of this type of truth, the object of its dearest affections' (118). The reason is simple: while philosophical language was at times 'even more obscure than the ideas it was intended to convey' (122), the language of 'the exact sciences' has always been precise and clear. In this light, she is optimistic about the fallacies of the human mind, in the wider context of what I configure as process epistemology—an approach to knowledge which embraces open-ended inquiries and recognises the provisional nature of knowledge, always subject to revision and refinement:

> One might have expected to go ashtray; and yet the errors of the human mind, seemingly inexhaustible, have all approached certain truths, and have not been as numerous as the defects of their methods might lead one to suppose. This is because the feeling of truth has never abandoned the authors of all these systems. This happy feeling has not been enough to preserve them from arbitrary and forced assumptions, but it has kept their imagination within certain limits (140).

Given the clarity of the language of the exact sciences and their consecutive prevalent position in seeking 'the truth', it is no wonder that Germain made the study of science in general, and the mathematical sciences in particular, central to her philosophical propositions. Her historical account of the working of the human mind inevitably brings to mind Comte's law of the three different theoretical states [états]: 'the theological, or fictitious; the metaphysical, or abstract; and the scientific, or positive' that all branches of knowledge and principal conceptions pass through on the plane of his *Philosophie Positive* (1835, 3). Parallels between her work and Comte's positive philosophy were identified by Stupuy in his introduction of her philosophical work (1896), but also in a number of reviews that followed.[11] But it is clear that Germain does not follow Comte's paradigm, and as Stupuy has commented, 'she does not distinguish between the logical processes which are specific to each category of knowledge; she does not indicate [...] the different destination of art and science, and her work is not exempt from all metaphysics' (1896, 54).

Germain probably did not even have time to read Comte's work, as the first volume of the *Philosophie Positive* appeared at the end of 1829, and by that time, she was already suffering from cancer. Even considering the time earlier in the 1820s, we should bear in mind that Germain was moving in the mathematical circles of the Academy of Sciences in Paris, and although she worked with many renowned mathematicians of her time, as we have seen in previous chapters, there is no evidence that she ever collaborated with Comte. In this light, Germain's philosophical ideas not only precede the formulation of positive philosophy, but they also belong to a different strand in the history of thought, as I will further argue.

What is therefore important to consider while reading her work is its social, cultural, and philosophical context, which included Comte and the rise of positivism in France, but also Hegel and German Idealism. Paul Ritti, the editor of her posthumously published philosophical work, has actually argued that since Germain had studied Kant, she should have read Hegel's philosophical work as well, and that she had probably met him during his visit to Paris in 1827, when the philosopher got to know many French intellectuals (1890, 354). There is no evidence for this claim and there is no explicit reference to Hegel's philosophy in Germain's work, other than deploying the abstract notion of being, which is 'the key to the vault of the Hegelian system', as Ritti has commented (354).

Indeed, the abstract notion of being recurs as an 'absolute reality' in Germain's work: 'will it be doubted that the archetype of being has an absolute reality, when we see the language of calculations springing forth from a single reality it has seized, encompassing all realities linked to the first by a common essence?' she asks in her treatise (1896, 130). In thus following the route of mathematical logic, she finally reaches the conclusion that 'of all our ideas, the most abstract is that of being; for the idea of nothingness is entirely negative. Being belongs to us, it penetrates our intelligence and illuminates it with the torch of truth' (138). In Ritti's reading, then, 'you have to be a metaphysician to be able to demonstrate that this being, which is, and which is not, becomes; and how becoming is a conciliation, of the two terms which seemed to exclude each other' (1890, 353).

It is precisely Germain's conception of being, its conditions, and its laws that 'have a scent of German metaphysics and especially Hegelian', Pierre Lafitte had argued in a critical reading of her work (cited in Ritti 1890, 354). For some of her commentators, then, Germain's Hegelian connections take her thought away from the field of positive philosophy.[12] As Ritti has further pointed out, Comte himself had acknowledged Germain's influence from German Idealism and it is no wonder that in devising his positivist calendar,[13] he placed her next to Hegel in the month named after Descartes and devoted to the modern philosophical trends of his time (352).

In the light of these debates and given that Germain's work weaves into different philosophical trends of her time, what I therefore argue is that her take on the dynamic nature of being and reality situates her philosophical work in the wider field of process philosophies. In mapping the history of process philosophy, Johanna Seibt (2023) has identified 'pockets of process thought' in the wider area of Western substance metaphysics: 'a special branch of process thought opened up in late 18th and early 19th century German Idealism, when Johann G. Fichte, Friedrich W. J. Schelling, and Georg W. F. Hegel responded to Immanuel Kant's system of a transcendental idealism' (2023). All these philosophers, Seibt notes, 'focused on the process by which the world of knowable appearances, including reflective reasoning, is generated' and Hegel in particular 'postulated that reality is the self-unfolding of dynamic structures or templates' (2023).

Germain's critique of Kantian epistemology and ontology does not therefore come out of the blue, nor does it simply follow the trend of her times, as Bucciarelli and Duorsky have commented (1980, 111). What I argue is that her treatise can be included in 'the pockets of process thought' that Seibt has identified in the wider area of Western substance metaphysics. While considering the context of Hegelian dialectics, within which her work was read, even by the positivist philosophical circles of the nineteenth century, what I argue is that there is an unacknowledged genealogy between her take on a process approach to cognition and Whitehead's process philosophy (1985). In doing so, I follow Seibt's apt remark that Hegel's dialectics is 'the hallmark of speculative process metaphysics' (2023) and that Whitehead's philosophy of the organism is the most rounded take on this speculative process metaphysics' (2023). Here, I have to clarify that given its unfinished and fragmented nature, Germain's philosophical thought can only be considered in terms of its processual aspects, namely her dynamic take on reality and the human mind.[14]

'The actual world is a process, and process is the becoming of actual entities', Whitehead has famously written in his major philosophical work *Process and*

Reality (1985, 22). As we have seen in Chapter 4, process is a fundamental fact of experience for Whitehead and 'involves the notion of a creative activity belonging to the very essence of each occasion' (1968, 151). As Steven Shaviro has perceptively pointed out, Whitehead's understanding of reality as process moves the analytical interest from the philosophical question of 'why is there something rather than nothing' to the more sociologically driven one of 'how is that there is always something new?' (2012, x)

This transition from the 'why' to the 'how' has also been highlighted in Germain's philosophical work: 'doubtless, the impression produced by reading an imaginary work does not resemble that which results from the study of a treatise on geometry' (1896, 80), she has written in considering connections between mathematics and literature. And yet, she goes on to observe, 'let us not hasten however to conclude that there is no common bond between works which initially appear so different' (80–1). It is only when we consider the process of their creation and follow the different phases of their composition that 'it will become evident that the highest literature, like the discoveries that enrich science have been inspired by a feeling of order and proportion which is the regulator of all intellectual movement' (81). In his commentary on her work, Stupuy has underlined Germain's interest in the 'how', tracing its genealogy to Diderot's thought: 'to seek the how and no longer the why, this is, in fact, what marks the philosophical progress outlined by the school of Diderot'(56), he has emphatically noted. In thus seeking 'the how' and not 'the why', Germain's process epistemology rejects static and fixed views of knowledge and understanding and emphasises instead that knowledge is dynamic, constantly evolving as well as shaped by ongoing processes of wonder, exploration, and discovery. In this light, she also embraces open-ended inquiries and acknowledges the provisional nature of knowledge, as we have already seen above.

But there is another important term in Germain's philosophical focus in her reference to 'a feeling of order and proportion' above, the notion of *feeling* [sentiment], which is recurrent in her *Considérations* as well as her *Pensées*. What is important to highlight here is that the main thrust of Germain's take on feelings is very different from the common understanding of feelings as affects or emotions. Throughout her treatise, she refers to feelings as the precursor of human understanding, the first stone in the long process of reasoning and knowledge production as well as the path to the ethics and aesthetics of human existence. Whether she refers to 'a profound feeling of order and proportions' (78), which is the regulator of all intellectual movement either in the works of high literature or the rich discoveries in science (81), 'a feeling of continuity', 'a feeling of analogy' (112), or 'a feeling of freedom' (135), Germain configures a universal feeling, which corresponds to a universal type of truth and has given rise to the creations of the human mind, as we have already noted above.

'As a matter of fact, whenever we get pleasure from a stroke of genius or a touch of eloquence in sciences, fine arts, or literature, it is because through them we can discern previously unseen relations and we are transported to a realm where we discover 'an unexpected order of ideas or feelings', she wrote (81).

Feelings also emerge in her *Pensées* connected to movement and power: 'force is in the body the faculty to move and to move others; it is in us *the feeling* of power' (209, emphasis added). Feelings are, in short, connected to the focus of Germain's philosophical work, namely investigating the 'how' of intellectual processes, but they are also framed within her process epistemology in terms of recognising the embodied nature of knowing. While emphasising feelings, Germain argues that knowledge is entangled with our experiences, emotions, and affects.

Indeed, the conception of *feeling* as a pre-cognitive understanding—in terms of sensing the true as a starting point in the long process of investigating, formulating, and understanding—is what defines her approach. Recall how in the first chapter of the *Considérations*, all intellectual movements either in literature, science, or the fine arts start from a chaotic exploration in the world of abstract ideas, unmodified hypotheses, or blurring forms: 'ideas crowd the poet's imagination; he remains uncertain for a while; a multitude of different sources seem capable of giving life to his composition; he follows its development, then abandons it. He makes a new choice, his mechanism becomes more complicated; he is not happy with it, he stops, he retraces his steps' (82).

In the same backdrop of 'a chaos of thoughts' (83), the mathematician eventually grasps 'the fruitful idea' as already discussed above. We know next to nothing of these intellectual adventures, Germain notes, as 'the poet will not give us an account of the subtle discussions that preceded the adoption of the emblems he has chosen' (88). Similarly, 'the genius, who has glimpsed one of the secrets of the natural order, will not tell us either how many times his imagination has wandered around the path, which was to lead him to the certain knowledge of a truth, which he is now able to demonstrate' (88–9). It is precisely the concealment of this process, argues Germain that has led to the historical separation of imagination from reason (89).

In addressing the 'what and the how' of knowledge, Whitehead initiated the philosophy of process, as we have already noted. Very much in accordance with Germain's principle—although never mentioning her work in his citations—Whitehead's thesis was that the world is one reality within which material and mental interrelations emerge and that this reality 'is the process' (1967, 72). What I have shown in this section is that despite its brevity, as well as its fragmented and unfinished state, Germain's philosophical work is an unrecognised trace of processual approaches to epistemology in the nineteenth century, by highlighting the dynamic, holistic, embodied, and relational nature of knowledge and thus embracing open-ended and provisional inquiries as well as pointing to the pragmatic orientation of knowledge in guiding action and facilitating further research. Her processual epistemology is further a rare exemplar of transdisciplinary thought, at a time when sciences were defining and defending their borders. In the following section, I will offer a glimpse into the literary ways in which Mary Somerville and Ada Lovelace engaged with and traversed scientific boundaries, each in her own distinctive manner.

Parallax, imagination, and poetry: Mary Somerville and Ada Lovelace

> The parallax of a celestial body is the angle under which the radius of the earth would be seen, if viewed from the centre of that body; it affords the means of ascertaining the distances of the sun, moon, and planets. When the moon is in the horizon at the instant of rising or setting, suppose lines to be drawn from her centre to the spectator and to the centre of the earth; these would form a right-angled triangle with the terrestrial radius, which is of a known length; and as the parallax or angle at the moon can be measured, all the angles and one side are given; whence the distance of the moon from the centre of the earth may be computed (Somerville 1849, 61–2).

The eighth section of Somerville's widely celebrated book *On the Connexion of the Physical Sciences* (1849) introduces the concept of parallax, a visual effect in astronomy where the apparent position of a celestial body shifts depending on the observer's vantage point. This shift highlights the relativity of perception in astronomy, as the measurement of parallax depends on the observer's position relative to both the Earth and the celestial body. In the passage, Somerville provides the mathematician's approach to defining parallax: it is measured as the angle under which the Earth's radius would appear if viewed from the celestial body. Using this parallax angle, astronomers can then apply geometry to calculate distances to celestial bodies like the Moon. For example, when the Moon is on the horizon, a right-angled triangle is formed between the Moon's centre, the observer, and the centre of the Earth. By knowing the Earth's radius and measuring the parallax angle, astronomers can use trigonometry to calculate the Earth-Moon distance accurately. Somerville's explanation thus illustrates how local, relative observations can be transformed into universal, objective measurements, bridging the gap between the observer's perspective and the objective reality of the cosmos.

In her exploration of Somerville's groundbreaking work *On the Connexion of the Physical Sciences* (1849) within the context of Victorian intellectual and cultural life, Michelle Boswell notes that parallax emerged as a powerful metaphor in nineteenth-century physical science, much like relativity and uncertainty did in the early twentieth century (2017, 732). Additionally, Anna Henchman has argued that parallax, along with other concepts from nineteenth-century astronomy, provided literary writers with rich metaphors and perspectives for representing subjective and multiple points of view in their narratives: 'Large-scale narrative works need to be understood as complex spatial systems in motion that change over time. In the nineteenth century a universe with *many* centres became the prevailing astronomical model' (2014, 1, emphasis in the text).

Somerville's use of stellar parallax as a key concept in the theory of vision—one that depends on both the observer's perspective and their movement—reflects her awareness of its broader influence on philosophical and literary discourses. Earlier, in the fifth section of her book, she references Francis Bacon's insightful speculation on the finite speed of light and the distinction between 'true' and 'apparent' phenomena in astronomy. Although Bacon's ideas were framed philosophically rather than empirically, they foreshadowed later scientific discoveries, particularly the realisation that light takes time to travel across space. This means that celestial bodies are observed not as they are at the moment of observation, but as they were in the past, a revelation that has profound implications for both science and philosophy:

> It produces in me, says the restorer of true philosophy, a doubt whether the face of the serene and starry heavens be seen at the instant it really exists, or not till some time later ; and whether there be not, with respect to the heavenly bodies, a true time and an apparent time, no less than a true place and an apparent place, as astronomers say, on account of parallax. For it seems incredible that the species or rays of the celestial bodies can pass through the immense interval between them and us in an instant, or that they do not even require some considerable portion of time (Bacon, cited in Somerville 1849, 37).[15]

On the Connexion of the Physical Sciences (1834), Mary Somerville's second book, became an immediate success, going through ten editions, selling over 15,000 copies, and being translated into German and Italian (Patterson 1983, 193). Published during a period of increasing specialisation in science, the book aimed to counter this trend by demonstrating the interconnectedness of various fields such as astronomy, physics, chemistry, geography, meteorology, and electromagnetism. Somerville's interdisciplinary approach resonated widely and received praise from prominent scientists of the time. By linking scientific disciplines, she positioned science as a global and collaborative endeavour, highlighting contributions from explorers, astronomers, and physicists across Europe and beyond. Her emphasis on shared discovery marked a shift towards viewing science as a collective human enterprise rather than an isolated pursuit.[16]

Somerville's epistemology in the book is rooted in descriptive explanation and analogy, avoiding technical jargon to make complex concepts accessible to a wider audience. Richard Holmes (2014, 433) observes that her clear and logical prose occasionally expands into sublime passages that underscore the aesthetic and universal significance of scientific principles. For example, her description of gravity's omnipresence—'in the descent of a raindrop, as in the falls of Niagara; in the weight of the air, as in the periods of the moon' (Somerville 1849, 1)—elevates scientific understanding to a universal experience.

In analysing the literary influences on Somerville's work, Boswell contrasts Somerville's optimistic embrace of scientific reasoning with the existential despair found in Lord Byron's *Cain* (1990 [1821]). While Byron's protagonist views humanity's insignificance in the cosmos with dread, Somerville celebrates human intellect and imagination as tools for engaging with the universe (Boswell 2017, 734). Somerville even invokes Byron in her discussion of parallax, quoting a line from Act II, scene 2 of *Cain* without direct attribution:

> In the milky way and the other starry nebulae, some of the stars that seem to us to be close to others, may be far behind them in the boundless depth of

> space; nay, may be rationally supposed to be situate many thousand times farther off; light would therefore require thousands of years to come to the earth from those myriads of suns, of which our own is but 'the dim and remote companion (Somerville 1849, 62).

Boswell's examination of Somerville's manuscripts at the Bodleian Library reveals that her personal notebook (Dep c. 352, fol. 5)[17] contains further lines from *Cain*, highlighting Somerville's engagement with existential themes. Boswell argues that these poetic references illustrate how Somerville's scientific work sought to address and resolve the very fears of the infinite that Byron's work expressed (2017, 734). This intersection of science and literature reflects Somerville's belief in the interplay between reason, observation, and the sublime, framing knowledge as a means of intellectual and aesthetic enrichment.

There are different aspects in what I would call Somerville's 'deep epistemology' that emerge from her 'popular' book but actually pervade the whole corpus of her scientific writings. Central to her epistemology is what Katherine Neely describes as her 'cosmic platform' (2001, 8)—a perspective from the heavens—that invites readers to imagine the universe not just in terms of observable phenomena, but also the countless unseen bodies that shape the cosmos. As Somerville writes, 'myriads of bodies may be wandering in space unseen by us, of whose nature we can form no idea, and still less of the part they perform in the economy of the universe' (1849, 446). This fusion of reason and imagination in Somerville's epistemology reflects a belief in human intellect's capacity to understand the laws of the universe, turning existential questions into opportunities for intellectual and aesthetic delight.

Somerville's refusal to use technical jargon and her ability to engage a broad audience demonstrate her commitment to democratising knowledge. Boswell (2017, 728) highlights that her work appeals to both specialists and lay readers, aligning with Enlightenment ideals of universal education while adapting them to the Victorian context, which increasingly emphasised interdisciplinary learning (see Lightman 1997). Somerville's interdisciplinary epistemology blends mathematics, astronomy, and literary aesthetics, revealing her nuanced understanding of how science and art can mutually enrich one another. Her references to Byron's *Cain* (1990 [1821]) are emblematic of a broader intellectual milieu in which science was seen not as a purely technical field, but as deeply connected to humanistic inquiry.

Moreover, Somerville's emphasis on the interconnectedness of physical sciences suggests an ethical dimension to her epistemology. By advocating for harmony and order within the universe, she presents science as a moral and spiritual endeavour rooted in humility and reverence for nature's laws. This framework reflects her optimism in the potential of human intellect to uncover the mysteries of nature, while acknowledging the provisional nature of knowledge and the ethical responsibilities that come with scientific understanding.

Ultimately, *On the Connexion of the Physical Sciences* exemplifies a relational epistemology that emphasises synthesis, accessibility, and the dynamic interplay of reason and imagination. Knowledge, in this view, emerges through the integration of ideas, disciplines, and discoveries, reflecting Somerville's broader philosophical outlook that sees interconnectedness as a fundamental property of reality. Her approach synthesises empirical observation, mathematical reasoning, and a unifying vision of nature's laws, embodying an optimistic belief in humanity's capacity to understand the universe. This vision aligns with the broader Enlightenment ethos, tempered by a Victorian sensibility that recognises the provisionality of knowledge and the moral responsibility inherent in scientific progress.[18]

While Mary Somerville engaged with Byron's literary figures to promote optimism in scientific exploration, Ada Lovelace—Byron's daughter, who had been reportedly discouraged from pursuing poetry by her mother's strict education—forged her own vision of mathematics as 'a poetical science' (LB 44, f.210)[19], as discussed in Chapter 5. Lovelace famously regarded mathematical principles not as cold, abstract formulas but as a form of artistic expression, where the elegance of logic and the beauty of imagination intersected. As noted in Chapter 3, her work on Charles Babbage's Engines demonstrated how mathematical concepts could be infused with creativity, envisioning a future where machines could perform not only calculations but also creative tasks, akin to a musician crafting a symphony.

In this context, her essay on imagination offers a particularly revealing outline of her epistemological outlook. Lovelace begins by questioning the common understanding of imagination: '*What* Is Imagination?' (LB 175, f.199)[20] she asks, observing that while the discourses around imagination are polyvalent, the meaning we have of it is rather obscure. 'We talk *much* of Imagination. We talk of the Imagination of Poets, the Imagination of Artists &c; I am inclined to think that in general we don't know very exactly what we are talking about' (f. 199). Her task, then, is to clarify the meaning of imagination by defining its grounds and premises. She therefore identifies imagination as a twofold faculty.

First, it is the '*Combining* Faculty' Lovelace explains: 'It brings together things, facts, ideas, conceptions, in new, original, endless, ever varying, Combinations. It seizes points in common, between subjects having no very apparent connexion, & hence seldom or never brought into juxtaposition' (f. 199). Seen this way, imagination synthesises disparate ideas, facts, and concepts into novel combinations, fostering new perspectives. This reflects Lovelace's appreciation for interdisciplinarity and her recognition of the value of cross-pollination between seemingly unrelated fields.

Second, she describes imagination as what I would call the Projective Faculty, which allows the mind to conceive and bring into presence the invisible or distant—whether spatially, temporally, or metaphysically. 'It *conceives* & brings into *mental* presence that which is far away, or invisible, or which in short does not exist within our physical & conscious cognizance', she writes (f. 199). As a 'noble faculty' (f. 199), it is closely tied to faith and the divine, suggesting that imagination bridges the human and the transcendent. It is 'God-like' (f. 199), enabling humans to transcend immediate realities: 'It renders Earth tolerable (at least should do so); it teaches us to *live*, in the tone of the eternal' (f.199).

The essay delves further into Lovelace's reflections on mortality, linking imagination to existential and spiritual concerns. She suggests that imagination has the power to reconceptualise death as a form of rebirth: 'Imagination teaches us that *Death* is *Birth*' (f. 199). She criticises dark or despairing uses of imagination, advocating instead for a hopeful, sanctifying vision that illuminates the human experience. 'We look to the Imagination to sun up our spirits within us', she writes (f. 199). For Lovelace, imagination is a source of light and sanctity, rendering life bearable while elevating the mind and spirit to a divine plane.

Lovelace also critiques what she perceives as the misuse of imagination by figures such as her father, Lord Byron. She argues that great imaginations have a responsibility to inspire and elevate rather than darken or despair. This reflects her belief that imagination should align with ethical and spiritual aspirations. She writes: 'My Father's Imagination was not always true to itself, as great, as glorious, as divine, as it might have been. It taught of desertion rather than hope. It sometimes dismissed its light rather than brightened' (f. 199). Her critique here resonates with Somerville's response to Byron's *Cain*, emphasising the ethical dimensions of imaginative expression. Finally, her metaphysical reflections transition seamlessly into her interest in the relationship between imagination, knowledge, and science. She asserts:

> Imagination is the *Discovering* Faculty, pre-eminently. It is that which penetrates into the unseen worlds around us, the worlds of Science. It is that which feels & discovers what is, the real which we see not, which exists not for our senses. Those who have learned to walk on the threshold of the unknown worlds, by means of what are commonly termed par excellence *the exact sciences*, may then with the fair white wings of Imagination hope to soar further into the unexplored amidst which we live (f. 199).

By proposing that imagination is a discovering faculty central to scientific inquiry, Lovelace challenges the view of science as purely empirical or mechanical. Instead, she emphasises the profound imaginative capacity required to grasp the unseen relationships underlying mathematical and scientific truths. Mathematics, as 'the exact sciences', becomes a bridge between the tangible and intangible. Understanding these relationships demands not just logical rigour but also an imaginative leap to intuit connections beyond sensory experience.

> Mathematical Science shows what *is*. It is the language of unseen relations between things. But to use & apply that language we must be able fully to appreciate, to feel, to seize, the unseen, the unconscious. Imagination too shows what *is*, the is that is beyond the senses. Hence she is or should be especially cultivated by the truly Scientific—those who wish to enter into the worlds around us! (f. 199, emphases in the text)

Lovelace celebrates mathematics as the 'true science' par excellence, a motif that runs through the philosophical reflections of women mathematicians. For those engaged in the exact sciences, imagination serves as the wings enabling them to transcend the boundaries of the known and explore the 'unexplored amidst which we live' (f. 199). This metaphor underscores her view of imagination as a navigational tool for traversing the liminal spaces between known scientific truths and the mysteries yet to be uncovered.

In Lovelace's essay, then, imagination emerges as both epistemological and ontological. Epistemologically, it facilitates the discovery of truths that are not immediately accessible to the senses, bridging the gap between perception and reality, intuition and logic. Ontologically, by connecting humanity to the divine and the eternal, imagination shapes not only what we know but also how we conceive of existence itself.

Overall, Lovelace's epistemology, as expressed in this essay, presents imagination as a cornerstone of human understanding. It is not merely a tool for creativity but a profound faculty that enables discovery, reconciles the visible with the invisible, and elevates the human spirit. Her reflections resonate with her larger contributions to mathematics and computing, where her imaginative insights anticipated the possibilities of what she termed 'poetical science' (LB 44, f.210). Through imagination, Lovelace bridges the realms of the empirical and the transcendent, positioning it as central to both scientific progress and the human experience.

Ada Lovelace's rare essay on imagination, partly transcribed and published in Betty Alexandra Toole's book *Ada, The Enchantress of Numbers* (1992, 136–7), remains largely hidden in the manuscripts of the Bodleian Libraries. Despite its limited availability, her reflections on imagination have been extensively highlighted and analysed in the rich body of literature surrounding her work. While Lovelace criticised her father's dark and morbid view of imagination, her own perspective is deeply rooted in the Romantic philosophical tradition. This influence is evident in her synthesis of creativity, logic, and science, as I will discuss next.

Lovelace grew up and was educated in the wake of Samuel Taylor Coleridge's *Biographia Literaria* (2014 [1817]), an influential text of the period. In this work, Coleridge distinguishes between two forms of imagination: primary and secondary. The primary imagination, according to Coleridge, is 'the living power and prime agent of all human perception, and as a repetition in the finite mind of the eternal act of creation in the infinite I AM' (205). This faculty is universal, enabling humans to perceive the world and create order from sensory input. The secondary imagination, however, is a higher-order faculty, described as 'an echo of the former, co-existing with the conscious will, yet still as identical with the primary in the kind of its agency, and differing only in degree, and in the mode of its operation' (206). It dissolves and reshapes the material of perception, striving to unify and recreate.

Lovelace's conception of imagination aligns closely with Coleridge's secondary imagination. We have good reasons to believe that she was influenced by Coleridge, since one of her first tutors, William Frend, had previously taught not only her mother but also Coleridge himself (see Lewis 1995, 390). Lovelace then viewed imagination not merely as passive perception but as an active, transformative force. This perspective echoes Coleridge's belief in imagination as a generative power capable of transcending the given and creating the new. Coleridge's

Romantic vision emphasises the unification of opposites—the natural and the supernatural, reason and feeling, science and poetry. Similarly, Lovelace's imagination sought to integrate logic, mathematics, and the arts, demonstrating a holistic and interdisciplinary perspective akin to Coleridge's synthesis of ideas. For both figures, imagination served as a bridge to the sublime and a means to engage with metaphysical concepts beyond ordinary comprehension. Lovelace's musings on imagination reflect a similar fascination with the infinite, contemplating its potential to expand human intellectual and creative capacities in unprecedented ways.

In her article 'The Enchantress of Numbers and the Magic Noose of Poetry: Literature, Mathematics, and Mysticism in the Nineteenth Century', Imogen Forbes-Macphail (2013) situates Lovelace within a broader Romantic and Victorian context. She demonstrates how Lovelace's mathematical insights were deeply intertwined with the imaginative and mystical traditions championed by figures such as Coleridge. At the heart of this interplay is the notion that the world might be read as a divine text or formula, created by a higher power and decipherable through symbolic systems like language or mathematics: 'if the world is conceived of as the Logos of God, or as a Divine formula, there must be a totalizing means of expressing experience in a semiotic system that is at once mathematical and poetic' (139). Forbes-Macphail highlights how this theme resonates with the works of both Lovelace and Coleridge, who each sought to bridge the symbolic and material worlds in unique ways.

Coleridge's writings, Forbes-Macphail observes, reveal a fascination with the analogy between language and algebra, seeing both as tools for organising and understanding reality (144). Lovelace extended this analogy in her vision of a 'poetical science' (LB 44, f.210), blending Romantic ideals with technological foresight. Forbes-Macphail presents a compelling argument for the persistent nineteenth-century desire to bridge art, science, and mysticism. Figures like Lovelace and Coleridge exemplify this yearning, using imagination to explore the intersections of language, mathematics, and reality. Their work reflects a shared belief in the transformative power of symbolic systems, guided by a Romantic sensibility that celebrates creativity and the search for unity.

Forbes-Macphail (2013) offers a compelling examination of the interplay between poetry and mathematics, presenting them as parallel symbolic systems that strive to articulate the truths of the world. While these disciplines may appear distinct, their shared reliance on structured forms and their ability to distill complex ideas into elegant, meaningful expressions underscore their deep connections. Poetry infuses mathematics with emotion and human resonance, while mathematics imbues poetry with structural precision and logic. This harmonious synthesis reflects the Romantic and Victorian aspiration for unity—a desire vividly embodied in the works of figures like Coleridge and Lovelace. In the following chapter, I will delve further into this entanglement of poetry and mathematics, examining its manifestation in Sofia Kovalevskaya's literary writings.

Re-imagining knowledge as process

This chapter has highlighted the profound contributions of women mathematicians to philosophical discourse, with a specific focus on their process-oriented epistemologies. Through their writings, Du Châtelet, Agnesi, Germain, Somerville, and Lovelace demonstrated a shared commitment to an understanding of knowledge as a dynamic, evolving process. Their works transcend disciplinary boundaries, integrating mathematics, philosophy, and creativity in ways that resonate deeply with contemporary theoretical frameworks.

Central to their epistemological approaches is the emphasis on the iterative and transformative nature of knowledge creation. Du Châtelet framed hypotheses as essential scaffolds for advancing scientific inquiry, championing speculative reasoning as a necessary step in the pursuit of truth. Her work exemplifies a methodological rigour that bridges empirical observation and metaphysical speculation. Agnesi, meanwhile, advocated for the clarity and precision of mathematical reasoning as the foundation for philosophical inquiry, coupling her belief in the universality of mathematics with an early feminist critique of gendered educational barriers.

Germain, in her philosophical reflections, placed symmetry and analogy at the heart of her epistemology, offering a unifying framework that linked mathematical beauty with philosophical insight. Her unfinished writings reveal an intellectual vision deeply attuned to the interrelations between reason, creativity, and the limits of human understanding. Somerville, on the other hand, advanced an interconnected vision of science, viewing the universe as a web of interdependent truths, a perspective that underscores the power of synthesis in knowledge-making. Finally, Lovelace captured the imaginative dimension of epistemology, blending analytical precision with creative foresight. Her concept of a 'poetical science' transcended traditional divides between reason and imagination, demonstrating how technological innovation and abstract thought could harmonise to generate new possibilities for understanding and creation.

Together, these women's philosophical works illuminate the deeply processual nature of their epistemologies. They resisted static conceptions of knowledge, instead embracing the unfolding, interconnected, and relational character of intellectual pursuits. By crossing traditional disciplinary boundaries, they forged transdisciplinary approaches *avant la lettre*, which continue to inspire and inform contemporary discussions in the history of ideas. Their writings invite us to rethink the very nature of knowledge itself—not as a finished product, but as an evolving process shaped by imagination, logic, and the interplay of diverse perspectives. This dynamic understanding of knowledge reflects their shared commitment to the potentiality of ideas and the transformative power of intellectual inquiry. Their legacies remind us that the pursuit of knowledge is not only a scientific or philosophical endeavour but also a profoundly creative and human one.

Notes

1 Germain's philosophical work was discussed at the time of its posthumous publication, mostly among the positivist philosophy circles (see Stupuy 1896). For contemporary engagements with her philosophical work, see Musielak 2020, particularly Chapter 10, Glesser, Suceavă, and Vajiac 2020; Muller 2023, and Tamboukou 2023, , 2025.
2 Newton's exact statement was: 'if I have seen further, it is by standing on the shoulders of giants'. See https://discover.hsp.org/Record/dc-9792/Description#tabnav [Accessed 10 December 2024].
3 The phrase 'Hypotheses non fingo' was used by Newton in the essay 'General Scholium', which was appended to the second edition of the *Principia,* published in 1713. He stated: 'I have not as yet been able to discover the reason for these properties of gravity from phenomena, and I do not feign hypotheses. For whatever is not deduced from the phenomena must be called a hypothesis; and hypotheses, whether metaphysical or physical, or based on occult qualities, or mechanical, have no place in experimental philosophy' (Newton 1999, 943).
4 For an elaborate discussion of 'the principle of sufficient reason', see Pruss 2006.
5 There is a rich collection of essays examining the principle of sufficient reason from different angles in Du Châtelet's work in Hagengruber's Époque Émilienne (2022). See, in particular, Hagengruber's Introduction to the collection, Carus 2022, Dunlop 2022, Hecht 2022, Lyssy 2022, Paganini 2022, and Reichenberger 2022.
6 See https://dcpm.hwps-digital-edition.uni-paderborn.de/dcpm/ [Accessed 20 December 2024].
7 Gilles Ménage (1613–92) was a French scholar particularly noted for his work on the history of women philosophers. His book, *Historia Mulierum Philosopharum* (History of Women Philosophers), published in 1690, is one of the earliest known works dedicated to this subject. In it, Ménage compiled biographies and contributions of various women philosophers from antiquity to his contemporary period (see Zedler 1984).
8 See https://historyofwomenphilosophers.org/agnesi/ [Accessed 20 December 2024].
9 For a comprehensive overview of Stoic philosophy, see Durand, Shogry, and Baltzly 2023.
10 It goes without saying that 'transcendental idealism' is a crucial concept in Kant's philosophy, with an important body of scholarship around it, which cannot be treated within the limitations of this chapter. See Stang 2023 for a recent overview of this important philosophical notion.
11 See the annexes of Germain's *Œuvres Philosophiques* (1896, 358–93).
12 See Goering, cited in *La Neue Freie Presse,* published in Vienna on 22 August 1888. An extract of this article is included in the Annexes of the *Œuvres Philosophiques* (Germain 1896, 3738).
13 See Comte's calendrier at https://gallica.bnf.fr/ark:/12148/bpt6k21868f.texteImage and Sarton 1952 for a discussion of Comte's positivist calendar.
14 For a further elaboration between Germain and Whitehead, see Tamboukou 2025.
15 This is an extract from Bacon's *Novum Organum Scientiarium* (1620). The translation of the extract that Somerville cites is slightly different from the translation by Joseph Davey (1901, 238).
16 For a detailed exposition of the book and its reception, see Patterson 1983, particularly Chapter 7. See also Neeley 2001, particularly Chapter 3.
17 Oxford Bodleian Libraries, Mary Somerville Collection, Dep. c. 369-372; Dep. b. 233; Dep. b. 206b; Dep. b. 157. Thereafter MSC followed by Department, Folder and Folio numbers.
18 See Houghton 1957, as a foundational text that explores the intellectual and cultural assumptions of the Victorian period, including their Enlightenment roots.
19 Oxford Bodleian Libraries, Archive of the Noel Byron and Lovelace Families, Dep. Lovelace Byron. In the text and thereafter, LB followed by the folder and folio number.
20 All emphases in the extracts from the essay are underlined words in the manuscript.

References

Agnesi, Maria Gaetana. 1738. *Propositiones philosophicae quas crebris disputationibus domi habitus coram clarissimis viris explicabat extempore, et ab objectis vindicabat Maria Cajetana de Agnesiis mediolanensis* [Philosophical propositions which Maria Gaetana Agnesi of Milan frequently explained extemporaneously and defended against objections in discussions held at home before the most illustrious men]. Milan: Malatesta.

Algarotti. 1737. *Il newtonismo per le dame, ovvero Dialoghi sopra la luce I colori* [Newtonianism for the Ladies, or Dialogues on Light and Colours]. Napoli [but Milan].

Anzoletti, Maria Luisa. 1900. *Maria Gaetana Agnesi*. Milan: Cogliati.

Bacon, Francis. 1901 [1620]. *Novum Organum or True Suggestions for the Interpretation of Nature*. Edited by Joseph Davey. New York: Collier.

Badiou, Alain. 2019. Malebranche. Translated by Jason E. Smith and Susan Spitzer. New York: Columbia University Press.

Barad, Karen. 2007. *Meeting the Universe Halfway: Quantum Physics and the Entanglement of Matter and Meaning*. Durham, NC: Duke University Press.

Boswell, Michelle. 2017. 'Poetry and Parallax in Mary Somerville's *On the Connexion of the Physical Sciences*'. *Victorian Literature and Culture* 45 (4): 727–44.

Brading, Katherine. 2019. *Émilie Du Châtelet and the Foundations of Physical Science*. New York and London: Routledge.

Bucciarelli, Luis, L., and Nancy Dworsky. 1980. *Sophie Germain: An Essay in the History of the Theory of Elasticity*. Boston, MA: Dordrecht, 1980.

Byron, George and L. Gordon. 1990 [1821]. *Three Plays: Sardanapalus; The Two Foscari; Cain*. Oxford and New York: Woodstock.

Carus, Clara. 2022. 'Du Châtelet's Contribution to the Concept of Time. History of Philosophy Between Leibniz and Kant'. In *Époque Émilienne*, edited by Ruth Hagengruber, 113–28. Cham: Springer.

Coleridge, Samuel Taylor. 2014 [1817]. *Biographia Literaria*. Edited by Adam Roberts. Edinburgh: Edinburgh University Press.

Comte, Auguste. 1835. *Cours de Philosophy Positive, La philosophie astronomique et la philosophie de la physique*. Tome Deuxième, Trente-Deuxième Leçon. Paris: Bachelier.

Costa, Emanuele. 2024. 'Education, Equality, and Proto-Feminism in Maria Gaetana Agnesi'. *Hypatia, First View*, 1–17. https://doi.org/10.1017/hyp.2024.6

De Gournay, le Jars, Marie. 2002 [1641]. 'The Equality of Men and Women'. In *Apology for the Woman Writing and Other Works*, edited by Richard Hillman and Colette Quesnel, 75–96. Chicago: University of Chicago Press.

Detlefsen, Karen. 2017. 'Du Châtelet on Women's Minds and Education'. In *What Makes a Philosopher Great?*, edited by Stephen Hetherington, 124–47. New York: Routledge.

Du Châtelet, Emilie. 1740. *Instutitions de Physique* [Foundations of Physics]. Paris: Prault.

Du Châtelet, Émilie. 2009. 'Foundations of Physics'. In *Selected Philosophical and Scientific Writings*, edited by Judith P. Zinsser and translated by Isabelle Bour and Judith P. Zinsser, 105–200. Chicago: The University of Chicago Press.

Du Châtelet, Émilie. 2024. 'Institutions de physique. The Paris Manuscript BnF Fr. 12265'. In *A Critical and Historical Online Edition (2021–2024)*, edited by Ruth E. Hagengruber, Hanns-Peter Neumann, Aaron Wells, Jil Muller, and Pedro Pricladnitzky. Accessed 19 December2024. https://dcpm.hwps-digital-edition.uni-paderborn.de/dcpm/.

Dunlop, Katherine. 2022. 'The Significance of Du Châtelet's Proof of the Parallelogram of Forces'. In *Époque Émilienne,* edited by Ruth Hagengruber, 85–111. Cham: Springer.

Durand, Marion, Simon Shogry, and Dirk Baltzly. 2023. 'Stoicism'. In *The Stanford Encyclopedia of Philosophy* (Spring 2023 Edition), edited by Edward N. Zalta and Uri Nodelman. Accessed 20 December2024. https://plato.stanford.edu/archives/spr2023/entries/stoicism/.

Findlen, Paula. 2003. 'Becoming a Scientist: Gender and Knowledge in Eighteenth Century Italy'. *Science in Context* 16 (1/2): 59–87.

Findlen, Paula. 2011. ‘Calculations of Faith: Mathematics, Philosophy, and Sanctity in 18th-century Italy (New Work on Maria Gaetana Agnesi)’. *Historia Mathematica* 38: 248–91.
Fontenelle, Bernard le Bovier de. 1990 [1686]. *Conversations on the Plurality of Worlds*. Translated by H. A. Hargreaves. Introduction by Nina Rattner Gelbart. Berkeley: University of California Press.
Forbes-Macphail, Imogen. 2013. ‘The Enchantress of Numbers and the Magic Noose of Poetry: Literature, Mathematics, and Mysticism in the Nineteenth Century’. *Journal of Language, Literature and Culture* 60 (3): 138–56.
Frenkel, Edward. 2013. *Love and Math: The Heart of Hidden Reality*. New York: Basic Books.
Frisi, Antonio Francesco. 1799. Elogio storico di D^{A} Maria Gaetana Agnesi, Milanese Dell’Academia Dell’ Instituto Delle Scienze, e Lettrice Onoraria di Matematiche nella Università di Bologna. Milan: Presso Giuseppe Galeazzi.
Germain, Sophie. 1833. *Considérations Generales Sur l’Etat Des Sciences Et Des Lettres Aux Différentes Époques de leur Culture*. Paris: Lachevardiere.
Germain, Sophie. 1896 [1879]. *Œuvres philosophiques de Sophie Germain, suivies de pensées et de lettres inédites. Et précédées d’une notice sur sa vie et ses œuvres par H^{te} Stupuy (Nouvelle Édition)* [Philosophical works of Sophie Germain, followed by Thoughts and unedited letters. Preceded by a Notice on Her Life and Works by H^{te} Stupuy (New Edition)]. Paris: Firmin-Didot.
Glesser, Adam, Bogdan D. Suceavă, and Michaela Vajiac. 2020. ‘The Infinite is the chasm in which our thoughts are lost: Reflections on Sophie Germain’s essays’. *Memoirs of the Scientific Sections of the Romanian Academy* XLIII: 215–23.
Hagengruber, Ruth, ed. 2012. *Emilie du Châtelet: Between Leibniz and Newton*. London and New York: Springer
Hagengruber, Ruth, ed. 2022. *Époque Émilienne*. Cham: Springer.
Hecht, Hartmut. 2022. ‘Three French Newtonians and Their Leibnizian Background’. In *Époque Émilienne*, edited by Ruth Hagengruber, 175–96. Cham: Springer.
Henchman, Anna. 2014. *The Starry Sky Within: Astronomy and the Reach of Mind in Victorian Literature*. Oxford: Oxford University Press.
Holmes, Richard. 2014. ‘In Retrospect: On The Connexion of the Physical Sciences’. *Nature* 514: 432–33.
Houghton, Walter E. 1957. *The Victorian Frame of Mind, 1830–1870*. New Haven: Yale University Press.
Kölving, Ulla et Andrew Brown, eds. 2018. *La Correspondance d’ Émilie Du Châtelet*. 2 vols. Ferney-Voltaire: Centre international d’étude du xviiie siècle.
Lewis, Judith, S. 1995. ‘Princess of Parallelograms and her daughter. Math and Gender in the Nineteenth Century English Aristocracy’. *Women’s Studies International Forum* 18 (4): 387–94.
Libri, Guglielmo. 1832. ‘Sciences Mathématiques-Mlle Germain’. *Journal Des Débats*, 18 May 1832, pp. 1–2.
Lightman, Bernard, ed. 1997. *Victorian Science in Context*. Chicago: Chicago University Press.
Lyssy, Ansgar. 2022. ‘Leibnizian Causes in a Newtonian World—Émilie Du Châtelet on Causation’. In *Époque Émilienne*, edited by Ruth Hagengruber, 197–218. Cham: Springer.
Mamiani, Maurizio. 1998. ‘La rivoluzione incompiuta: i confini tra matematica e fisica da Newton a d’Alembert, in L’età dei Lumi’ [The Unfinished Revolution: The Boundaries Between Mathematics and Physics from Newton to d’Alembert, in The Age of Enlightenment]. In *Saggi sulla cultura settecentesca* [Essays on Eighteenth-century Culture], edited by Antonio Santucci, 39–48. Bologna: il Mulino.

Masotti, Arnaldo. 1940. '"Maria Gaetana Agnesi". *Rendiconti del Seminario Matematico e Fisico di Milano* [Reports of the Mathematical and Physics Seminar in Milan] 14: 89–127.

Maylender, Michele, 1926–1930. *Storia delle accademie in Italia,* 5 vols.Bologna: Capelli.

Mazzotti, Massimo. 2007. *The World of Maria Gaetana Agnesi, Mathematician of God.* Baltimore: Johns Hopkins University Press..

Minonzio, Franco. 2006. *Chiarezza e Metodo: L'Indagine Scientifica di Maria Gaetana Agnesi* [Clarity and Method: The Scientific Investigation of Maria Gaetana Agnesi]. Milano: Lampi di Stampa.

Muller, Jil. 2023. 'Space and Time: Mathematical and Moral Thoughts in Sophie Germain and Blaise Pascal'. In *Exploring the Contributions of Women in the History of Philosophy, Science, and Literature, Throughout Time. Women in the History of Philosophy and Sciences*, edited by C. C. Harry and G. N. Vlahakis, vol. 20, 85–99. Cham: Springer.

Musielak, Dora. 2020. *Sophie Germain: Revolutionary Mathematician.* 2nd ed. Cham: Springer.

Neeley, Kathryn A. 2001. *Mary Somerville: Science, Illumination, and the Female Mind.* Cambridge: Cambridge University Press.

Newton, Isaak. 1999 [1723]. *Philosophiae Naturalis Principia Mathematica*, General Scholium. 3rd ed. Translated by I. Bernard Cohen and Anne Miller Whitman. Berkeley: University of California Press.

Paganini, Gianni. 2022. 'Émilie Du Châtelet's Epistemology of Hypotheses'. In *Époque Émilienne*, edited by Ruth Hagengruber, 21–56. Cham: Springer.

Patterson, Elizabeth Chambers. 1983. *Mary Somerville and the Cultivation of Science 1815–1840.* The Hague: Martinus Nijhoff.

Poullain de la Barre, François. 2005 [1673]. *Three Cartesian Feminist Treatises.* Edited by Marcelle Maistre Welch and translated by Vivien Bosley. Chicago: University of Chicago Press.

Pruss, Alexander R. 2006. *The Principle of Sufficient Reason: A Reassessment.* Cambridge: Cambridge University Press.

Reichenberger, Andrea. 2012. 'Leibniz's Quantity of Force: A Heresy? Émilie Du Châtelet's Institutions in the Context of the Vis viva Controversy'. In *Emilie du Châtelet: between Leibniz and Newton*, edited by Ruth Hagengruber, 157–71. London and New Yourk: Springer.

Reichenberger, Andrea. 2022. 'The Reception of Émilie Du Châtelet in the German Enlightenment'. In *Époque Émilienne*, edited by Ruth Hagengruber, 129–45. Cham: Springer.

Ritti, Paul. 1890. 'Un discours de M Hippolyte Stupuy sur Sophie Germain'. *La Revue occidentale, philosophique, sociale et politique*, July 1st, 1890, 350–57.

Rossi, Elena. 2015. 'Introduzione' [Introduction], translated in English by Karen Pennau Fronduti'. In *Proposizioni Filosofiche, con testo latino a fronte a cura di Elena Rossi* [with Latin text on the facing page] by Maria Gaetana Agnesi, edited and translated in Italian by Elena Rossi, 11–31. Sansepolcro: Biblioteca del Centro Studi 'Mario Pancrazi'. . Accessed 11 December 2024. https://www.centrostudimariopancrazi.it/proposizioni-filosofiche.html.

Seibt, Johanna. 2023. Process Philosophy'. In *The Stanford Encyclopedia of Philosophy* (Winter 2023 Edition), edited by Edward N. Zalta and Uri Nodelman. https://plato.stanford.edu/archives/win2023/entries/process-philosophy

Shaviro, Steven. 2012. *Without Criteria: Kant, Whitehead, Deleuze and Aesthetics.* Cambridge, MA: The MIT Press.

Somerville, Mary. 1849 [1834]. *On the Connexion of the Physical Sciences.* 8th ed. London: J. Murray.

Souriau Étienne. 2009 [1956]. *Du mode d'existence de l'œuvre à faire. In Les différents modes d'existence, suivi de De l'œuvre à faire.* Presented and introduced by Isabelle Stengers and Bruno Latour, 195–217. Paris: Presses Universitaires de France.

Stang, Nicholas F. 2023. 'Kant's Transcendental Idealism'. In *The Stanford Encyclopedia of Philosophy* (Winter 2023 Edition), edited by Edward N. Zalta and Uri Nodelman. https://plato.stanford.edu/archives/win2023/entries/kant-transcendental-idealism.

Stengers, Isabelle. 2011. *Thinking with Whitehead.* Translated by Michael Chase. Cambridge, MA: Harvard University Press.

Stupuy, Jean-Léon-Hippolyte. 1896 [1879]. 'Étude sur la Vie et les œuvres de Sophie Germain'. In *Œuvres philosophiques de Sophie Germain*, suivies de pensées et de lettres inédites, 1–73. Paris: Firmin-Didot.

Tamboukou, Maria. 2025. 'Fragments of thought: Considering Sophie Germain's Process Epistemology', *Hypatia*: A Journal of Feminist Philosophy, 40 (2), 350-365, doi.org/10.1017/hyp.2024.45.

Tamboukou, Maria. 2023. 'Feeling Happiness, Feeling Science: Diffractive Readings of Émilie Du Châtelet's and Sophie Germain's Philosophical Writings'. *Journal of the History of Women Philosophers and Scientists* 2: 68–88.

Toole, Betty Alexandra. 1992. *Ada, the enchantress of numbers.* Moreton-in-Marsh: Strawberry Press.

Van Schurman, Anna Maria. 2013 [1641]. 'A dissertation on the Natural Capacities of Women for Study and Learning'. In *The Equality of Sexes: Three Feminist Texts of the Seventeenth Century*, edited by Desmond Clarke. Oxford: Oxford University Press.

Watson, James. 2001. The Double Helix: A Personal Account of the Discovery of the Structure of DNA . Guildford: Touchstone.

Whitehead, Alfred North. 1967 [1925]. *Science and the Modern World.* New York: Free Press.

Whitehead, Alfred N. 1968 [1938]. *Modes of Thought.* New York: Free Press.

Whitehead, Alfred North. 1985 [1929]. *Process and Reality* [Corrected Edition]. Edited by David Ray Griffin and Donald W. Sheburne. New York: The Free Press.

Zedler, Beatrice. 1984. *The History of Women Philosophers.* Translated from the Latin with an introduction by Beatrice H. Zedler. Lanham: University Press of America.

Zen, Stefano. 2017. *Maria Gaetana Agnesi: tra Newton e Leibniz: scienza, religiosità e autonomia femminile nell'Italia del Settecento* [Maria Gaetana Agnesi: Between Newton and Leibniz: Science, Religiosity, and Female Autonomy in 18th Century Italy]. Roma: Aracne editrice.

Zinsser, Judith P. 2009. 'Volume Editor's Introduction'. In *Selected Philosophical and Scientific Writings*, edited by Judith, P. Zinsser and translated by Isabelle Bour and Judith, P. Zinsser, 105–8. Chicago and London: The University of Chicago Press.

7 Narrative Rhythmanalysis

Listening to the Echo of the Subject

> I was twenty-two years old when I moved to Petersburg. Three months earlier I had graduated from a university abroad and returned to Russia, doctoral degree in hand. After five years of isolated, cloistered existence in a small university town, life in Petersburg immediately enveloped and, as it were, intoxicated me. Putting aside for a while the consideration of analytic functions, space and the four dimensions, which had so recently obsessed me, I threw myself into new interests. I made acquaintances left and right. I tried to penetrate the most varied circles. With greedy curiosity I turned my attention to all the essentially empty but initially so engaging manifestations of the complex hubbub that we call life in Petersburg (Kovalevskaya 2001 [1892], 3).

In the opening pages of her autofictional novella, *The Nihilist Girl*, Sofia Kovalevskaya crafts a vivid and multifaceted portrait of life in Russia shortly after earning her doctoral degree in mathematics in 1874. As noted in previous chapters, Kovalevskaya was the first woman to receive such a high-level university degree in mathematics and later became the first woman to hold a chair in mathematics in modern Europe. Her novella, written during a period of recalibration as she sought to re-engage with the Russian society she had left behind in 1869 to access higher education in Europe, is imbued with the tensions and resonances of a subjectivity caught between borders—cultural, intellectual, and existential. Among her varied literary endeavours, which include poetry, an autobiographical account of her early years in Russia, theatrical plays, and several unfinished novels, *The Nihilist Girl* is her only completed novella.[1]

Kovalevskaya's literary works were highly regarded during her lifetime and beyond. The editor of the journal *Russkaya Starina* famously claimed that her *Russian Childhood* was worthy of standing alongside Tolstoy's *Childhood* (Kochina 1985, 224–5). Such claims, however exaggerated, underscore the vibrant reception of her literary output. Yet, as discussed in earlier chapters, much of the attention on Kovalevskaya has centred on her extraordinary biography and mathematical achievements. In this chapter, I move away from those well-trodden paths to explore her literary writings through the modality of what I have configured elsewhere as *narrative rhythmanalysis* (Tamboukou 2021), thus attuning to their

DOI: 10.4324/9781003533993-8

spatio-temporal rhythms and the flows of desire, affect, and becoming that they unleash.

The chapter unfolds in four sections. Following this brief introduction, I explore key philosophical treatments of rhythm, mapping what I conceptualise as a rhythm *assemblage*. Next, I turn to Kovalevskaya's literary writings, reading them through the lens of *narrative rhythmanalysis*. This approach reveals a dynamic interplay where life, literature, poetry, and mathematics converge, forming *assemblages* that resist fixed identities and linear narratives. What I argue is that Kovalevskaya's subjectivity emerges not as a stable unity but as a 'rhythmic knot' (Lacoue-Labarthe 1989)—a dynamic nexus where intensities fold, unfold, and reconfigure across temporal and spatial dimensions. By attuning to the rhythms, pulsations, and echoes within these knots, *narrative rhythmanalysis* amplifies both the resonances and disruptions in her work. It traces the interplay of continuity and rupture, repetition and novelty, emphasising her resistance to disciplinary and biographical constraints. Rather than a singular narrative, what therefore unfolds is a multiplicity of becomings, where rhythm operates as both a method and a creative force, illuminating the intricate entanglements between Kovalevskaya's intellectual and imaginative pursuits. By way of conclusion, I highlight the role of 'the echo of the subject' in the entanglement of the figures of the poet and mathematician within the textual construction of Kovalevskaya's subjectivity, and I show how the philosophical components of the *rhythm assemblage* illuminate different aspects of Kovalevskaya's inscription as a 'rhythmed' subject.

Mapping a *rhythm assemblage*

In recent years, the concept of rhythm has gained significant traction across various disciplinary fields, informing a wide range of analyses. However, as Jonas Rutgeerts[2] succinctly notes, this surge of interest in rhythmanalysis has often left the notion of rhythm itself vague or 'enigmatic' (2019, 85). Elsewhere, I have employed the framework of *narrative rhythmanalysis* (Tamboukou 2021) to explore oral stories of forced displacement, drawing primarily on Henri Lefebvre's (2004) rhythmanalysis, which he developed through Gaston Bachelard's earlier insights, as well as Gilles Deleuze and Félix Guattari's (1988) notion of the refrain (*ritornello*). In this chapter, however, my focus shifts to literary narratives, examining how the rhythm of poetic order stages an encounter between the narrator, her protagonist, the reader/researcher, and the pulsations of the natural and sociocultural rhythms that shape subjects and their worlds.

To establish a conceptual foundation, I trace rhythm's rich philosophical genealogy, emphasising its role in shaping the writing subject. The etymology of the word 'rhythm' is widely acknowledged to stem from the Greek term *rhythmós* [ῥυθμός], derived from the verb 'to flow' [ῥεῖν] and the suffix *-thmos* [-θμός], which signifies a measured flow or movement. As Jūratė Baranova, Laura Junutytė, and Lilija Duoblienė succinctly summarise, rhythm 'designates the movement or variation characterised by the regular recurrence or alternation of different quantities or conditions' (2016, 10). Yet, there is disagreement among linguists about its precise origins. Émile Benveniste, in his influential essay 'The Notion of 'Rhythm' in Its Linguistic Expression', argues that *rhythmós* in the works of Presocratic philosophers, Herodotus, and the tragedians referred to form, schema, or figure, rather than deriving from the verb 'to flow' (1973, 282).

Benveniste draws a distinction between *schema* [σχῆμα], which he describes as a fixed form, and *rhythmós*, which he defines as 'the form in the instant that it is assumed by what is moving, mobile, and fluid' (285–6). Rhythm, in this sense, captures the fleeting, improvised, and changeable nature of form. This nuanced understanding underscores the entanglements between rhythm, flow, the moment, and form, themes that will be further developed in this chapter's *rhythm assemblage*.

Despite differences in linguistic interpretations, rhythm's philosophical history is equally rich and varied. As Peter Cheyne, Andy Hamilton, and Max Paddison note in their introduction to *The Philosophy of Rhythm* (2019), rhythm's conceptual journey begins with Heraclitus' fragment 'everything flows' [τὰ πάντα ῥεῖ], imbuing rhythm with a cosmic resonance that persists in process philosophies. Plato introduced a socio-cultural strand by defining rhythm as 'order within movement', emphasising its structural and regulative dimensions (1–2). Rutgeerts further highlights that by the early twentieth century, rhythm had become a key concept across philosophy, science, and social theory, often framed as a mechanism underpinning all movement (2019, 100n21).

By tracing these genealogical lines, I aim to situate my own perspective within this complex terrain. This chapter maps an *assemblage* of theoretical approaches, drawing on the twentieth-century debate on rhythm between Henri Bergson and Gaston Bachelard. While their conceptualisations of rhythm diverge significantly, notable intersections exist, particularly in their shared interest in the philosophy of time, as explored in subsequent scholarship.[3] I read these entanglements diffractively through Deleuze and Guattari's (1988) notion of the refrain (*ritornello*) and Philippe Lacoue-Labarthe's (1989) interpretation of rhythm as an *intermezzo* between beat and figure, articulated in his configuration of the 'echo' as constitutive of subjectivity. Here, as in previous chapters, my approach aligns with feminist affirmative methodologies, which advocate for reading theoretical insights 'through one another' to rework foundational concepts within their respective traditions (Barad 2003, 811). By employing this diffractive lens, I explore rhythm not as a static or universal concept but as a dynamic interplay of flows, interruptions, and reconfigurations that shape the writing subject and the narratives they produce.

Bergson's work on time, duration, and memory has had a profound and lasting influence, shaping philosophical thought at the dawn of the twentieth century and continuing to resonate in contemporary process philosophy and new materialisms. Central to Bergson's philosophy is the concept of rhythm, which plays a pivotal role in his early essay *Time and Free Will* (2001 [1889]) and is further developed in his influential work *Matter and Memory* (2002 [1896]). Bergson challenges the compartmentalised view of time as a homogeneous sequence of discrete instants, proposing instead the image of *duration [durée]*—a dynamic, qualitative flow where past, present, and future intermingle within lived experience: 'In a word, pure duration might well be nothing but a succession of qualitative changes, which melt into and permeate one another, without precise outlines, without any tendency to externalize themselves in relation to one another, without any affiliation with number: it would be pure heterogeneity' (2001, 104). Bergson frequently employs music as a metaphor to illustrate the experience of duration, with rhythm serving as a crucial aesthetic tool. Rhythm, he suggests, evokes the fluidity of duration by disrupting ordinary sensations and ideas, creating a profound emotional resonance: 'In music, the rhythm and measure suspend the normal flow of our sensations and ideas [...] and they take hold of us with such force that even the faintest imitation of a groan will suffice to fill us with the utmost sadness' (14–15).

However, as Rutgeerts points out, in *Matter and Memory*, rhythm transcends its role as an aesthetic metaphor to become the defining characteristic of all entities and phenomena. For Bergson, rhythm is not merely a representation of duration but an intrinsic expression of it: each entity does not simply *have* a rhythm—it *is* its rhythm, Rutgeerts pithily notes (2019, 91, emphasis in the text). This rhythm reflects the force of duration that animates and permeates all existence, causing everything to move, change, or 'vibrate' (91).

Bergson's reality, then, is fundamentally polyrhythmic, characterised by countless rhythms coexisting and interacting. These rhythms vary in their rates of vibration, which determine the uniqueness of entities and their positions within the 'scale of being'. Writes Bergson: 'In reality there is no one rhythm of duration; it is possible to imagine many different rhythms which, slower or faster, measure the degree of tension or relaxation of different kinds of consciousness, and thereby fix their respective places in the scale of being.' (2002, 207)

This diversity of rhythms also explains the challenges of perception. To perceive something, we must often impose a dominant rhythm onto the slower, subtler rhythms of the phenomena around us, effectively immobilising them. As Bergson notes, 'to perceive is to immobilize' (208). Yet, this immobilisation comes at a cost: it simplifies and reduces the complexity of reality, turning our backs on the deeper, more nuanced knowledge available through intuition (199). For Bergson, true knowledge lies not in imposing our rhythms onto the world but in surrendering to the rhythms of duration. Intuition, for him, becomes the mode of aligning oneself with these rhythms—a way of coinciding with the 'extra-intellectual matter of knowledge' (2005 [1907], 389) by adopting its rhythm and movement. Through this alignment, one grasps reality from within, experiencing its fluid and interconnected nature.

While influenced by Bergson, Bachelard critically diverged from both the concept of duration and the image of rhythm as flow. On the opening page of his book *Intuition of the Instant* (2013 [1932]), a philosophical response to Gaston Roupnel's drama *Siloë* (1927), Bachelard foregrounds Roupnel's central thesis: 'time has but one reality, the reality of the instant' (Bachelard 2013, 6). This notion, that time exists only as solitary instants, recurs throughout the book, effectively stripping past and future of any ontological reality, as Rutgeerts has noted (2019, 93). However, Bachelard also grapples with the Bergsonian challenge of explaining why, despite this, we experience time as something that unfolds continuously. To address this, he introduces rhythm as a central concept in his philosophy of time, intertwining it with his understanding of subjectivity. He writes, 'The individual being, insofar as it is complex, corresponds first of all to a simultaneity of instantaneous actions. It rediscovers itself only to the degree that it resumes these simultaneous actions' (2013, 39). For Bachelard, the subject is not just a sum of its qualities but also its potentialities—its *becoming*. This becoming is shaped by a harmony of temporal rhythms, which serve as the medium through which the paradox of 'discontinuous continuity' is understood (39). Rhythm, in this framework, bridges the gaps between instants, much as the subject traverses the temporal vacuum separating them (39).

Thus, Bachelard rejects the Bergsonian concept of duration as an ontological force, instead portraying it as a subjective construct—a grouping of instants shaped by rhythms. As he asserts, 'There is but one reality: the instant. Duration, habit, and progress are only groupings of instants—the simplest among the phenomena of time. None of these temporal phenomena can have an ontological privilege' (51). Habits thus occupy a foundational role in Bachelard's philosophy of time and being, shaping the subject through 'more or less accurate repetitions, more or less detailed reflections' (41). Past memories also contribute, as the individual strives to trace the patterns of today upon the rhythms of yesterday. Yet, since 'the past is as empty as the future' and 'the future as dead as the past' (28), the past's presence in the now is merely an echo: 'in us the past is a voice that has found an echo' (31), Bachelard observes. We will return to the notion of the echo later on in this section.

In *The Dialectic of Duration* (2000 [1936]), Bachelard expands on rhythm as the means of connecting solitary instants into a continuous refrain, while also situating rhythm within a dialectical process involving states of action and existence. He writes, 'there is a fundamental heterogeneity at the very heart of lived, active, creative duration, and [...] in order to know or use time well, we must activate the rhythm of creation and destruction, of work and repose' (29). Rhythm, as Rutgeerts remarks, is not simply a sequence of instants but an alternation or interaction between opposing forces (2019, 96). Furthermore, Bachelard's dialectical view sees rhythms as interdependent and layered, conditioned by one another—a perspective that echoes Bergson's notion of a polyrhythmic reality. In this way, Bachelard critiques and reconfigures rhythm, not as a passive flow but as a dynamic interplay of forces that shape time, being, and creativity. His philosophy presents rhythm as both the connective tissue of temporal experience and the driving force behind the active dialectics of existence.

As we have seen so far, rhythm plays a central role in how both Bergson and Bachelard conceptualise temporal existence. It is not only constitutive of reality and the self but also serves as a tool we use to position ourselves in the world and attune to it. As Rutgeerts succinctly observes, for Bergson, rhythm operates through perception, where 'we impose the fast rhythm of our thinking onto the slow rhythm of material things, thus immobilizing them and making it possible for us to use them' (2019, 96). In contrast, for Bachelard, rhythm is embedded in our habits—repetitions, refrains, and echoes of the past—through which we feel at home in the world. While Bergson views rhythm as an expression of pure duration, Bachelard approaches it as a means of constructing and interpreting the continuity of the discontinuous. It is within this tension that Deleuze and Guattari's (1988) concept of 'the refrain' and Lacoue-Labarthe's idea of 'the echo of the subject' (1989) emerge as crucial components in the *rhythm assemblage* mapped in this chapter. These notions further illuminate the ways rhythm mediates between self, time, and the world, offering a dynamic framework for understanding existence.

For Deleuze and Guattari, rhythm serves as a force that organises otherwise chaotic and immeasurable moments, emerging in the transitions between milieus or environments. They describe rhythm as 'the milieu's answer to chaos', emphasising its capacity to 'tie together critical moments' (1988, 313). This perspective acknowledges Bachelard's view of rhythm as a linkage between active moments, operating on a distinct plane. As Deleuze and Guattari point out, 'Bachelard is right to say that "the link between truly active moments (rhythm) is always affected on a different plane from the one upon which the action is carried out"' (313).[4] On this analytical plane, Deleuze's and Guattari's concept of the refrain [*ritournelle*] (312) plays a pivotal role in exploring themes across philosophy, art, culture, and the relationships between human and non-human entities.

The refrain refers to any repetitive, patterned activity or mode of being that introduces a sense of order or structure within the chaos Deleuze and Guattari describe as *chaosmos*. As they explain, 'what chaos and rhythm have in common is the in-between—between two milieus, rhythm-chaos or the chaosmos' (313). Beyond mere repetition, the refrain embodies becoming and transformation: 'the refrain is a prism, a crystal of space-time [that] acts upon that which surrounds it, sound or light, extracting from it various vibrations, or decompositions, projections or transformations' (348). Notably, while rhythm and the refrain are interconnected, they remain distinct. Rhythm arises from refrains in music and beyond, as Deleuze and Guattari write: 'Music is pervaded by every minority, and yet composes an immense power. Children's, women's, ethnic, and territorial refrains, refrains of love and destruction: the birth of rhythm' (300). Deleuze's and Guattari's configuration of the refrain also intersects with Bachelard's notion of rhythm as tying together 'critical moments' (313). They argue that 'it is of the nature of the refrain to become concentrated by elimination in a very short moment' (349), thereby 'fabricating time' (349) and aligning with Bachelard's view of the instant as the only temporal reality.

While Bergson's influence on Deleuze is well-documented and extensively analysed in both Deleuze's own work (1991) and secondary literature[5], the concept of the refrain also gestures towards Bergson's idea of vibration as an expression of duration's force. This force, as discussed earlier, gives rise to the singularity and uniqueness of entities. In Deleuze and Guattari's refrains, vibrations extracted from the *chaosmos* similarly mark the transformative power of duration. I will return to these theoretical entanglements with rhythm in my analysis of Kovalevskaya's literary writings in the next section. For now, I turn to Lacoue-Labarthe's (1989) concept of the 'echo of the subject', which forms the final component of the *rhythm assemblage* I map in this chapter.

In his exploration of the relationship between language, subjectivity, and literature, Lacoue-Labarthe engages deeply with the concept of the subject, examining its entanglements with language, representation, and the political sphere. In his essay 'The Echo of the Subject' (1989), he specifically traces connections between autobiography and music, using rhythm as a central motif. Lacoue-Labarthe asks, 'What is it that ties together autobiography, that is to say, the autobiographical compulsion [Zwang] (the need to tell, to confess, to write oneself), and music—the haunting by music or the musical obsession?' (140). This question anchors his investigation into the writing subject—'the subject that writes itself [s'écrit]: that writes about the subject, that is written about, that is written'—a subject that exists only insofar as it is inscribed (141). Indeed, for Lacoue-Labarthe, the act of inscription is foundational to the subject's becoming. The subject is 'one' only in the act of writing or being written—a dynamic process that resists fixed identity. This conceptualisation of subjectivity as constituted through inscription provides a crucial framework for understanding Kovalevskaya's self-fashioning through her literary work, as I will explore in the next section.

In thus following psychoanalytic and philosophical trails in his explorations, Lacoue-Labarthe situates music as the catalyst for the autobiographical gesture (151), an idea that resonates intuitively: the familiar experience of a song or melody evoking memory and reflection. Since 'music is prime' (151), the subject is inherently configured through sounds, refrains, and echoes. Rhythm, though elusive—it cannot be seen, only heard—operates as a 'pre-specular' and 'pre-figural' force, defined through its connections to resonance, echo, and reverberation, which emerge from repetition (196). For Lacoue-Labarthe, there is no rhythm without repetition—or shall we say *refrains*? Rhythm then becomes 'the spacing and the division in the Same, the repeated difference-from-itself of the Same' (196). This understanding positions rhythm as essential to subjectivity. Without rhythm, the subject becomes unmoored, unable to find the reassuring echo that differentiates and affirms its existence. As John Martis notes, the loss of rhythm signals the loss of the subject itself (2005, 44). Conversely, rhythm constitutes the condition of possibility for subjectivity: 'we are constituted by this rhythm', Jacques Derrida observes in his famous introduction to *Typography* (1989, 31).

Lacoue-Labarthe's insights draw heavily on Paul Valéry's conception of the self as fundamentally relational and reflective, rather than autonomous. Valéry's assertion that the self is 'nothing but an echo' (cited in Lacoue-Labarthe 1989, 139) underscores the subject's dependence on external resonances, where rhythm and relationality continuously shape its formation. In this view, the self is not static but a dynamic process—a perpetual reconfiguration through the echoes of encounters. Lacoue-Labarthe's deployment of the notion of 'the echo' further highlights how meaning and subjectivity emerge through the interplay of sound, language, history, and culture. The subject becomes entangled in this web of echoes, perpetually mediated by rhythm and inscription: 'every soul is a rhythmic knot. We ("we") are rhythmed', Lacoue-Labarthe notes (202), invoking Mallarmé's *La Musique et les Lettres* (140). Rhythm, in his work, transcends music and representation, existing 'between beat and figure' (202). It is, as he poetically articulates, 'a rhythm in which "I" seek desperately to recognize "myself"' (203). This perspective resonates with both Bergson and Bachelard. As Rutgeerts has aptly put it, for Bergson, 'each entity is its rhythm' (2019, 91), while for Bachelard, the self is 'the integral sum of rhythms' (2013, 41). Together, these philosophies, along with Lacoue-Labarthe's nuanced take on rhythm as the interplay of inscription, music, and subjectivity, create a fertile framework for examining how rhythm shapes the self's becoming, both in life and in art.

Despite the diverse philosophical approaches to rhythm, a significant body of scholarship has emerged around its role in literature.[6] As Rebecca Wallbank succinctly notes, 'rhythmic auditory-imagery is prevalent within one's experience of literature, even if one is not attentive to it, and it can alter the tone, tenor, and mood of an author's writings in a non-voluntary, non-foregrounded manner' (2019, 371). It is within this context that I examine the spatio-temporal rhythms in Kovalevskaya's literary writings, attuning to her existential need—and indeed her desire—to navigate two intertwined forces in her life: continuity, as shaped by her identity as an aristocratic Russian woman, and discontinuity, marked by rupture and change.

Indeed, Kovalevskaya's life unfolded across different space/time planes, punctuated by instants, breaks, and durations. There was the Bachelardian 'intuition' of becoming a mathematician—a rupture from the expectations of her aristocratic upbringing that deterritorialised her from her homeland, Russia. Her cosmopolitan wanderings across Europe signified further dislocations, juxtaposed with the continuity and intensity of her mathematical pursuits. Simultaneously, her fervent engagement with radical politics and the struggle to establish herself as a professor in mathematics created rhythms of tension and perseverance, all while managing the demands of being a single mother. Her literary writings illuminate the intricate spatial and temporal dimensions she navigated, uncovering a rhythmic interplay of rupture and flow, permanence and change, continuity and disruption. These rhythms echo Bergson's insight that there is 'no unique rhythm of duration', but rather a multiplicity of 'different rhythms' (2002, 207), as previously discussed. Let us now trace the tracks of Kovalevskaya's spatio-temporal rhythms within her literary corpus, exploring how these interwoven patterns shape her narrative and offer insights into her life's dynamic interplay of stability and transformation.

Writing, reading, and sensing rhythm

> Palibino lay quite close to a wood, which gradually became denser and denser, and at last merged into the enormous imperial forest, which extended for hundreds of miles. In its thickets you never heard the sound of an axe, except perhaps at night, when a peasant was bold enough to steal a little crown-wood. Many curious stories were told about this forest. Of course, superstition peopled it with supernatural beings: fairies, trolls and gnomes (Kovalevskaya 1895a, 242).

In this excerpt from the first version of her autobiography—written in the third person and framed as a novel, as explored in Chapter 3—Kovalevskaya vividly renders the landscape of her childhood home at Palibino, the family estate where she moved at the age of eight. The thick, almost tangible materiality of the woods envelops the scene in a profound silence, a stillness that is suddenly ruptured by the dissonant sound of an axe. This sharp, intrusive noise invokes time in its Bachelardian conception as an instant, a singular moment that interrupts the flow of duration (2013, 6). As we have discussed in the previous section, for Bachelard, time's reality lies not in continuity but in the fragmentary, fleeting nature of the instant, which disrupts and reconfigures the rhythm of existence. In this sense, the axe's sound serves as a temporal fissure, crystallising the tension between the permanence of the landscape and the transience of human activity. Kovalevskaya's depiction mirrors Bachelard's insight into how such instants create a dynamic interplay between stillness and movement, materiality and temporality, underscoring the profound rhythm of lived experience.

The evocative depiction of the Palibino countryside in Kovalevskaya's *A Russian Childhood*, written in the first person and first published in Russia in 1890, provides a compelling counterpart to the third-person version of her autobiography discussed above. These slightly different texts not only enrich the narrative but also illuminate the entanglements of refrain and rhythm, as theorised by Deleuze and Guattari's notion of the *ritornello* (1988) and Lacoue-Labarthe's concept of 'the repeated difference-from-itself of the Same' (1989, 196). The nuanced variations between the two versions invite readers to tune into the rhythms of Kovalevskaya's prose while also acknowledging that some of these differences may stem from the complexities of translation—from Swedish to English, or Russian to English—over the years, as outlined in Chapter 3.

> One side of our estate abutted upon wooded land almost to its very edge. At first this wood was kept cleared and looking like a park, but little by little it had grown denser and more impassable, until at last it merged with the vast State forest. This latter extended over an area of hundreds of versts. No axe had ever sounded in it in the memory of man except clandestinely, perhaps, in the hands of some poaching peasant come to make off with government wood by dark of night. Strange legends about this forest circulated among the peasants, and it was hard to determine where truth left off and myth began. Like all Russian forests, it was of course inhabited by a variety of evil spirits such as wood-demons and river-nymphs (Kovalevskaya 1978, 83).

In these forceful descriptions of the forest, space, time, and the history of the Russian Empire are deeply entangled. The Palibino woods, overgrown with time, become saturated with folklore, myths, and the collective imagination, embodying a continuity that accumulates density over generations. Yet, this continuity is disrupted by the dissonance of a peasant's clandestine axe—an act that reclaims, or perhaps resists, an imperial imposition of property rights. This soundscape of silence and intrusion encapsulates a rhythmic interplay of continuity and discontinuity, framed within the historical context of the 1861 Emancipation Reform, which abolished serfdom shortly before Kovalevskaya's family moved to Palibino (see Koblitz 1988).

This interplay of rhythm between landscape and history—of space (*extension*) and time (*inextension*)—reflects a literary manifestation of Bergson's philosophy, wherein space and time are no longer radically distinct but exist as extremes on a continuum, as Rutgeerts notes (2019, 91). Rhythm, in this framework, emerges as 'the specific outcome of the concrete interplay between the forces of extension and inextension that takes place in each phenomenon and characterizes it' (91). Readers are drawn into this rhythm, captivated by the vibratory interplay of space and time that pulses through Kovalevskaya's descriptions of Palibino, wherein her own echo as a rhythmed subject can be discerned.

Immersed in Bergson's image of rhythm, we encounter a world of incessant movement and vibrations. Within this polyrhythmic reality, stability arises only through the imposition of our intense rhythm of duration upon the slower rhythms of our surroundings. As we have seen in the previous section, for Bergson, 'to perceive means to immobilise' (2002, 208). This act of stabilising through rhythmic imposition finds poignant expression in Kovalevskaya's depiction of the rhythms of her childhood estate, such as the house clock's measured ticking—an enduring refrain that imposes order upon the flux of time and space. It is to this moment of rhythmic stability that we now turn.

> The clock on the classroom wall struck seven. Those seven repeated strokes reached me through sleep and gave rise to the sad certainty that now, this minute, Dunyasha, the chambermaid, would be coming in to wake me up. But sleeping was still so sweet that I tried to convince myself that those seven repulsive strokes of the clock were only in my imagination (Kovalevskaya 1978, 98).

In this evocative passage from *A Russian Childhood*, Kovalevskaya's recollection of her childhood self-captures a poignant tension between two temporal modes: the objective, measured time of the clock and the subjective, fluid experience of duration. This interplay echoes Bergson's concept of dual perception, where he asks, 'do we not sometimes perceive in ourselves, in sleep, two contemporaneous and distinct persons, one of whom sleeps a few minutes, while the other's dream fills days and weeks?' (2002, 207). The seven harsh strokes of the clock, signifying the external, linear passage of time, are juxtaposed with the child's experience of the Bergsonian *durée*—an infinite, subjective moment of bliss where the boundaries of time dissolve.

Indeed, the strokes of the clock serve as an example of Bergson's 'amelodic images', which, as Christopher Corbier observes, illustrate the intuition of duration (2012, 296). The monotone and staccato repetition of the strokes becomes the backdrop against which self-forgetfulness can emerge, allowing the child to temporarily disregard the rigidity of measured time and immerse herself in the flow of duration. As Rutgeerts notes, this self-forgetfulness creates the conditions for the experience of real duration—the child's decision to dismiss clock time and savour the sweetness of sleep (2019, 91). However, this passage also resonates with Bachelard's concept of the instant, particularly the fleeting bliss of 'the last moment of sleep'. The stroke of the clock, while a measure of linear time, paradoxically anchors the child's indulgence in the fullness of the instant. It evokes the Bachelardian insight that even within the flow of duration, moments of temporal rupture allow for the intensification of experience.

Rather than viewing duration and the instant as oppositional, Kovalevskaya's literary recollection reveals their entanglement. The child, immersed in the Bergsonian *durée*, rewrites herself as someone who resists the demands of clock time and instead lingers in the rhythm of the final moments of sleep. This rhythm, oscillating between repetition and refusal, allows her to inhabit a state where duration and the instant coexist—a dynamic interplay that resonates throughout Kovalevskaya's literary writings. Ultimately, this short passage becomes a literary microcosm of Kovalevskaya's negotiation with time, rhythm, and self-inscription, illustrating how measured and lived temporalities converge in the act of remembering and writing.

As Kovalevskaya matured and nurtured her dream of becoming a mathematician, time in her writings transforms into what Étienne Souriau (2009) conceptualises as the *œuvre à faire*, a project-in-the-making. This notion of time as a creative endeavour aligns with her sense of becoming, which can be viewed through the lens of Bachelard's rhythmanalysis. Bachelard posits rhythmanalysis as a method to 'discover new opportunities for creating rhythms' (2000, 148), emphasising the generative potential of rhythm in shaping human experience and creativity. For Bachelard, rhythmanalysis is not merely descriptive but transformative, offering a means to reconfigure one's relationship to time and existence. This perspective aligns seamlessly with Souriau's *œuvre à faire (2009)*, in which creation is always in process, shaped by the rhythms of the subject's engagement with the world.

Here, it is crucial to recognise the origins of rhythmanalysis in the work of the Portuguese philosopher Lúcio Alberto Pinheiro dos Santos (1931), who first articulated the study of rhythms as a way of understanding life, nature, and thought. For Pinheiro, rhythmanalysis involved examining the dynamic interplay of patterns and movements that govern human existence, proposing rhythm as an essential framework for grasping the temporal structures of experience. His pioneering ideas significantly influenced Bachelard, who incorporated rhythmanalysis into his explorations of temporality, poetics, and the phenomenology of creative acts.

In her literary oeuvre, Kovalevskaya crafts temporalities that resonate with Bachelard's view of rhythmanalysis as a creative act. In this context, her life and work can be seen as a rhythmic negotiation between the *œuvre à faire* and her temporal experiences, as captured in her literary writings. Bachelard's rhythmanalysis provides a lens through which we can interpret this interplay, as it involves both the lived rhythms of daily life and the imagined rhythms of creative becoming. For Kovalevskaya, the rhythm of her becoming a mathematician—her *œuvre à faire*—required navigating between periods of stability and rupture, repetition and novelty. These rhythms shaped not only her intellectual pursuits but also her identity as a writer, scientist, and individual. Pinheiro's (1931) emphasis on the interrelation between natural and human rhythms further enriches this understanding. By situating human experience within broader rhythmic patterns—biological, ecological, and cosmic—Pinheiro's rhythmanalysis underscores the inseparability of personal and universal temporalities. Kovalevskaya's writings reflect this tension: her mathematical endeavours represent a disciplined, almost metronomic rhythm, while her literary work captures the fluid, unpredictable rhythms of memory, emotion, and creativity.

Thus, Kovalevskaya's becoming, viewed through the prisms of rhythmanalysis and the *œuvre à faire*, emerges as an intricate dance of temporalities. Her writings become spaces where rhythms are not only experienced but actively created, weaving together the scientific precision of mathematics with the poetic fluidity of life's unfolding—a harmonious yet dynamic interplay of time, rhythm, and creative potential. It is thus in her poetry, published posthumously, that we find traces of such interplays. Heidelberg, the little German town where she first studied mathematics, thus appears as a heterotopic space, both real and imagined,[7] in the verses of her poetry:

> April 13
> Now unfolds a different image
> In the changing scene I'm viewing
> Of a tidy German village
> With a castle, now in ruins.
> All the mountains outlined softly
> And the chestnuts lush and verdant
> Are enveloped in a misty,
> Hazy blue transparent curtain
> Green, as boundless, as the ocean,
> And the sky of vivid azure
> I see whimsically patterned,
> With great arches and with towers
>
> (in Coleman 2021, 55)

In the rhythm of Kovalevskaya's verses, nature and culture merge in dynamic interplay, weaving landscapes and timescapes into poetic assemblages. Each space and time evoked in her stanzas carries its own rhythm, forming a layered and multifaceted composition. In the poem above, the fusion of distinct rhythms generates a tapestry of sensations, allowing readers to experience the vibrancy of her poetic imagination. Through rhythmanalysis, this interplay can be understood as more than mere literary technique—it becomes a method of articulating and inhabiting complex emotional and temporal landscapes.

As a poet, Kovalevskaya transforms her feelings into vivid images, which are then translated into words that follow the 'laws of rhythm', as Bergson conceives them (2001, 15). Her poetic rhythm is not only a structural device but also a conduit for conveying the affective forces that underpin her creative process. These forces resonate with Bergson's assertion that rhythm enables readers to feel the emotional charge behind poetic images. According to Bergson, 'we should never realize these images so strongly without the regular movement of the rhythm by which our soul is lulled into self-forgetfulness, and, as in a dream, thinks and sees with the poet' (15). Rhythm thus bridges the gap between the poet's internal world and the reader's imaginative engagement, establishing a shared temporal and affective space.

For Bergson, the power of rhythm lies in its ability 'to lull the imagination, to rock it [bercer] to and fro between like and like with a regular see-saw motion, and thus prepare it submissively to accept the vision suggested' (2008, 34). This oscillation between repetition and variation mirrors the rhythmic structure of Kovalevskaya's poetry, where the alternation of natural and cultural elements creates a dynamic balance. Through this rhythm, her verses not only evoke sensory and emotional impressions but also invite readers to enter the temporal flows and spatial dimensions she portrays.

By applying rhythmanalysis to her work, we can uncover the layers of meaning embedded in her poetic rhythms. These rhythms function as temporal and affective forces, mediating between the poet/mathematician's inner experiences and the external world she seeks to depict. Kovalevskaya's poetic rhythm, then, is not merely an aesthetic choice but a method of engaging with the interplay of memory, imagination, and perception—a rhythmic articulation of her lived experience that transcends the boundaries of individual sensation to resonate with universal patterns of feeling and thought.

What I therefore argue is that Bachelard's emphasis on the creative potential of rhythm as a bridge between moments intertwines with Bergson's focus on fluidity and the oscillation of imagination, while Pinheiro's universal framework situates these rhythms within broader patterns of existence. Extending these insights into the *narrative assemblage* of Kovalevskaya's literary writings, I explore how rhythms both shape and are shaped by processes of subjectivity and becoming. These interwoven perspectives form the foundation of my own configuration of *narrative rhythmanalysis*, revealing how Kovalevskaya's poetry enacts a dynamic interplay of discontinuity and continuity, imagination and memory, nature and culture—an intricate *assemblage* of time, space, and subjectivity in the making.

By writing poetry, Kovalevskaya materialised Karl Weierstrass' idea that 'the highest point within our science is accessible only to one who is to a certain extent a poet and has prophetic vision and a sense of beauty' (cited in Kochina 1985, 118), but she did it through the support of a network of significant others, including Gösta Mittag-Leffler, who first invited her to Stockholm University, as we have seen in previous chapters. It was in Stockholm that Kovalevskaya met Mittag-Leffler's sister, Carlotta, with whom she co-wrote a theatrical play, *The Struggle for Happiness*, first published in Swedish in 1887 as *Kampen för Lyckan* and later in Russian in 1892 (Kovalevskaya 1974). The concept, plot, and characters originated from Kovalevskaya, who narrated the story to Leffler, who would then draft the text. The play is divided into two contrasting parts: *How It Was* and *How It Might Have Been*, exploring different outcomes for the same characters based on their adherence to societal norms versus following their personal desires.

The play was performed in Sweden, and the second part was staged in Moscow at the Korsh Theatre (see Kochina 1985, 237). In *How It Was*, Karl, an engineer, avoids proposing to Alisa, a sensitive and idealistic young Swedish aristocrat, due to societal and financial pressures, leading Alisa to marry her cousin Hjalmar, with tragic consequences. In *How It Might Have Been*, Alisa defies convention, divorces Hjalmar, modernises the factory she had inherited, and ensures workers' security by making them shareholders. Kochina has observed that while idealistic, the play is significant as one of the first to depict workers on stage (237), reflecting Kovalevskaya's progressive views and her engagement with social and economic themes through personal and societal conflicts. Here, in the opening scene of the first drama, we see the protagonist, Alisa, returning home from university studies, full of dreams but already feeling the pressure of her family to fulfill her purpose as a dutiful daughter of the aristocracy.

> MDM SELEN (*emerges from the castle, turns to Aunt Amelia and points towards Alisa)*
>
> What a lovely picture! Everyone's so happy to be welcoming our precious Alisa back home again.
>
> AUNT AMELIA (*seems to be thinking before speaking in a meek, mournful voice)*
>
> Of course- It's wonderful! Our precious girl has always brought such life to this house! She has so many plans- plans to organize schools, public lectures, and other projects to help the workers.
>
> MDM SELEN
>
> I heard she performed brilliantly on the university exams.
>
> AUNT AMELIA
>
> Yes, she has a good head on her shoulders. She's like her father. And since the good Lord hasn't seen fit to give my dear brother a son, it's a good thing that at least his daughter …
>
> MDM SELEN
>
> So, she will be continuing with classes at Uppsala?
>
> AUNT AMELIA (*smiling and casting a glance at Yalmar who is coming towards her)*
>
> Oh, I am hopeful that something very different could be happening in the near future.
>
> (in Coleman 2021, 75)

The extract above reflects the interplay of continuity and discontinuity within the narrative, as dreams of the young clash with the expectations of the older generation in the aristocratic space of Gerrrgamra Castle. The mansion serves as a rhythmic focal point, imbued with the past's permanence yet disrupted by the youthful energy of Alisa's ambitions. Aunt Amelia's cautious optimism, contrasted with

Alisa's idealistic plans for schools and lectures, captures a rhythm of intergenerational tension—one rooted in the aristocracy's traditions yet challenged by Alisa's modern vision. Bachelard's concept of the *instant* helps us understand these clashes as concentrated moments where the rhythm of the past meets the potential for rupture and transformation. The mansion becomes a temporal node where these forces intersect, reflecting the larger narrative rhythm of continuity and discontinuity. Through the lens of *narrative rhythmanalysis*, the dialogues and settings reveal a rhythmic negotiation of time and space, shaped by tensions between personal aspirations and societal constraints. This dynamic resonates with both Bachelard's (2013) notion of the 'solitary instance', but erupt nevertheless in the Bergsonian durée (2001) of 'the struggle for happiness':

> ERNEST.
>
> There seems to be some kind of magnetism in the air tonight. On a night like this, I think it would be impossible for any person with a drop of blood in his vein not to be a little bit enamoured.
>
> ALISA. (*runs up to them and embraces Paula*).
>
> What a wonderful night! Even the air! How intoxicating! How perfectly beautiful the sky and the hills, there in the distance! Oh! Paula! It's so good-so completely wonderful just to be alive!
>
> (in Coleman 2021, 88)

The second extract highlights an embodied immersion into Bergsonian *durée*, where Alisa and Ernest attune themselves to the rhythm of the intoxicating night. The beauty of the sky and hills, combined with the visceral sensation of being alive, creates a rhythmic interplay between external nature and internal emotion. As Bergson suggests (2001, 90), comprehension of life requires relaxing one's own rhythm and diving into the polyrhythmic reality of duration—a method achieved through intuition. In this scene, Alisa and Ernest abandon their mundane concerns and align themselves with the rhythms of the natural world, experiencing fleeting moments of existential joy. These moments, while transient, offer a glimpse into the harmony of life's continuous flow.

Rutgeerts observes (2019, 88) that for Bachelard and Bergson, concrete duration arises from the relationship between dynamic forces (youthful dreams, love, nature) and resistance (tradition, societal expectations). The narrative rhythm of *The Struggle for Happiness* embodies this dynamic, as characters navigate the tension between their aspirations and the constraints of their environment. *Narrative rhythmanalysis*, in this context, reveals how these forces shape the characters' experiences of time and space, allowing us to trace the oscillation between moments of resistance and release. Both extracts highlight how *The Struggle for Happiness* uses rhythm as a narrative device to explore the interplay between continuity and disruption, past and future, resistance and possibility. The mansion becomes a site where temporal rhythms collide, and the beautiful night offers a fleeting respite from these tensions. By situating the characters within these rhythmic flows, the play mirrors Bachelard's and Bergson's insights into the polyrhythmic nature of existence, where life unfolds as a dynamic interplay of forces that momentarily align in the rhythm of happiness.

The closing scenes of the first play in *The Struggle for Happiness* bring together space, time, and music in a rhythmic crescendo, as Alisa and Yalmar envision the *Waterfall Symphony*. This musical narrative, inspired by the waterfall on their estate—a symbol of both creative force and failed ambition—represents their attempt to reconcile the emotional and financial collapse of their arranged marriage. The waterfall, once a source of energy for the factory, becomes the metaphorical and literal foundation for their imagined symphony, blending the rhythms of nature, industry, and human aspiration:

ALISA

A musical narrative—'The Waterfall'—our waterfall, of course! It's a great idea! First you would need to portray it in primitive times, flowing freely in a wide, rough stream from the region of eternal snows-from the glaciers [...]

ALISA (continuing)

So, then, the construction of the factory begins—dams are constructed and the waterfall is forced to flow through a certain channel.

YALMAR

Yes, of course, and a deafening grumble from the waterfall!

ALISA

And the scratch of a saw, sharply and disharmoniously rushing to join the powerful roar of the water!

YALMAR

A saw?? Yes, it seems right! You are a genius (*He embraces her)* What wonderful discords could be created on this theme!

[...]
YALMAR

Then the resistance of our waterfall will end, as all the other streams merge into a powerful chorus, audible from everywhere-from the ends of the earth! Then our waterfall will not want to be left behind. Its voice will be heard and will sound, at last, with increasing force, until its roar finally covers the entire orchestra with its high silvery tone. Wouldn't this be an excellent finale?

(in Coleman 2021, 138)

The *Waterfall Symphony* encapsulates what *narrative rhythmanalysis* reveals as the interplay of continuity and rupture, harmony and discord. Drawing on

Bergson's and Bachelard's philosophies, the symphony becomes a framework for understanding how space and time interact dynamically in the characters' lives. Bergson famously critiqued the spatialisation of time, rejecting the idea of discrete, compartmentalised moments and proposing instead that time unfolds as a *durée*—a qualitative, continuous flow akin to the experience of melody: 'we have here to do with a synthesis which is, so to speak, qualitative, a gradual organization of our successive sensations, a unity resembling that of a phrase in a melody' (2001, 111). In this context, Alisa and Yalmar's symphony reflects an attempt to synthesise their fragmented past into a harmonious temporal narrative. The sounds of the waterfall—its primitive roar, industrial hum, and eventual symphonic crescendo—become metaphors for the multifaceted rhythms of their lives, shaped by the estate, the factory, and the struggles that defined their relationship.

Bachelard further extends this idea in *The Dialectic of Duration* (2001), where he emphasises that continuity in time is constructed through rhythm. For Bachelard, rhythm and melody are not immediately apprehended but are gradually recognised as a 'reflux of impressions flowing from the present to the past' (110). Alisa and Yalmar's symphony, with its interplay of individual elements and collective orchestration, mirrors Bachelard's concept of rhythms as layered and interdependent (111). The symphony is not just a reflection of their personal history but also a rhythmic narrative of the estate's fortunes and misfortunes, encapsulating the continuity of the family's legacy.

The dynamic interplay of space and time in the *Waterfall Symphony* challenges traditional notions of their separation. While Bergson critiques space for distorting time, the waterfall—a spatial feature tied to the estate's identity—embodies both dynamism and synthesis. In Kovalevskaya's narrative, space becomes a living force, shaping and being shaped by temporal rhythms. The waterfall's evolving role—from natural wonder to industrial tool to artistic inspiration—demonstrates how space and time converge in the existential rhythms of the characters' lives. The rhythm of the symphony also reflects Alisa and Yalmar's shifting emotional states. Initially constrained by societal expectations and financial pressures, they find fleeting harmony in their collaborative creative vision, immersing themselves in the rhythms of memory and imagination. The symphony becomes a temporal *assemblage*, where past struggles and future aspirations coalesce in the present moment of creative expression.

The *Waterfall Symphony* finally echoes elements of Kovalevskaya's own life, particularly her 'fictitious' marriage, initially arranged to secure her freedom to study in Europe. Like Alisa, Kovalevskaya experienced moments of personal and professional harmony—achieving academic success and welcoming the birth of her daughter—before her marriage ended in ruins and tragedy with her husband's suicide. Later, during her years as a professor, she found new love with Maxim Kovalevskii, whose villa on the Riviera inspired her unfinished novella, *A Fragment of Romance on the Riviera*.

> With a sharp and shrill whistle, the train took off from a tunnel near Genoa and rushed along the coast of the Mediterranean Sea towards Ventimiglia. It was the end of January; the morning was lovely and warm. On the blue background of the sky white lambs roamed, and the whole vicinity was constantly changing character and colour; when the sun went behind a cloud, the sea suddenly darkened and twitched with a grey haze; but in a minute it sparkled again and took off; rocks monotonously steel and surrounding mountains took on a pink hue. Grayish green shimmered with silver. The sea was stained suddenly with countless different colours. Along the banks, where the ground was sandy, there was a pale bluish stripe; a little further away it turned into emerald green. Even further, the presence of an underwater rock was detected with a whole row of small white scallops, and behind a large sail long, glittering, golden ripples stretched across a ship (Kovalevskaya 1951, 168).

This evocative passage from Kovalevskaya's novella captures the rhythms of space and time as the train races along the Mediterranean coast. The spatio-temporal fluxes of the landscape—the interplay of shifting sunlight, the shimmering sea, and the changing hues of rocks and vegetation—merge with the mechanical rhythm of the train, creating a layered narrative that invites rhythmanalytical interpretation. Through the protagonist Marya Nikolaevna's journey, Kovalevskaya weaves together personal, natural, and mechanical rhythms into a dynamic tapestry of sensations and reflections.

Indeed, the monotone and staccato repetition of the train engine acts as an amelodic rhythm, a steady backdrop that induces a state of self-forgetfulness in Marya, allowing her to immerse herself in the unfolding scenery. This echoes Bergson's idea of rhythm as a 'vector of hypnosis' (Corbier 2012, 298), where repetition lulls the individual into experiencing time as continuous *durée*. Recall that, for Bergson, rhythm facilitates a synthesis of successive impressions, turning discrete moments into a unified melody of perception. In this passage, the rhythmic motion of the train carries both Marya and the reader into a state where the external world and internal emotions align, creating a rhythm of interconnectedness.

As readers, we too are drawn into this rhythmic flow. The train's steady beat, combined with the vivid descriptions of the Mediterranean landscape, blurs the boundaries between individual strokes of the engine and the fluid continuity of the journey. This rhythmic interplay mirrors the melodic continuity Bachelard associates with lived experience, where temporal rhythms are not immediately apprehended but gradually recognised as impressions flow from the present into the past. In thus following the rhythm of the train, we trace connections between Alisa, Marya, and also the author herself, in the dreams of well-educated young women to be of service to the world, while also falling in love and feeling intensely the vibes of nature. In the extract below, the protagonist of the novella meets a well-known figure of the Russian intelligentsia, Mikhail Mikhailovich Zvantsev, on the train, and they start a conversation about her plans, but also about the beauties of the Riviera:

Oh, how good it is here! What a wonderful place! —constantly exclaimed Marya Nikolaevna. 'You know, I'm even scared that I'm spoiling myself here? But after that I will have to go to the wilderness, to the Saratov province, to take up my position as a country teacher! she suddenly said.

Yes, Saratov, of course, is not like the Riviera. Only why are you taking a position this far! Really you did not find a place in Petersburg? asked Zvantsev naively, wanting to show his interest.

Marya Nikolaevna looked at him and blushed all over.

I am not going out of necessity, she said, rushing and confused. I finished first the courses and, of course, I could have stayed in Petersburg. And besides, I do not need a position anyway. But it seems to me that every intelligent person can be more useful in the provinces. Here I am with my friend Yulia Ivanovna Rumyantseva, and we are to open a rural school in the Saratov province. There are enough intelligent people in St Petersburg without us.

(Kovalevskaya 1951, 174)

Marya's journey is not merely physical but also symbolic of her aspirations and identity as an educated young woman seeking to make a meaningful contribution to society. Her reflections and her dialogue with Zvantsev reveal her inner rhythm, shaped by both personal desires and social ideals. Her intention to leave the cosmopolitan comforts of St. Petersburg for the rural Saratov province underscores her commitment to aligning her rhythm with that of the wider world, emphasising a duty to serve and uplift others. This rhythm of self-transformation resonates with the broader theme of *narrative rhythmanalysis*. The repetitive yet forward-moving rhythm of the train mirrors Marya's internal struggle to reconcile the expectations of her class with her aspirations for independence and social impact. Kovalevskaya's portrayal of Marya's decision to open a rural school with her friend Yulia Ivanovna reflects the rhythmic negotiation between continuity (her aristocratic roots) and rupture (her embrace of progressive ideals), embodying what I would call the rhythm of becoming.

The character of Zvantsev, a liberal thinker with echoes of Kovalevskaya's lover Maxim Kovalevskii, introduces another layer of rhythm: the interplay of personal relationships and intellectual commitments. Maxim's support for Kovalevskaya's literary work, particularly his role in publishing *The Nihilist Girl*, highlights the dual rhythms that Kovalevskaya inhabited as both a mathematician and a poet. These dual rhythms align with Bachelard's (2001) view of the scientist and the poet as complementary figures, with each embodying distinct yet interdependent modes of engagement with the world. Kovalevskaya's attunement to these dual rhythms speaks to the synthesis of space, time, and subjectivity in her work. The rhythmic intermingling of the train's movement, the changing Mediterranean landscape, and Marya's reflections become entangled components of *narrative rhythmanalysis*, revealing how life emerges through the interplay of diverse rhythms—natural, mechanical, emotional, and social.

Overall, Kovalevskaya's narrative in this unfinished novella intertwines rhythmic layers that go beyond the train's mechanical motion or the beauty of the Riviera. It constructs an *assemblage* of space, time, and subjectivity, where rhythms are both experienced and created. The train journey becomes a metaphor for the larger journey of becoming, as Marya navigates her role in a changing world. Through *narrative rhythmanalysis*, we can trace the connections between the protagonist, the landscape, and Kovalevskaya herself, revealing the profound interplay of continuity, rupture, and transformation that defines both the novella and its author's legacy.

Echoes of the subject: the mathematician and the poet

Bachelard's philosophy of science and imagination intricately weaves together the figures of the scientist and the poet, showing how both realms rely on the mind's capacity to construct worlds within rules, whether logical or aesthetic (see Smith 2016). The scientist is the one who breaks with tradition and engages in rigorous and objective inquiry, which, in the case of Kovalevskaya, was her passionate work in mathematics that eventually led to one of the highest mathematical awards of her times: the Prix Bordin that she received from the French Academy of Sciences in 1888, as we have seen in Chapter 4. But as we have already discussed in this chapter, mathematics went hand in hand with poetry and imagination both for Kovalevskaya and Bachelard.[8] 'Being a poet means multiplying the temporal dialectic and refusing the easy continuity of sensation and deduction; it means refusing catagenic repose and welcoming a repose that vibrates, a psyche that vibrates', Bachelard wrote in *The Dialectic of Duration* (2001, 119). In further making a distinction between 'thought poetry' and 'poetry that is spoken', Bachelard discussed their interrelation, pointing out that 'the echo will reveal the voice that lies deep' (119). We listen to the echo of this hidden voice through our bodily rhythm, since 'the rhythm we think is our starting point for organising the rhythm we hear, and not the reverse', he argued in line with his phenomenological position (119).

This idea is mirrored in the reflections of Valéry, whose conception of the self aligns with Bachelard's focus on the self as a constantly shifting entity. For Valéry, the self is shaped by the reverberations of experience, and time, rather than being a fixed external measurement, transforms the self through its rhythms, pulses, and echoes: 'Perhaps there is in us a slow, periodic memory, deeper than the memory of impressions and objects, a long-term memory or *echo* of ourselves which carries us back and unexpectedly restores to us our former inclinations, powers, and even hopes of long ago' (1985, 133). Both Bachelard and Valéry then place significant emphasis on how external forces—whether rhythm, sound, or the flow of time—impact subjectivity, with Valéry viewing these forces as internalised echoes that form the very substance of the self. This relational view of the self, perpetually shaped by the external, invites us to reconsider the traditional separation between the arts and sciences.

Valéry himself bridges this divide, remarking that the poet and the mathematician are united by the same intellectual and emotional labour, each seeking beauty through their respective mediums, whether through a line of verse or a mathematical proof. 'The transition from the non-mathematical to the mathematical has always excited me—it is the passage between any clear idea and *mental action*, that is, between conceiving or seeing and being able to act', he writes in the *Cahiers* in the section on mathematics (1974, vol. 2, 814, emphasis in the text). This passage reveals his fascination with the interplay between abstraction and action in the realm of mathematical thought. He describes the transition from the 'non-mathematical to the mathematical' as an exciting process that hinges on *mental action*—the ability not only to conceive an idea but to act upon it, transforming it into something operable. This broadens our understanding of creativity, uniting disciplines that are often seen as disparate. Valéry thus positions the poet and the mathematician not as separate figures but as two manifestations of the same imaginative and disciplined mind.

Valéry's perspective provides a compelling viewpoint for understanding Kovalevskaya's sense of herself as both a mathematician and a writer. Her ability to navigate between the logical rigour of mathematics and the imaginative freedom of literature reflects the interplay Valéry describes. Kovalevskaya, like Valéry's mathematician, moved beyond mere ideas to construct elegant systems—whether mathematical proofs or literary narratives—that reveal the creative unity of thought across disciplines. Her own reflections reinforce this connection between poetic and mathematical thought. This is what she wrote to the Russian writer Aleksandra Stanislavovna Shabelskaya in the autumn of 1890, in response to a fan letter she had received from her:

> I understand that you are surprised that I can pursue both literature and mathematics at the same time. Many people who have never been presented with an opportunity to know more about mathematics confuse it with arithmetic and believe it to be a dry and arid science. In essence, though, it is the science that demands the greatest imagination, and one of the most respected mathematicians of our century says completely correctly that you cannot be a mathematician without at the same time being a poet in your heart. Only, naturally, in order to understand the truth in this pronouncement, you need to rid yourself of your old prejudice that a poet has to compose something imaginary, that imagination and invention are one and the same. I think that a poet should see what others don't see, see more deeply than others. Which is what a mathematician should also do (Kovalevskaya 1951, 314).

Kovalevskaya's words reveal a profound alignment between the roles of the poet and the mathematician, as both must perceive the world differently, through a deeper, more imaginative lens. In her literary works, Kovalevskaya exemplifies the soul of a poet, seeing what others cannot. This oscillation between the figures of the poet and the mathematician positions her as a 'rhythmic knot' (Lacoue-Labarthe 1989, 202), she is indeed rhythmed between mathematics and literature. As she wrote to Shabelskaya in the letter above, throughout her life she was unable to decide whether she had more aptitude for mathematics or literature:

> Just when my head starts getting tired over purely abstract speculations, it immediately begins gravitating towards observations of life, to stories and, on the contrary, at other times everything in life suddenly starts seeming petty and uninteresting, and only the eternal, immutable laws of science appeal (Kovalevskaya 1951, 314).

Ultimately, Kovalevskaya emerges as a figure who bridges philosophical insights from Bergson, Bachelard, Deleuze and Guattari as well as Lacoue-Labarthe and Valéry, her life and work embodying the dynamic construction of meaning through rhythms, echoes, and refrains. Indeed, her mathematical achievements reflect the structured elegance Valéry attributes to the mathematician's mind, transforming abstract ideas into profound creations. Simultaneously, like Bachelard's poet, she embraced the vibrant interplay of intuition and formalism, disrupting linear thought to cultivate imaginative insights. Reflecting Lacoue-Labarthe's conception of the subject, Kovalevskaya moves within the resonant interplay of rhythm and reverberation, navigating the tension and synergy between her roles as mathematician and poet. Deleuze's and Guattari's notion of the refrain deepens this understanding, as Kovalevskaya's work acts as a repeated yet transformative motif, carving out territories of meaning while remaining open to new possibilities. And through Bergson's metaphysics of time and rhythm, we see her creative endeavours as expressions of durée—the lived, dynamic, and ever-evolving flow of time, in which intuition and rhythm merge to shape the self.

Thus, the rhythms of Kovalevskaya's literary writings become essential to hearing her voice, for without them, her echoes risk being lost amid the myths and dominant discourses that often obscure historical figures in the archives and beyond (see Tamboukou 2022). Lacoue-Labarthe reminds us that rhythm is fundamental to the self, its absence threatening the subject's coherence (1989, 195). In Kovalevskaya's case, her sense of self oscillates between the poetic and the mathematical, hesitating and undecidable. Her writings—both literary and mathematical—offer a unique means to attune to this oscillation, allowing us to hear the subtle interplay of these roles. Through *narrative rhythmanalysis*, we can trace the echoes of Kovalevskaya's self, emerging not from a fixed identity but from the dynamic interaction of her creative pursuits. Despite Valéry's caution that 'there is always some fraction of meaning, some delicate or faraway echo that escapes us' (1985, 276), the interplay of rhythm, refrain, and durée allows us to glimpse Kovalevskaya's voice as it resonates through her work. Her echoes emerge not from a fixed position but from the dynamic convergence of her creative endeavours, where mathematics and literature blend to form a harmonious and evolving expression of the self. Through these entangled perspectives, Kovalevskaya's life becomes a testament to the rhythmic and temporal forces that underpin human creativity, illuminating the dynamic interplay between science, art, and philosophy.

Notes

1 See Kovalevskaya 1951, 1974, 1978 [1895]; Coleman2021.

2 Rutgeerts (2019) offers a very comprehensive overview of the 'rhythmanalysis' literature.
3 See Rutgeerts 2019 for a critical overview of Bergson's and Bachelard's take on rhythm. See also Rizo-Patron 2017; Worms and Wunenburger 2008; Smith 2016.
4 The quote from Bachelard is from *La dialectique de la durée,* cited in Deleuze and Guattari 1988, 547n4).
5 See amongst others, Moulard-Leonard 2008; Robinson 2009; Pearson 2002.
6 See Wallbank 2019 for a comprehensive overview of this literature.
7 I draw here of course on the Foucauldian notion of 'heterotopia' that I have discussed at length, elsewhere in my work (see Tamboukou 2000, 2012).
8 Here it is important to note that there is a rich body of literature exploring links and entanglements between poetry and mathematics, most notably in the philosophical work of Alain Badiou and Barbara Cassin (see Hickmott 2020). The genealogy of the poetry/mathematics entanglement goes back, of course, to the Romantic tradition and mostly in Novalis (see Taylor 2024).

References

Bachelard, Gaston. 2000 [1936]. *The Dialectic of Duration.* Translated by Mary McAllester Jones. Manchester: Clinamen Press.

Bachelard, Gaston. 2013 [1932]. *Intuition of the Instant.* Translated by Eileen Rizo-Patron. Evanston: Northwestern University Press.

Barad, Karen. 2003. 'Posthumanist Performativity: Toward an Understanding of How Matter Comes to Matter'. *Signs: Journal of Women in Culture and Society* 28 (3): 801–31.

Baranova, Jūratė, Laura Junutytė, and Lilija Duobliene. 2016. 'Introduction'. In *Rhythm and Refrain: In Between Philosophy and Arts*, 9–22. Vilnius: Lithuanian University of Educational Sciences.

Benveniste, Emile. 1973. 'The Notion of "Rhythm" in Its Linguistic Expression'. In *Problems in General Linguistics,* edited by Miami Linguistic, vol. 8, 281–8. Miami: University of Miami Press.

Bergson, Henri. 2001 [1889]. *Time and Free Will: An Essay on the Immediate Data of Consciousness.* Translated by Frank, L. Pogson. Dover Publications.

Bergson, Henri. 2002 [1896]. *Matter and Memory.* Translated by N. M. Paul and W. S. Palmer. Zone Books.

Bergson, Henri. 2005 [1907]. *Creative Evolution.* Translated by Arthur Mitchell. New York: Cosimo Classics.

Bergson, Henri. 2008 [1911]. *Laughter: An Essay on the Meaning of the Comic.* Translated by Cloudesley Brereton and Fred Rothwell. Rockville: Arc Manor.

Cheyne, Peter, Andy Hamilton, and Max Paddison. 2019. 'Introduction: Philosophy of Rhythm'. In *The Philosophy of Rhythm*, edited by P. Cheyne, A. Hamilton, and M. Paddison, 1–12. Oxford: Oxford University Press.

Coleman, Sandra DeLozier. 2021. *Mathematician with the Soul of a Poet: Poems and Plays of Sofia Kovalevskay.* Bohannon Hall Press.

Corbier, Christopher. 2012. 'Bachelard, Bergson, Emmanuel: Mélodie, rythme et durée'. *Archives de Philosophie* 75 (2): 291–310.

Deleuze, Gilles. 1991. *Bergsonism.* Translated by Hugh Tomlinson and Barbara Habberjam. New York: Zone Books.

Deleuze, Gilles, and Felix Guattari. 1988. *A Thousand Plateaus.* Translated by Brian Massumi. The Athlone Press.

Derrida, Jacques. 1989. 'Introduction'. In *Typography: Mimesis, Philosophy, Politics*, edited and translated by Christopher Fynsk, 1–42. Cambridge, MA: Harvard University Press.

Hickmott, Sarah. 2020. *Music, Philosophy and Gender in Contemporary French Thought: Nancy, Lacoue-Labarthe, Badiou.* Edinburgh: Edinburgh University Press.

Koblitz, Ann Hibner. 1988. 'Science, Women, and the Russian Intelligentsia: The Generation of the 1860s'. *The History of Science Society* 79 (2): 208–26.

Kochina, Pelageya. 1985. *Love and Mathematics: Sofia Kovalevskaya*. Translated by Michael Burov. Mir Publishers.

Kovalevskaya, Sofia. 1951. *Воспоминания и письма* [Memories and Letters], edited and annotated by Solomon Yakovlevich Shtraikh. Moscow: USSR Academy of Sciences. [2nd ed. 1961].

Kovalevskaya, Sofia. 1974. *Воспоминания* повести [Recollections, Novellas]. Edited by Pelageya Kochina. Moscow: Nauka.

Kovalevskaya, Sofia. 1978. *A Russian Childhood*. Translated by Beatrice Stillman. New York: Springer-Verlag.

Kovalevskaya, Sofia. 1895a. 'Reminiscences of Childhood [Life in Russia: The Sisters Rajevski]'. In *Sonia Kovalevsky. Biography and Autobiography*, translated by Louise Von Cossel, 167–317. London: Walter Scott.

Kovalevskaya, Sofia. 2001 [1892]. *Nihilist Girl*. Translated by Natasha Kolchevska and Mary Zirin. New York: The Modern Language Association of America.

Lacoue-Labarthe, Philippe. 1989. 'The Echo of the Subject'. In *Typography: Mimesis, Philosophy, Politics*, edited and translated by Christopher Fynsk, 139–207. Cambridge, MA: Harvard University Press.

Lefebvre, Henri. 2004. *Rhythmanalysis: Space, Time and Everyday Life*. Translated by Stuart Elden and Gerald Moore. London: Continuum.

Martis, John. 2005. *Philippe Lacoue-Labarthe: Representation and the Loss of the Subject*. New York: Fordham University Press.

Moulard-Leonard, Valentine. 2008. *Bergson Deleuze Encounters*. New York: SUNY.

Pearson, Keith Ansell. 2002. *Philosophy and the Adventure of the Virtual*. London: Routledge.

Pinheiro dos Santos, Lúcio Alberto. 1931. *Ritmo e Sincronismo: Ensaio de Ritmanálise Geral [Rhythm and Synchronism: Essay on General Rhythmanalysis]*. Companhia Editora Nacional.

Rizo-Patron, Eileen. 2017. *Adventures in Phenomenology: Gaston Bachelard*. New York: Suny.

Robinson, Keith. 2009: *Deleuze, Whitehead, Bergson: Rhizomatic Connections*. New York: Springer.

Roupnel, Gaston. 1927. *Siloë*. Paris: Librairie Stock.

Rutgeerts, Jonas. 2019. 'Revisiting Rhythmanalysis: How Rhythm Operates in the Work of Gaston Bachelard and Henri Bergson'. *Parrhesia* 31: 85–102.

Smith, Roch C. 2016. *Gaston Bachelard: Philosopher of Science and Imagination*. Albany: Suny Press.

Souriau Étienne. 2009 [1956]. 'Du mode d'existence de l'œuvre à faire'. In *Les différents modes d'existence*, suivi de *De l'œuvre à faire*, presented and introduced by Isabelle Stengers and Bruno Latour, 195–217. Paris: Presses Universitaires de France.

Tamboukou, Maria. 2021. 'Narrative Rhythmanalysis: The Art and Politics of Listening to Women's Narratives of Forced Displacement'. *International Journal of Social Research Methodology* 24 (2): 149–62.

Tamboukou, Maria. 2022. 'Traces in the Archive: Re-imagining Sofia Kovealevskaya'. *Life Writing* 19 (3): 341–56.

Valéry, Paul. 1974. *Cahiers* [Notebooks] Vol. 2. Edited by Judith Robinson. Gallimard.

Valéry, Paul. 1985 [1958]. *The Art of Poetry*. Translated by Denise Folliot, with a Introduction by T. S. Eliot. Princeton University Press.

Taylor, Charles. 2024. *Cosmic Connections: Poetry in the Age of Disenchantment*. Cambridge, MA: Harvard University Press.

Wallbank, Rebecca. 2019. 'The Not-So-Silent Reading: What Does it Mean to Say that We Appreciate Rhythm in Literature?' In *The Philosophy of Rhythm*, edited by P. Cheyne, A. Hamilton, and M. Paddison, 362–73. Oxford: Oxford University Press.
Worms, Frédéric, and Jaen-Jacques Wunenburger. 2008. *Bachelard et Bergson: continuité et discontinuité.* Paris: Presses Universitaires de France.

Conclusion

Future Pasts: Charting Assemblages

When I began the project of writing a feminist genealogy of automathographies, my knowledge of the first women mathematicians was surprisingly limited, apart from Hypatia, whose prominence had featured in my classical education. I initially attributed this gap to my lack of scholarly focus on mathematics or its history. However, as I began presenting my work to audiences of historians, cultural theorists, sociologists, and even mathematicians, I discovered that this lack of knowledge was not unique to me.

I recall my first archival visit to the Ambrosiana Biblioteca in Milan in November 2022, and a subsequent dinner in Turin, a city with a rich mathematical tradition. To my surprise, none of the Italian mathematicians present, including a woman professor, were familiar with Maria Gaetana Agnesi, despite her historical connections with mathematics and philosophy professors in Turin. Similarly, during my visits to France over the years, I have found Émilie Du Châtelet largely unknown among French mathematicians, although Sophie Germain's name has held a more familiar resonance. In British circles, Ada Lovelace is somewhat recognised, but her mentor and friend Mary Somerville remains far less well known. Meanwhile, Sofia Kovalevskaya's name has remained familiar among Russian mathematicians, yet she is largely unrecognised more broadly, including in the Nordic countries, where her academic career had flourished. One particularly memorable moment occurred at the Irish Sociological Association's graduate conference in Limerick in February 2024. During my keynote, I introduced Wang Zhenyi, an extraordinary yet little-known Chinese mathematician, as we have seen, only to have Chinese graduate students approach me afterwards to confess that they, too, were unaware of her existence or contributions.

These encounters reveal the pervasive mechanisms of erasure that render women mathematicians invisible across temporal and geographic contexts, even within the very disciplines ostensibly most attuned to their contributions. They underscore the necessity of a genealogical endeavour—a critical excavation of silenced narratives, an unearthing of subjugated knowledges, and a rewriting of intellectual history to disrupt dominant epistemic frameworks. Such a project does not simply restore forgotten figures but interrogates the conditions under which their exclusion was produced, situating these stories within the broader matrix of power, knowledge, and history.

DOI: 10.4324/9781003533993-9

The surprises continued to unfold as I delved deeper into this project. Immersed in the extensive bodies of literature surrounding the individual lives and works of the six women mathematicians, I began to discern a striking absence: even scholars who had rigorously engaged with Du Châtelet, for instance, had rarely, if ever, considered tracing connections with Agnesi. These gaps were not confined to personal relations—which, perhaps unsurprisingly, were absent—but extended to intellectual linkages, comparisons of their work, or analyses of the cultural and geographical contexts of the distinct Enlightenments they emerged from. When I wrote my first paper on the educational trajectories of Du Châtelet and Agnesi (Tamboukou 2025), I was surprised to discover that apart from a graduate thesis in German (Martini 2017), there was no other published literature exploring their connections.

The same silences pervaded the scholarly treatment of Ada Lovelace and Mary Somerville. Despite the rich academic focus on their individual lives and their profound intellectual relationship, no study had explored their extensive correspondence. This omission was particularly surprising given how illuminating their letters are, as I demonstrate in this book and in a related article (Tamboukou 2023a).

Accessibility posed yet another challenge. The auto/biographical, literary, and philosophical writings of these women mathematicians remain largely untranslated in English, with Wang Zhenyi's works unavailable in any European language. As I have discussed at length in Chapter 3, this lack of translation severely limits their accessibility to international audiences, perpetuating their marginalisation. Furthermore, Germain's philosophical writings remain underexplored in contemporary discussions of gender, philosophy, and science, and no comparative study exists examining the philosophical oeuvres of Germain and Du Châtelet. Addressing this gap, I published an article on their treatment of feelings in epistemology (Tamboukou 2023b), yet the broader absence remains glaring.

These gaps are not merely incidental but symptomatic of the broader mechanisms of exclusion and invisibility that have shaped the epistemic frameworks of intellectual history. The omissions highlight the urgent need for genealogical projects that excavate, connect, and recontextualise these women's lives and works. By bringing their writings into dialogue and tracing the intricate networks of ideas, relations, and cultural contexts that shaped them, we not only recover subjugated knowledge but also disrupt the disciplinary boundaries and power structures that have long obscured their contributions.

The greatest surprise I encountered in this project was the process of writing this book. After years of genealogical excavation in the archives of women mathematicians and immersing myself in the extensive literature surrounding their lives and work, I initially believed the book would almost write itself. It did not. While studying each woman mathematician individually provided a solid foundation, and drawing bilateral comparisons between their lives and contributions offered important insights, bringing their lives and works together required a fundamentally different conceptual approach—one that could situate their multiplicities on what Deleuze and Guattari call a *plane of consistency*, 'which knows only speeds and affects' (1988, 262). The 'plane of consistency' refers indeed to a conceptual space where *assemblages* of diverse elements—ideas, events, bodies, affects, and intensities—coexist and interact without reducing their uniqueness or imposing hierarchical structures. On a *plane of consistency* then, connections emerge not through fixed identities or linear relationships but through dynamic interactions that generate new possibilities and open pathways for becoming.

In this light, writing the book was no longer about drawing intersecting lines between distinct historical moments or conditions of possibility, such as the differing Enlightenments of Du Châtelet and Agnesi or Lovelace and Somerville's disruptions of British mathematical traditions. Nor was it merely about situating Germain and Kovalevskaya's achievements in the context of their prestigious prizes from the French Academy of Sciences. Instead, bringing these figures together meant constructing an *assemblage*—a dynamic configuration of mathematics, philosophy, and the literary arts—that revealed not only their historical contexts but also the untapped potentialities of their intellectual trajectories.

In Deleuze's and Guattari's (1988) image of thought, *assemblages* are traversed by relations of interiority and exteriority. Internally, they are configured by the ways their components connect—whether through shared ideas, practices, or material conditions. Externally, they form unexpected alliances and connections with other *assemblages*, sparking new intensities and unanticipated *lines of flight*. By suggesting the idea of 'the exteriority of relations', Deleuze has argued that things are conditioned, but not determined by their relations, that relations may change without their constituent components changing and that, finally, entities and subjectivities are never fully realised within their relations (Deleuze and Parnet 2002, 41). Thus, to assemble the lives and works of these women mathematicians on a *plane of consistency* meant opening a new world where past possibilities and future potentialities coexist, allowing their multiplicities to resonate without subsuming them into a singular narrative.

What unfolded was not a neatly ordered history, but a dynamic cartography of intersecting forces—a network of connections that transcends disciplinary boundaries and invites fresh perspectives on the entanglements of mathematics, philosophy, and art. Each chapter in this book has been conceived as a *plane of consistency*, weaving together themes that have emerged from assembling lines of thought drawn from the life documents of women mathematicians, alongside literary and philosophical writings. In this conclusion, I aim to trace some of these lines that traverse the becomings of women mathematicians, moving beyond genealogy to explore the transformative potential of the *assemblage*.

Assemblages of class and gender

Thinking with the idea of forming *planes of consistency* in the overall complexity of the themes, differentiations, and variations that have emerged from this study, all women mathematicians came of age during the rise of 'public science' and benefited from the educational opportunities afforded to daughters of the upper classes to which they belonged, albeit with important differentiations. Du Châtelet, Lovelace, and Kovalevskaya were aristocrats; Agnesi and Germain were daughters of wealthy merchants; and only Somerville hailed from the minor gentry, what we might call the middle class today. Yet, it was in how they positioned themselves within and beyond their class privileges that a paradigmatic network of relations of exteriority emerged, shaping their unique contributions to mathematics, philosophy, and the literary arts.

Émilie Du Châtelet, born into the French aristocracy, was educated and groomed for court life, excelling in its entertainments and intrigues. Yet, despite her role as a courtier, she defied conventions by becoming a leading intellectual of the French Enlightenment. Her engagement with mathematics, physics, and philosophy extended far beyond the expected pursuits of a salonnière, and she leveraged her privileges to secure some of the finest mathematical tutors of her era, while maintaining scientific correspondence with prominent mathematicians across Europe. Her ability to connect with the broader intellectual currents of her time exemplifies relations of exteriority, as she moved beyond the confines of her social position to contribute to a wider *assemblage* of scientific knowledge.

Moreover, the ethical norms of love and marriage among the aristocracy in eighteenth-century France provided Du Châtelet with the latitude to live her relationship with Voltaire openly. Their partnership was not merely personal but also deeply intellectual, encompassing literature, drama, and science. This interplay between her aristocratic milieu and her intellectual pursuits highlights how relations of exteriority allowed her to transform her class position into a platform for engaging with and influencing the broader Enlightenment *assemblage*.

Lovelace's aristocratic status similarly enabled her to pursue her passion for mathematics, even within the constraints of Victorian Britain's restrictive ethics. Unlike Du Châtelet, who had greater autonomy within the freer social norms of her era, Lovelace faced stringent financial controls imposed by her mother and husband, which limited her ability to fully realise her intellectual ambitions. Despite her title as Countess, she was frequently infantilised by those closest to her. Her mother and husband referred to her affectionately but patronisingly as their 'bird', as reflected in the 'Dear bird' salutations in her husband's correspondence. The shadow of her father, the Romantic poet Lord Byron, added another layer of constraint, his dark legacy shaping public perceptions of her and weighing heavily on her psychosocial status. This legacy seemed to stifle her creative potentials, serving as both an inspiration and a burden she struggled to reconcile throughout her life.

Yet, Lovelace's enduring intellectual relationship with Charles Babbage demonstrates the power of relations of exteriority in her life. Operating within and against the confines of her social position, she forged a connection that allowed her to contribute to one of the most visionary scientific projects of her time. The very fact that she cultivated and sustained such a relationship in a society that sought to circumscribe her intellectual pursuits speaks to her determination, *agonistic stance*, and passion for mathematics. These external connections became the means by which she transcended the restrictive *assemblage* of her aristocratic and domestic roles, reinventing herself as a thinker and innovator.

Kovalevskaya, a daughter of the Russian aristocracy, was deeply entangled with the radical generation of the 1860s, a cohort envisioning transformative futures for social and gender relations (see Koblitz 1988). Her marriage functioned as a pragmatic arrangement—a radical contract designed to circumvent societal constraints and enable her immersion in the pursuit of mathematics. Operating within the destabilised ethical framework of the Bakhtinian *carnivalesque*—a concept that subverts and temporarily overturns established hierarchies and norms—she used the relational flexibility afforded by her role as a 'wife' to explore new configurations of becoming within and beyond the established norms of filial duty, marriage, and motherhood.[1]

Kovalevskaya's life unfolded through the interplay of relations of interiority—her aristocratic lineage and its associated expectations—and relations of exteriority, including her engagement with radical movements, international academic networks, and intellectual exchanges. These relations were not stable or fixed but instead dynamic and productive, continually reshaping her trajectory. While not fully embodying the radicalism of her heroine in *The Nihilist Girl*, Kovalevskaya engaged with the political movements of her time, participating in the shifting *assemblages* of social and gender equality that defined the intellectual and political milieu of the nineteenth century. Her academic career, marked by her position as the first tenured woman academic in Europe, was the effect of these complex entanglements. As a single mother in Sweden, she disrupted conventional expectations of motherhood, even within the progressive intelligentsia, demonstrating the fluidity of the *assemblages* through which she navigated.

Kovalevskaya's life cannot be reduced to her aristocratic origins or her radical aspirations. Indeed, she remained an 'outsider within' the Russian aristocracy, resisting full alignment with its values while never entirely escaping its influence. Her subjectivity emerges as a site of ongoing reconfiguration, where relations of exteriority allowed for connections and transformations that deterritorialised her position within her inherited position. Always on the move between *assemblages*—aristocracy, revolutionary politics, mathematics, the literary arts, and cosmopolitan intellectual circles—she embodied a multiplicity of becomings that resisted fixed trajectories or singular narratives.

Agnesi was not born into the elite circles of aristocracy; her family's wealth originated in the silk trade of Milan, a foundation rooted in material success rather than inherited privilege. Her father, however, harboured aspirations that extended beyond material accumulation. He envisioned social ascent not through wealth alone but through the cultivation of intellectual distinction. His daughters became instruments of this ambition, their talents leveraged for the realisation of his vision. Within this *assemblage*, Agnesi's role was defined by relations of exteriority: her intellectual abilities were projected outwards, fashioned into public performances to create a spectacle of brilliance for societal consumption. Music became the domain of her sister, while Agnesi was cast as an intellectual prodigy. Her achievements were thus externalised, transformed into markers of her family's imagined trajectory, leaving little space for her to develop an interiority independent of these demands.

Yet, within the exteriority imposed upon her, Agnesi began to reshape the relations constituting her *assemblage*. The public display of her intellect, initially a spectacle, became the means through which she negotiated a new interiority. Rather than rejecting the exterior forces shaping her life, she worked within them, carving out spaces for intellectual and spiritual exploration. It was in this reconfiguration—this boundary between exterior spectacle and interior retreat—that Agnesi turned to mathematics. No longer a mere tool for external validation, mathematics became her refuge, a means to engage deeply with questions of truth, divinity, and existence beyond the gaze of societal expectation.

In Deleuze's and Guattari's terms, Agnesi's journey illustrates the dynamics of *becoming imperceptible* (1988, 232). Her intellectual pursuits marked a *line of flight*—a movement away from the rigid *assemblage* of social performance and towards a fluid space of thought and being. This shift was not a simple withdrawal from the world but a transformation of her relational *assemblage*. Mathematics allowed her to navigate the tensions between the interior and exterior, creating a space where she could think and act beyond the roles imposed upon her. As Tamsin Lorraine notes, in becoming imperceptible, 'you allow lines of flight to release you both from the central reference of a self as well as the organizing focus of coherent objects' (2000, 191). For Agnesi, this meant transcending the externalised image of the prodigy and embracing a life defined not by visibility but by quiet, profound engagement with her chosen path.

Following her father's death, Agnesi gained the freedom to determine the trajectory of her life. Yet, despite her intellectual prominence and the recognition she had garnered, she lived her later years in poverty, dedicated to serving the destitute. Her death, marked by burial in an unmarked grave, became the ultimate expression of her *becoming imperceptible*. This act of literal and metaphorical disappearance did not signify resistance but a dissolution into a realm of exteriority beyond societal recognition. Agnesi's life traced a delicate interplay of interiority and exteriority, revealing how one might reconfigure an *assemblage* not through rejection, but through transformation—moving beyond the constraints of imposed visibility to exist quietly, fluidly, and profoundly unmarked.

Germain, the daughter of a wealthy merchant, belonged to the liberal and educated bourgeoisie of late eighteenth-century France. As we have seen in previous chapters, her father, Ambroise-François Germain, was a prominent figure in the political landscape, serving as Deputy of the Third Estate for Paris and later as a member of the National Constituent Assembly (Stupuy 1896, 3). Positioned at the intersection of privilege and revolutionary ideals, one might assume that Germain's path to mathematics would be unobstructed in the newly proclaimed era of *liberté, égalité, fraternité*. Yet, the French Revolution, despite its sweeping declarations, failed its women. Germain's family, the post-revolutionary educational system, and the newly established academic structures collectively denied her formal access to the mathematical world.

Within this *assemblage* of exclusion, Germain's journey highlights the tension between relations of interiority and exteriority within and between *assemblages*. Interiority in Germain's case refers to the systems—family, education, and academia—that sought to define her capacities from within their rigid frameworks, excluding her on the basis of gender. Exteriority, by contrast, captures her engagement with spaces and connections beyond these constraints, enabling her to reconfigure her position within the mathematical *assemblage*. To bypass the barriers erected by her milieu, Germain assumed the male pseudonym 'Monsieur LeBlanc' to access lectures at the newly formed École Polytechnique. This act of gender masquerade was not merely subversive but indicative of her effort to escape the interior forces that sought to delimit her intellectual potential.

Despite living in Paris, at the heart of revolutionary change and the burgeoning scientific community, Germain remained institutionally and psychosocially marginalised. Yet, her persistence allowed her to form alliances within the French Academy—mathematicians like Legendre and Fourier who advocated for her work—and connections with external networks, reaching beyond national boundaries to figures such as Gauss in Germany and Libri in Italy. These interactions illustrate the interplay of relations of interiority and exteriority in Germain's *assemblage*: while the institutional systems constrained her, her ability to connect with external *assemblages* of mathematicians enabled her to transcend those limitations.

Her *agonistics of knowledge*, as discussed in previous chapters, enabled her to bridge these two realms. Through her correspondence with Gauss, initiated under the guise of 'Monsieur LeBlanc', Germain not only found recognition but began to etch her name into the mathematical discourse of her time. Her work on number theory and elasticity unfolded through a delicate balance between the restrictive interiors of Parisian academia and the expansive exteriors of transnational intellectual exchange. In the end, Germain's journey is not merely one of *agon* and perseverance but a demonstration of the dynamics of *assemblages*. She operated within the folds of an *assemblage* that sought to marginalise her, yet reconstituted it by forming new relations that transcended its limitations. In doing so, she charted a path to 'the unreachable star',[2] securing her place in the constellation of French mathematics—not as an isolated figure but as one whose *agonistics* and connections redefined the boundaries of inclusion and recognition in a deeply stratified intellectual milieu.

Mary Somerville's social origins were rooted in the minor gentry, a class that occupied a distinct position in the British social hierarchy. The minor gentry, while not part of the peerage or the wealthiest echelons of the landed aristocracy, held a privileged status above commoners due to land ownership, familial connections, or professions aligned with the elite. This standing granted them access to networks, education, and opportunities unavailable to the lower classes, yet without the full societal influence and wealth of the greater gentry or aristocracy. Somerville's familial connections illustrate the nuances of this status. Through her mother, she was distantly related to the Earl of Minto, while her father claimed ties to the Barons Fairfax of Cameron in the Scottish peerage. Though her immediate family was not particularly wealthy or prominent within the peerage, her father's knighthood—awarded for his role in the 1797 Battle of Camperdown—bestowed the title of 'Sir', elevating their social standing. This knightly status, combined with her eventual marriage to William Somerville, a physician who ranked among baronets and knights, placed Somerville in a position of modest but meaningful privilege (see Stenhouse 2021a, 26–7).

What is significant and unusual in Somerville's case is the critical role her second husband, William Somerville, played in her scientific career, as Brigitte

Stenhouse has insightfully demonstrated (2021b). At a time when women were excluded from formal scientific institutions, Somerville relied on informal networks and William's active mediation to engage with the intellectual community. His multifaceted contributions—as her chaperone, representative, secretary, and literary agent—enabled her to overcome societal barriers and pursue her ambitions in ways that challenge traditional understandings of patriarchy and heteronormative relations.

Within the *assemblage* of their partnership, William's contributions reveal a dynamic interplay between relations of interiority and exteriority. His domestic roles—managing correspondence, sourcing essential books, and coordinating the logistical aspects of her publications—illustrate interior relations, where the structure of their household was a resource integral to Somerville's intellectual work. Simultaneously, William's external activities as her representative highlight relations of exteriority. His memberships in prestigious societies like the Royal Society, and his connections with influential figures such as Laplace and Babbage, allowed him to bridge the divide between Somerville and the male-dominated institutions that formally excluded her (see Stenhouse 2021b).

As her chaperone, William expanded Somerville's physical and social mobility, enabling her to attend gatherings and travel across Europe, where she engaged directly with leading thinkers. As her literary agent, he managed negotiations with publishers and ensured that her works, including the landmark *Mechanism of the Heavens*, reached a wider audience. These efforts were instrumental in positioning Somerville as a central figure in the circulation of advanced mathematical ideas in Britain, as detailed in earlier chapters.

The nuanced approach of the *assemblage* sheds light on how their partnership defies reductive interpretations of patriarchy. William's active support did not diminish Somerville's intellectual force but rather enabled her to navigate the systemic exclusions of her time. Their relationship, rooted in collaboration, challenges the binary view of heteronormative marriage as inherently restrictive for women. Instead, it underscores how such partnerships can operate as *assemblages* of mutual reinforcement, where interior and exterior relations work in tandem to transcend societal limitations.

Somerville's achievements undeniably reflect her intellectual hard work within the *agonistics of knowledge*. However, they also reveal how relational dynamics within the *assemblage* of her marriage and her standing within the minor gentry created conditions for her success. While her social class provided a foundation for engaging with intellectual life, it was the rare and multifaceted support of her husband that enabled her to emerge as one of the foremost mathematicians and scientists of her era. Their partnership thus exemplifies how a nuanced understanding of interiority and exteriority within *assemblages* can offer fresh perspectives on gender, power, and collaboration in historical contexts.

William's role in Somerville's career stands out, both in its time and today, as an unusual example of spousal support that operated within and beyond traditional gender norms. Yet, his contributions connect to a broader *plane of consistency* across the lives of the women mathematicians discussed in this study: the critical support they received from men. These men—whether fathers, tutors, mentors, friends, lovers, or colleagues—were not merely passive enablers but active participants in *assemblages* that shaped the trajectories of these women's intellectual lives. By focusing on relations of interiority and exteriority within and between these *assemblages*, we can better understand how such collaborations disrupted the myth of homogeneous patriarchal structures, creating spaces for unexpected and transformative possibilities.

For Du Châtelet, the overlapping relationships with Voltaire and her tutor Maupertuis marked the beginning of a new phase in her intellectual life. These relationships, though stormy and fraught with disagreements, formed a plane of intensities that catalysed the emergence of new ideas, as seen in her *Discours sur le bonheur* (1961). The dynamic between collaboration and conflict did not merely sustain her work but enriched it, creating fertile ground for philosophical and scientific innovation.

Agnesi's intellectual development was profoundly influenced by the educational *assemblages* that she was entangled in. Her dedications to mentors such as the Theatine priest Augustino Tolotae in the *Oration* and Count Carlo Belloni in the *Philosophical Propositions*, along with her correspondence with Michele Casati, her philosophy tutor, reflect not only a deep sense of indebtedness to their guidance but also the warmth of friendship. As Louisa Anzoletti (1900, 128) has noted, the emotional tone of her dedication to Belloni offers rare glimpses of affect and gratitude within an otherwise formally composed dedication text. Furthermore, her pedagogical relationship with Ramiro Rampinelli, her mathematics tutor, and her subsequent correspondence with the Riccati family—connections fostered through Rampinelli—were pivotal in catalysing her monumental contributions to calculus, as explored in Chapter 2.

Germain's case illustrates how networks of male mathematicians, especially those associated with the Paris Academy, were indispensable to her mathematical advancement. Yet, her relationship with this male-dominated world was marked by ambivalence. As a single woman navigating these spaces, she encountered both opportunities and resistance. Germain was not merely a mentee; her famous revelation of her identity to Gauss, accompanied by her efforts to secure his safety during wartime, highlights her remarkable capacity to integrate personal care with scientific engagement. This act of protection offers a poignant reminder of how 'the personal and the scientific' are often deeply intertwined, as I have argued in Chapter 2.

Lovelace's relationships with Augustus De Morgan and Charles Babbage were similarly transformative. These men played pivotal roles in her mathematical development, but Lovelace actively reconfigured these dynamics. She shaped these relationships not only as avenues for mathematical synergy but also as platforms for philosophical reflection, speculation, and innovation, carving out a unique intellectual identity within the constraints of her time.

Kovalevskaya's life presents perhaps the most exceptional case in this spectrum of gender relations. From her early encouragement by her father and uncle to the pivotal decision of her fictitious marriage, Kovalevskaya's trajectory was shaped by unconventional alliances. Her doctoral supervisor, Karl Weierstrass, became a central figure in her intellectual journey, providing rigorous mentorship, but also poetic inspiration and warm friendship, while Gösta Mittag-Leffler's efforts to create academic opportunities for her were instrumental in enabling her career to flourish.

Throughout this study, these complex and dynamic gender relations illuminate a crucial insight: rather than reinforcing the homogeneity of patriarchal institutions like the family, state, or academy, these *assemblages* disrupted and reconfigured them. This does not imply the absence of patriarchy but reveals how focusing solely on institutional stratification misses the multiplicity of forces at work. As Deleuze and Guattari argue, societies should not be defined solely by their structures but by what escapes them—the untimely, the inventive forces that open new futures (1988, 216). These *assemblages* of support, conflict, and creativity exemplify how life's relational dynamics hold the potential to exceed societal constraints, forging unpredictable and transformative paths. But what about other women? How do they figure in the *assemblages* of women mathematicians' lives? This is what I consider next.

Rhizomatic connections and the politics of becoming

Mothers often occupied polarities within women mathematicians' lived experiences—either remote and absent or omnipresent and stifling—affecting the flows and intensities of their daughters' becomings in ways that were ambiguous or constraining. Sisters and daughters sometimes formed important nodes of relational continuity, but it was often the rhizomatic connections with women friends that proved the most transformative, creating planes of consistency that sustained and amplified their intellectual trajectories.

As I will further discuss, these kinship relationships and friendships were not static but dynamic, operating as *lines of flight* within patriarchal structures that sought to territorialise intellectual spaces. Women—whether relatives, friends, fellow mathematicians and scientists, or collaborators—formed *assemblages* of solidarity, redistributing affective and intellectual intensities and enabling the exchange of ideas. These networks were not merely supportive—they disrupted hierarchical systems, opening up new pathways for thought and creation. Through these *assemblages*, women mathematicians emerged as singularities within fields otherwise dominated by molar structures of exclusion. Women's networks, then, were catalytic, allowing for a becoming that transcended the limits imposed by gendered norms, fostering the unforeseen and the unthinkable. Yet, these relations were far from uniform and must be considered in their uniqueness and differentiations.

Du Châtelet's mother, Gabrielle Anne de Froullay, Baronne de Breteuil, appears as a distant figure in her life. Du Châtelet's correspondence contains only sparse references to her mother's visits, with no indication of intellectual exchanges or meaningful connections between them. Yet, as Du Châtelet's biographers note, Froullay was renowned not only for her beauty but also for being 'well-read, particularly in theology and astronomy' (Zinsser 2006, 20). Despite the apparent emotional distance, Du Châtelet ensured her elderly and widowed mother's care when

Froullay retreated to Créteil, on the outskirts of Paris. Their lives intersected most poignantly in their final years, with Froullay passing away in 1740, only nine years before her daughter's untimely death (39).

Despite the rarity of sources, one could assume that her mother likely played a significant role in Du Châtelet's enculturation into court life and its protocols. As Judith Zinsser has speculated, she may also have influenced her daughter's early biblical education, as suggested by an anecdote from a family memoir. At a dinner table discussion, young Émilie posed a provocative question to her mother: 'Which should one take as more certain, that Nebuchadnezzar turned into a beast, or Prince Chérie into a bird?' (27) Without hesitation, Froullay replied, 'Neither' (27). Undeterred and bordering on impertinence, Émilie pressed the point, insisting she had 'seen it in the Bible' (27), an authority not easily dismissed in their discussions. Froullay accepted the challenge and responded, 'You never saw anything like that in the Bible' (27). As Zinsser comments (27), rather than silencing her daughter, as might have been expected in a traditional noble household, Froullay instructed Émilie to fetch the Bible and show the passage in question. The little girl eagerly complied, returning to proudly point out the reference describing Nebuchadnezzar's madness—how he lived like a beast in the countryside, eating grasses, and letting his hair and nails grow. This moment illustrates, according to Zinsser (27), not only Émilie's persistent and questioning nature, which would later define her intellectual pursuits, but also her mother's unusual openness to having her authority tested. Such encouragement of critical inquiry and unorthodox behaviour must have been instrumental in shaping Du Châtelet's remarkable intellect and independence.

Kovalevskaya's relationship with her mother was equally remote, marked by emotional distance, reflective of her aristocratic upbringing. She grew up with nannies, governesses, and tutors, who looked after her upbringing and education. Unlike her uncle and father, who figure prominently in her early interest in mathematics, as we have seen above, her mother in the autobiography of her early years appears as an unapproachable and almost fairy-tale figure:

> When I recall my mother during the first phase of my childhood, I always picture her as a quite young and very beautiful woman. I see her always gay and elegantly dressed. Most often I remember her in a low-necked ball gown, with bare arms and wearing a mass of bracelets and rings. She is getting ready to go out somewhere to a party and has dropped in to say goodnight to us (Kovalevskaya 1978, 52).

This distant relationship with her mother pervades Kovalevskaya's autobiographical writings, including her letters, and there is no extant correspondence between them. Neither are there any references to her mother in the many memoirs that were written after her untimely death. The only evidence that we have from her autobiography is the fact that Kovalevskaya lived her childhood with an intense feeling that she was 'unloved'. As already discussed in Chapter 5, this feeling of 'not being loved' has become a refrain not only in Kovalevskaya's but also in the biographical tales, memoirs, and discourses that have unfolded around her after her untimely death. Her early years, marked by a profound sense of isolation, would leave an indelible imprint on her self-perception and the way her life story has been remembered, but also in the way her own relationship with her daughter unfolded, as we will see later in this section.

Ann Isabella Milbanke, Lady Byron, emerges as a crucial component within the *assemblage* of Lovelace's intellectual development. Far from being a singular, isolated influence, Lady Byron's role was entangled with broader social, educational, and familial forces, forming a dynamic and interconnected network that shaped Lovelace's trajectory as a mathematician. An aristocrat of the highest status and a woman of considerable intellect, Lady Byron's relationship with her daughter has been extensively documented, mythologised, and analysed (see Moore 1977, Stein 1985). Beyond these narratives lies the enduring reality of her central role in fostering Lovelace's passion for mathematics.

Lady Byron's engagement with mathematics and her understanding of contemporary pedagogical theories positioned her within the progressive movements of her time, including advocacy for the education of the working classes (Taylor 1998). Her role as a mother was mediated through the *assemblage* of nineteenth-century aristocratic norms, which dictated the delegation of daily child-rearing to nannies and governesses. While Lady Byron often managed these caregivers remotely through correspondence, she maintained a firm hand, ensuring that their influence aligned with her vision for Lovelace's upbringing. As Doris Langley Moore notes, 'the list of nurses, governesses and tutors who fell from grace, especially when they gained any influence over Ada, is an extensive one' (1977, 9).

When it came to Lovelace's education, however, Lady Byron adopted a more direct role, exemplifying her resistance to certain elements within the *assemblage* of caregiving practices. In a notebook from Lovelace's sixth year, written tellingly in her daughter's voice, Lady Byron declared her intention to oversee her daughter's education personally, seeking to avoid what she called 'the evil of Governesses' (Stein 1985, 23). This decision highlights her determination to guide Lovelace's intellectual formation on her own terms, shaped by her personal experience and values.

Lady Byron's turbulent relationship with Lord Byron introduced a significant dynamic to the *assemblage* of influences shaping Lovelace's intellectual development. Determined to distance her daughter from poetry and the cultural legacy associated with her father, Lady Byron consciously nurtured Lovelace's mathematical and scientific interests. This commitment is evident in their correspondence, which reflects Lady Byron's unwavering support for Lovelace's pursuits. Her dedication was further solidified when she enlisted her own mathematics tutor, William Frend, to guide Lovelace's studies, ensuring a rigorous foundation in the discipline (see Moore 1977, 18).

Lady Byron's influence extended beyond intellectual instruction, intersecting with the social and scientific networks of the time. She actively encouraged Lovelace's collaboration with Babbage, attending his scientific soirées and supporting her daughter's engagement with his revolutionary ideas. This support was recognised by contemporaries such as De Morgan, who, in a letter to Lady Byron on 21 January 1844, praised Lovelace's mathematical brilliance as 'utterly out of the common way for any beginner, man or woman' (LB, 339, f. 2)[3]. Judith Lewis captures this relational dynamic, observing that 'mathematics became an important part of the emotional connection between the two women' within the unique configuration of their single-parent household (1995, 387).

This *assemblage* of relational and material forces is further evident during Lovelace's work on her celebrated translation and commentary on Menabrea's paper. In a letter to her mother on 15 August 1843, Lovelace detailed the laborious process of revising mathematical formulae for publication: 'No one can estimate the trouble & *interminable* labour of having to revise the printing of *mathematical* formulae' (LB, 42, f. 86, emphasis in the text). Despite her fatigue, she expressed a forward-looking vision: 'This is a pleasant prospect for the future, as I suppose many hundreds & thousands of such formulae will come forth from my pen, in one way or another' (f.86).

Lady Byron's influence within this *assemblage* was both profound and paradoxical. While her guidance and encouragement formed an essential component of Lovelace's intellectual development, the dynamics of their relationship also allowed mathematics to become a means through which Lovelace asserted her independence. As Lewis notes, Lovelace's engagement with mathematics ultimately created a psychosocial distance from both her mother and her husband, enabling her to establish herself as a scientist and thinker (1995, 391). Within the assemblage that shaped her life, Lady Byron was a critical yet complex component, simultaneously nurturing Lovelace's genius and facilitating her emergence as a unique female figure in the history of mathematics.

Beyond the fact that Agnesi's mother, Anna Fortunata Brivio, died tragically young at thirty-three due to complications during childbirth when Agnesi was just fourteen (Mazzotti 2007, 157n11), little is known about their relationship or any influence Brivio might have had on her daughter. However, her mother's untimely death appears to have had a profound impact on Agnesi, intensifying the 'mysterious and persistent malady' that plagued her adolescence, as discussed in the Introduction (32). This loss, occurring during such a formative period, adds an emotional dimension to Agnesi's early years, hinting at the complex interplay between personal tragedy and her intellectual development.

Unlike Du Châtelet and Kovalevskaya, whose aristocratic mothers remained distant and uninvolved in their mathematical pursuits, or Agnesi, whose mother died before her engagement with mathematics began in earnest, Germain's mother, Marie-Madeleine Gruguelu, played a uniquely active role in her daughter's life, though primarily as a social chaperone rather than an intellectual influence, as was the case with Lovelace. Initially resistant to Germain's nocturnal studies, as we have seen in previous chapters, her mother eventually acquiesced to her daughter's passion for mathematics and became a constant presence in her public and social interactions. Invitations to Germain consistently included her mother, and visitors to their home would invariably be received in her mother's presence.

Dora Musielak (2020, 33) notes that even in Germain's adulthood, her acquaintances were required to obtain written permission from her mother to visit her. In one instance, when Germain was forty, Fourier's invitation to lunch was addressed first to Madame Germain, implying that Germain's attendance was contingent on her mother's approval and likely her accompaniment (13). While there is no evidence of her mother directly influencing Germain's education or scholarly activities, this dynamic undeniably shaped the environment in which Germain navigated her intellectual pursuits. For instance, it is likely that any early meetings Germain had with Lagrange were either accompanied by her mother or confined to brief, formal exchanges at home—settings that may have limited the scope for meaningful mentorship or collaboration (200). This persistent maternal oversight established a protective framework around Germain's social and professional interactions, yet it raises questions about whether this dynamic hindered her ability to build the broader networks necessary for advancing her mathematical career. Within this *assemblage* of familial oversight and intellectual ambition, Germain's life exemplifies a complex interplay of forces—her mother's vigilance serving as both a protective stratum and a territorialising constraint, shaping the flows through which Germain pursued her mathematical becoming.

The mother-daughter dynamics in the lives of Du Châtelet, Kovalevskaya, Agnesi, and Germain remain largely speculative and undocumented. In contrast, Lovelace's relationship with her mother was intense and well-documented until the end of her life. Similarly, the correspondence preserved in Somerville's papers at the Bodleian Libraries reveals a vibrant and multifaceted relationship with her mother, Margaret Charters Fairfax. Somerville frequently references her mother in her autobiography, though certain details were omitted in the published version. Notably, the first draft includes the statement: 'My mother, whose maiden name was Margaret Charters, learnt to read she scarce knew how, and had only three months' instruction in writing and accounts' (1872, 1–2). This omission may have been intended to protect her mother's memory, avoiding any perception that her limited formal education diminished her significance in Somerville's life.

Somerville credited her mother with teaching her to read the Bible (2001 [1873], 58), yet, as explored in previous chapters, Margaret, alongside her husband, initially opposed her daughter's nocturnal studies and engagement with mathematics. This early resistance forms a point of tension in their relationship, illustrating an *assemblage* of intersecting forces—maternal concern, societal expectations, and Somerville's emerging intellectual trajectory. Despite these initial obstacles, the bond between mother and daughter deepened over the years.

After Somerville's first marriage ended in widowhood, she returned to her parents' home with her two children. It was during this period that her passion for mathematics flourished. Her mother, once sceptical, grew to take immense pride in her daughter's scholarly accomplishments. Their correspondence reflects Margaret's transformation into a steadfast supporter of Mary's intellectual endeavours. In 1831, she wrote twice to enquire about the reception of Mary's first book: 'I am [...] anxious to know how your book is received by the public—write me soon, my dearest Mary' (cited in McMillan 2001, 27). Early the following year, she expressed joy at the book's success and relief at her daughter's good health (27). Somerville, in turn, kept her mother closely informed during her travels, sharing news of her encounters with Europe's intellectual elite. In a letter from Paris, dated 18 August 1817, during her first trip to France, she wrote to her mother: 'we were invited with much hospitality to the houses of the most celebrated philosophers of France, who made entertainments on purpose to make us acquainted with everyone who was eminent in science' (cited in Patterson 1983, 20). The Somerville mother-daughter relationship exemplifies the complex interplay of social, intellectual, and emotional forces that shaped Somerville's life. Far from being a static or hierarchical bond, it functioned as a dynamic *assemblage*, where early resistance gave way to profound support, enabling Somerville to navigate her intellectual pursuits while maintaining a deep personal connection.

The relationships between women mathematicians and their mothers often reflect a duality of constraint and empowerment, showcasing how familial networks can simultaneously hinder and amplify individual trajectories of growth and achievement. But what about their relationships with their own daughters? Here, too, the dynamics are multifaceted and nuanced, shaped by the interplay of social class, generational expectations, societal pressures, and personal ambitions. Yet, a recurring theme emerges: these relationships were often fraught with tension and difficulty, revealing the complexities of balancing intellectual pursuits with maternal roles.

In previous chapters, we have examined how Du Châtelet passionately defended women's right to education, as articulated in her translator's introduction to *The Fable of the Bees* (2009) and her *Discours sur le bonheur* (1961). However, what remains puzzling is her decision to dedicate the *Institutions de Physique* (1740) solely to her son, Florent-Louis, born in 1727, while omitting her daughter, Gabrielle-Pauline, born just a year earlier. Little is known about Gabrielle-Pauline's education, though evidence suggests she accompanied her mother at Cirey and even participated in the theatrical productions staged by Du Châtelet and Voltaire. Her formal education seems to have been fragmented, including intermittent convent schooling and home instruction in Latin, following her mother's scholarly pursuits (see Zinsser 2006, 27, 40).

Gabrielle-Pauline's trajectory aligns closely with the expectations and ethics of the eighteenth-century French aristocracy. At just sixteen, she married into a prominent Neapolitan aristocratic family, a union emblematic of the social and political alliances valued by her class. Even within this context, her mother maintained a distant but dutiful interest in her life. In correspondence with the mathematician François Jacquier, who visited Pauline in Naples, Du Châtelet wrote from Versailles on 17 December 1745: 'Tell me if you saw my daughter at Naples and if you were happy about it. She must have given birth now' (in Kölving and Brown 2018, vol. 1, 498). Such a detached inquiry about her daughter's well-being and maternal status underscores the limited personal closeness between them, reflecting the constrained familial bonds often shaped by aristocratic norms.

The absence of extant letters between Du Châtelet and Gabrielle-Pauline highlights a perceived distance in their relationship, though this perception is complicated by the loss of their correspondence. Despite Du Châtelet's prolific letter-writing, no surviving records document sustained communication with her daughter, leaving little evidence of an intellectual bond akin to those Du Châtelet cultivated with her male peers. Gabrielle-Pauline's early death in 1754, likely in childbirth at the age of twenty-eight, occurred only five years after her mother's untimely passing in 1749.

This dynamic must be understood within the *assemblage* of eighteenth-century French aristocracy, where family ethics and societal expectations often dictated the nature of relationships. For aristocratic women, marriage frequently severed ties with natal families, redirecting loyalties and obligations towards their husband's lineage. Du Châtelet's relationship with her daughter, shaped by these norms, reflects the broader tensions between personal affections, intellectual pursuits, and the demands of aristocratic decorum in an era where women's roles were rigidly prescribed.

Kovalevskaya's relationship with her daughter Fufa[4] was equally complex. When she first visited Stockholm for an unsalaried academic position, she understood that going there with her daughter would be a huge strain for both of them. Thus, she decided to leave Fufa with her godmother, Iulia Lermontova, in Moscow. When her extraordinary professorship was confirmed in the summer of 1884, the question of bringing her daughter to Sweden was raised again, but Kovalevskaya felt that things had not yet settled in Sweden and that it was in the child's best interest to stay in Russia. Although her friend Theresa Gyldén had written to warn her that there were rumours and gossip in Sweden regarding her separation from her daughter, Kovalevskaya's response was swift and determined. She explained that together with Iulia, they had decided that it would be better for Fufa to stay in Moscow, where the little girl had a sense of belonging:

> You must also think of how alone we are in the world, my little Sonya and me. Her birth was welcomed by a whole, happy family; only five years have passed and now she has neither a father, nor a grandmother or grandfather, she has no natural support other than me. In these circumstances, it is entirely understandable that the link connecting her to the Lermontov family is doubly precious, and that I am not acting frivolously in not daring not only to sever but even to weaken this link (Kovalevskaya 1951, 283).

Having carefully thought about the welfare of her daughter was not just a whim of the moment for Kovalevskaya. When Fufa eventually joined her in Stockholm in the fall of 1886, she found a new family among Kovalevskaya's circle of friends. In her 'Reminiscences' of her mother, Fufa has written at length about her impressions of her move to Sweden: 'At the end of August, after two and a half days by sea, we arrived in Stockholm just before sunset. The view of the beautiful Stockholm raid amazed even me, who still did not know how to appreciate pictures of nature.' (Kovalevskaya S.V. 1951, 362) Kovalevskaya spent a lot of time with her daughter in the beginning, reading her books and taking her to the market 'where, through visual learning, she significantly enriched my vocabulary of the Swedish language' (362). It was still summer, and Kovalevskaya's acquaintances were on holidays, but when they came back, little Fufa got to know the Lefflers and became particularly attached to the Gyldéns, who had a boy her age and three older children: 'I enjoyed going to them [...] they lived on the outskirts of the city, in a house with a tower, surrounded by a large garden. I began to live quite like a member of this family, particularly during the time of my mother's frequent absences and vacations' (363).

As we have seen in previous chapters, Kovalevskaya travelled a lot during the academic holidays, mostly to keep in touch with the mathematical circles in France and Germany, but also to have some time for her research writing. Fufa remembered that 'during the summer and Christmas holidays my mother always went to France or Germany, while I stayed in Sweden or at Lermontova's. We spent only one summer with my mother' (362). She also wrote that she had not kept early memories of her mother, as 'until the age of seven, I saw her for the most part only in fits and starts' (360), but she clearly remembered her life in her uncle's family as 'the happiest period of my early childhood.' (in Kochina 1985, 318). Throughout the several recollections of her life with her mother that she wrote,[5] it is obvious that they did not have a close relationship: 'my feelings for mother, were rather complicated and were not as intimate as those for Iulia Lermontova', she wrote, (320). Fufa could open up to Iulia as 'she loved me as I was' (320), despite the fact that she was always careful not to display her full affection for the child of her friend: 'no doubt, she loved me but thought it improper to give vent to her tenderness, and I was almost never caressed, apart from a kiss for the night' (320).

Kovalevskaya's role as a mother resists easy categorisation within conventional ideals of motherhood, prompting the broader question: can such an ideal even exist? Moreover, how are discourses of motherhood constructed, and under what conditions do they emerge? Ann Koblitz observes that Kovalevskaya's limited involvement in her daughter's upbringing conformed to the norms of her culture and social class: 'It was the custom for women of the continental European gentry and educated classes to leave the care of their children to nursemaids and other servants' (1989, 187). This approach to motherhood, shaped by the aristocratic *assemblage* of familial roles and societal expectations, was the same modality in which Kovalevskaya herself, as well as Du Châtelet and Lovelace, were raised—a model they, in turn, perpetuated with their own daughters.

In a different social context, Somerville's relationship with her daughters, Martha Charters (1815–79) and Mary Charlotte (1817–75), offers a distinct configuration of motherhood within an *assemblage* defined by extraordinary closeness and mutual interdependence. The three women lived together, with both daughters remaining unmarried and dedicating their lives to supporting their mother's intellectual pursuits. This familial constellation exemplifies a collective *assemblage*, where their roles extended beyond domestic assistance to active participation in sustaining and amplifying Somerville's scientific and intellectual legacy. Martha, in particular, played a pivotal role after Somerville's death, preserving her mother's legacy by gathering personal journals and letters, editing the first draft of Somerville's autobiography, and publishing it with the help of family friends. Through this effort, Martha ensured that Somerville's contributions to science and society were not only remembered but integrated into public and scholarly discourses.

Somerville's *Recollections* offer glimpses into the intellectual environment of their household, where knowledge and creativity were cultivated collectively. She wrote: 'My daughters became excellent linguists and wrote when necessary for me, as I never attempted to write any language but English. They showed a very decided taste for music at an early age and became good musicians which was more owing to their own industry and talent than to teaching' (1872, 155). Their proficiency in languages and music reflects the cultural richness of this *assemblage*, where the flow of intellectual and artistic pursuits was not confined to the mother but distributed across the family.

Sadly, both daughters passed away soon after their mother—Mary Charlotte in 1875 and Martha in 1879—marking the end of this remarkable familial *assemblage*. Their lives underscore the intricate interplay of personal devotion, intellectual ambition, and shared purpose within a relational network that transcended conventional roles of motherhood and daughterhood. Together, they forged a collective legacy that continues to resonate, illustrating how individual trajectories are shaped and sustained within the dynamic interplay of familial and cultural *assemblages*.

As already noted above, Agnesi's relationship with her mother has not been documented, and as a single woman, she did not have the experience of raising children. Her relationship with her sister Teresa,[6] however, was deeply intertwined, shaped by their shared upbringing in a household that staged their father's meticulously curated intellectual and artistic spectacles. Born just two years apart—Gaetana in 1718 and Teresa in 1720—the sisters grew up within the same cultural and educational milieu, though their trajectories diverged dramatically in adulthood, both in their pursuits and their life choices.

While Gaetana became renowned as a mathematician and philosopher, Teresa channelled her talents into music, becoming an accomplished composer, harpsichordist, and singer. The specifics of Teresa's education remain unclear due to the paucity of scholarship on her life, yet it is reasonable to assume that she benefited from some of the same tutors who shaped her sister's intellectual formation. This shared intellectual foundation reveals a relationship of interiority within their familial *assemblage*, where the same influences fostered their respective talents in distinct fields.

Teresa's career as a composer was deeply influenced by her dynamic engagement with external forces, particularly through the dissemination of her music to royal patrons in Vienna and Dresden (Mazzotti 2007, 133–4). These exchanges allowed her creative flows to deterritorialise beyond the confines of Milan, where her work was less celebrated, exemplifying a relationship of exteriority—her compositional *assemblage* shaped and transformed through interactions with the cultural and political rhizomes of Europe's courts. Her surviving compositions, including at least four operas, keyboard and vocal pieces, and works for small ensembles, reveal her capacity to navigate these *assemblages* with fluidity and force, even as much of her oeuvre has been lost.

Teresa's dedication of her compositions to figures like Maria Antonia Walpurgis, herself a composer, and Empress Maria Theresa signals her participation in the collective intensities that sought to affirm women's creative power during this period. As Mazzotti observes, these dedications also celebrated the figure of 'the strong woman', aligning Teresa with a broader set of cultural deterritorialisations centred on intellectual and moral equality between men and women (2007, 133). This motif, which coursed through the works of learned women in mid-eighteenth-century Milan, serves as a plane of convergence between the two sisters' lines of becoming.

Gaetana's celebrated oration on women's potential to engage in the liberal arts and sciences and her dedication of the *Instituzioni analitiche* (1801 [1748]) to Empress Maria Theresa resonate with Teresa's musical tributes. These expressions of their respective *assemblages*, operating through mathematics and music, demonstrate their shared commitment to destabilising existing strata and affirming the creative and intellectual capacities of women. Together, they generated a multiplicity of flows that not only traversed but reconfigured the cultural and intellectual *assemblages* of their time.

Despite these successes, Teresa's trajectory was sharply reterritorialised by her marriage to Pier Antonio Pinottini in 1752, just three months after their father's death—a union reportedly opposed by their father (Mazzotti 2007, 124). This event marked a significant bifurcation from her sister's line of becoming. While Gaetana's intellectual and spiritual flows continued relatively undisturbed—albeit deterritorialised away from science and mathematics into philanthropy and religious devotion—Teresa's *assemblage* became entangled in a plane of financial instability and hardship, exacerbated by her husband's irresponsibility.[7] The marriage introduced strata that constrained her creative production, yet Teresa continued to compose, navigating these limitations with resilience. By 1771, she was recognised in Milan as 'an authority in harpsichord' (149), signalling her capacity to sustain and redirect her creative flows despite the forces working to territorialise her life. This persistence reflects the multiplicity within Teresa's *assemblage*, where deterritorialised artistic intensities intersected with the weight of social and material constraints, generating a *line of flight* that allowed her to remain a force within Milan's musical landscape.

Teresa's life can be seen as a complex *assemblage* of relations of interiority and exteriority: the interior influences of her family and shared intellectual environment intersecting with the exterior forces of royal patronage, societal norms, and personal challenges. Her contributions to music stand as a testament to her resilience and ambition, even as they were shaped and, at times, constrained by external circumstances. In contrast, Gaetana's uninterrupted scholarly pursuits highlight a different negotiation of similar familial and societal *assemblages*. Ultimately, the two sisters' lives, while distinct, illuminate parallel yet divergent paths of becoming—Gaetana through intellectual rigour and Teresa through creative artistry, each navigating the intersections of shared origins and external pressures in their own ways. Gaetana's brief note, accompanying a small jewel she left to her 'much-loved sister Teresa' when she renounced her rights to the family inheritance shortly after their father's death, conveys the depth of their sisterly bond with touching simplicity: 'a small token that would remind you of me' (Mazzotti 2007, 145).

Agnesi's parallel creative life with her sister resonates with Kovalevskaya's relationship with her beloved sister, Anna Jaclard, affectionately known as Anyuta. While earlier in Chapter 5, I explored the love and admiration within Kovalevskaya's bond with her sister, here I focus on the ways their unique bond shaped their intellectual and personal journeys, spanning Russia and Europe. Anyuta, six years older than Sofia, was a literary talent who published her first stories in Fyodor Dostoyevsky's journal *Epokha* in 1864 (Krukovskaya 1864a, 1864b). She was not only admired by her sister as 'a raving beauty' (Kovalevskaya 1978, 135) but was pivotal in Kovalevskaya's intellectual development and early politicisation into the Enlightenment ideals flourishing in 1860s Russian society (see Kochina 1985, 30–6). It was Anyuta who first approached Vladimir Kovalevskii about entering into a fictitious marriage, but he stipulated that 'the bride' should be Sofia, not her. In a letter written shortly before her 'marriage', Kovalevskaya envisioned their future as Russia's new women:

> I am getting ready for my examination, writing my dissertation. Anyuta puts her travel notes in order; later I study on my own, still later we together set up a colony, and I go to Siberia. I find a mass of difficulties and disillusions there, but I can manage to be of help there nonetheless. Anyuta writes a wonderful paper; I succeed in making a discovery; we establish a high school for girls and another one for boys; I have a physics laboratory of my own (in Kochina 1985, 47).

Upon moving from her family house in Palibino to St. Petersburg with her newlywed 'husband', Kovalevskaya decided to focus on mathematics, confiding first in Anyuta: 'You know, my incomparable Anyuta, I all but made up my mind not to study medicine, but to enter a physico-mathematical department' (50–1). This decision stemmed from her lack of interest in practical fields and her deep engagement with mathematics, which she described as her only source of true happiness (51). When Kovalevskaya and her 'husband' eventually left for Europe in the beginning of 1869, Anyuta joined them.

After a brief stay in Vienna, however, the trio parted ways: Kovalevskaya went to Heidelberg to study mathematics, Vladimir remained in Vienna, and Anyuta moved to Paris to pursue her literary aspirations without informing her parents. However, the secret was short-lived; when her allowance was cut off, Anyuta began working as a typesetter in Paris. Through this work, she became involved in the radical workers' movement and met Victor Jaclard, a prominent revolutionary leader who became her lifelong partner and husband. Together, Anyuta and Victor actively participated in the Paris Commune of 1871.

Anyuta played a significant role during the Commune, serving on the Montmartre Committee of Vigilance and a committee supervising girls' education. She was instrumental in organising food supplies for the besieged city and acted as a representative of the Russian Section of the International. Her commitment extended to serving on a committee addressing women's rights (see Eichner 2004, 81, 103, 124). During this turbulent period, Kovalevskaya temporarily paused her doctoral studies in Berlin. She and Vladimir travelled to Paris to support Anyuta and Victor. Vladimir wrote to his brother Aleksander from Paris on 28 May 1871: 'Very many of our good friends have been wounded or killed' (in Koblitz 1993, 110). Though Anyuta and Victor survived, they faced persecution and imprisonment. Anyuta managed to escape to England, where she found refuge with Karl Marx's family, while Victor, who had been sentenced to death, managed to escape through the intervention of Kovalevskaya'sinfluential Russian family. Eventually, the couple reunited and found refuge as political exiles in Zurich.

The Paris Commune left an indelible mark on Kovalevskaya. According to her friend and co-author Carlotta Leffler, she planned to write a novel, *The Sisters Raevsky Under the Commune*, but her premature death left this and many other projects unfinished (Leffler 1895, 44). In 1874, Kovalevskaya and Anyuta returned to Russia, but in 1880, following a general amnesty for Commune participants, Anyuta returned to Paris, where she eventually passed away, as discussed in Chapter 4.

Kovalevskaya's relationship with Anyuta was one of profound intellectual respect and mutual support. Anyuta served as a confidante and emotional anchor for Kovalevskaya, while their shared commitment to challenging societal norms united them. Anyuta's political activism complemented Kovalevskaya's academic pursuits, and her early death in 1887 was a devastating loss for Kovalevskaya, who deeply mourned the passing of her beloved sister and trusted ally. The bond between Kovalevskaya and her sister, Anyuta, exemplifies the critical role of sibling relations in the *assemblage* of becoming. Anyuta's intellectual curiosity and shared defiance of societal conventions were instrumental in Kovalevskaya's early intellectual awakening. Together, they created an affective and intellectual milieu that supported Kovalevskaya's *lines of flight* from a stifling family environment and the restrictive expectations of Russian society. This sibling connection, imbued with shared ambitions and struggles, acted as an initial *plane of consistency* that sustained Kovalevskaya's later alliances with mentors and peers.

Like Agnesi, Germain lived her life as a single woman, but she had two sisters, Marie-Madeleine, five years her senior, and Angélique-Ambroise, two years younger. While there is no indication of an intellectual bond between Germain and her sisters, her family nonetheless formed a crucial *assemblage* in the preservation and dissemination of her legacy. Central to this effort was Marie-Madeleine's son, Jacques-Amant Lherbette, whose role as a practising lawyer and editor of Germain's philosophical work in 1833 marked the beginning of her posthumous recognition. Together with his mother, Lherbette facilitated the deposit of Germain's papers at the Paris Academy in 1859, an act that, as noted in Chapter 1, was instrumental in amplifying her philosophical and scientific contributions.

As the executor of Germain's manuscripts, Lherbette not only acknowledged his aunt's significance but also actively worked to ensure her ideas endured. This effort, far from being a mere display of familial loyalty, reflects a complex interplay of social relations within the family network. The *assemblage* of family, shaped by both gendered dynamics and intellectual commitments, highlights the inseparability of personal and professional spheres in Germain's legacy. Rather than drawing a strict division between the roles of men and women in Germain's life and career, the family *assemblage* captures the intricate entanglement of these relationships. This complexity underscores how the web of human relations—spanning generations, genders, and social roles—functions as a multiplicity, shaping and reconfiguring Germain's intellectual afterlife and ensuring her enduring place in the histories of philosophy and science.

Beyond family *assemblages*, where women mathematicians navigated roles as mothers, daughters, or sisters, their emergence in the public spheres of mathematics, philosophy, and science was intricately connected to extended networks of scholarly but also revolutionary women. In such *assemblages,* familial and friendship networks often intersected, with sisters contributing a foundational intensity upon which friendships with other women could build. These *assemblages* reveal that women's intellectual becomings were rarely isolated or linear but instead *rhizomatic*, composed of multifaceted and evolving relations that sustained creativity, resistance, and the emergence of new possibilities.[8]

Somerville's life and legacy exemplify the importance of such intellectual *assemblages* and *rhizomatic* connections among women. Through her extensive correspondence, she fostered a dynamic community of knowledge exchange and mutual support, shaping both her own work and the broader intellectual landscape. Somerville's correspondents included an impressive array of women scholars and writers, among them Lovelace, as discussed in Chapter 2; the Scottish dramatist and poet Joanna Baillie; suffragist and writer Frances Power Cobbe; Irish novelist and educationalist Maria Edgeworth; science writer Jane Marcet; astronomer Caroline Lucretia Herschel; and sociologist and philosopher Harriet Martineau. These relationships reflect the collective nature of intellectual progress, where ideas were not developed in isolation but through vibrant interactions within an interconnected network of women thinkers.

Notably, Somerville and Herschel broke new ground in the history of women in science by becoming the first female members of the Royal Astronomical Society in 1835. Frances Power Cobbe played a critical role in preserving Somerville's legacy by assisting Martha Somerville in editing her mother's autobiography, as we have seen in Chapter 1.

This network exemplifies the dynamics of women's intellectual *assemblages*—a fluid and relational configuration of shared interests, mutual empowerment, and intellectual exchange. Somerville's correspondence and interactions with these women were not merely acts of social engagement but deliberate efforts to create a collaborative intellectual society. These relationships were pioneering in their recognition of women's contributions to diverse fields, from literature and astronomy to education and sociology.

The *assemblage* framework captures the non-hierarchical and rhizomatic nature of this network. Each woman, through her distinct expertise and experiences, contributed to a collective force that transcended individual achievements. For example, Jane Marcet's popular science writings likely influenced Somerville's pedagogical approach, while her collaboration with Lovelace formed a rare mentorship between women mathematicians, as we have seen in Chapter 2. Frances Power Cobbe's editorial assistance not only ensured the preservation of Somerville's legacy but also underscored the practical dimensions of this *assemblage*, where intellectual labour and emotional support coalesced.

This *assemblage* of women intellectuals transcended the traditional boundaries of public and private spheres, forging a collective that challenged societal constraints while amplifying individual potential. Their collaborative endeavours represented a *becoming-woman* of intellectual life—an unfolding process that redefined the possibilities for women across disciplines and geographies. As an extension of these synergies, Somerville became the first signatory of John Stuart Mill's landmark 1866 petition for women's suffrage (see Pugh 1980), a bold assertion of women's right to participate fully in civic life. Her advocacy did not stop there; she also lent her support to the establishment of Girton College, the first women's college associated with the University of Cambridge, demonstrating her commitment to expanding women's access to education. In a symbolic and practical act of solidarity, Somerville donated her scientific library to the college, ensuring that future generations of women scholars would inherit both the tools and the inspiration to pursue intellectual endeavours. This network of women was not merely a community but an active, generative *assemblage* that reshaped the cultural and intellectual landscape of its time (see Somerville 2001, 346).

Somerville's politics and her unwavering advocacy for women's rights strongly resonate with Kovalevskaya's gendered politics, which shaped her life's decisions and directions. This alignment is evident in the networks of women that emerged around Kovalevskaya and the assemblages of shared ideals, struggles, and ambitions that connected them.

When Kovalevskaya moved to Europe with her husband, Anyuta joined them, as we have seen above. Soon, a dynamic and intellectually vibrant colony of radical Russian women formed in Heidelberg, each pursuing academic and political aspirations. The first to join was Iulia Lermontova, Kovalevskaya's close friend and eventual trailblazer in chemistry, becoming the first woman in the world to earn a PhD in the field in 1874 from the University of Göttingen—the same year Kovalevskaya completed her own doctorate. Lermontova's memoir captures the camaraderie and joy of their shared journey, describing how they would walk together in Heidelberg, and 'finding ourselves on a flat stretch of road, the two of us, Sofa and I, would take off at a dead run, exactly like two little children. Dear heavens! How much gaiety and happiness there is in these memories of the first period of our university life!' (Lermontova 1951, 382)

The network expanded further when Anyuta's friend Zhanna Evreinova joined them. Evreinova, who would later become the first Russian woman lawyer, was already a staunch advocate for women's equal rights, reflected in her writings and activism. Her escape from Russia was as daring as her ideals—crossing the border on foot 'without a passport and under the fire of the guards' (Kochina 1985, 58). In Heidelberg, she attended her first university lectures in law before continuing her studies in Leipzig. Kovalevskaya's cousin Natasha Armfeldt also joined the network. A woman of the same age as Kovalevskaya, she fled her parents' disapproval to study mathematics but quickly became immersed in revolutionary activities. Her path was fraught with danger: she was arrested multiple times and sentenced to hard labour in Siberia in 1879, where she married an exiled revolutionary. Tragically, the harsh conditions of exile took their toll, and she died of consumption in 1887 (59).

These women were part of a broader radical generation of the 1860s, united by their belief that scientific knowledge and intellectual freedom could dismantle the constraints of the past and pave the way for social revolution. According to Elizaveta Litvinova, another member of this network and the second woman to earn a doctorate in mathematics from the University of Bern in 1878, Kovalevskaya was 'the bright star that guided girls who wanted to study' (59). Kovalevskaya herself captured the spirit of these networks in her reflections:

> We were so enthusiastic about the new ideas; so sure that the present social state could not continue long. We pictured to ourselves the glorious period of liberty and universal enlightenment of which we dreamed, and in which we firmly believed. Besides this, we had the sense of true union and cooperation. When three or four of us met in a drawing-room among older people, where we had no right to advance our opinions, a tone, a glance, even a sigh, were sufficient to show one another that we were one in thought and sympathy (in Leffler 1895, 11).

Throughout her life, Kovalevskaya remained acutely aware of her place within a larger story of women's struggle for equality and emancipation. When she was offered a position at the newly established mathematics department at the University of Stockholm, her first question to Gösta Mittag-Leffler addressed the institution's stance on women in a letter dated 14 October 1880: 'I would be very grateful if you were so kind as to inform me clearly with respect to the attitude your University holds to us: does it open the doors for us without restraints, or does it admit us only in exceptional cases and a special favour, or does it refuse us completely?' (IML/SK-ML/L1)[9] Kovalevskaya's words reflect her own university experience and her alignment with the broader 'women's question' in science. Her overall thoughts on women and science have been recorded in Litvinova's reminiscences:

> My destiny, or if you wish, the main goal of my life, but I like more the word destiny, because the goal of my life is in myself, while destiny is of divine origin. I feel that my destiny is to serve the truth, that is, science, and to blaze the trail for women, because that means to serve justice. I am very glad that I was born a woman, because this gives me a chance to serve both truth and justice at the same time. But it is not always easy to follow your destiny (in Kochina 1985, 75–6).

Kovalevskaya's life and work unfolded within *assemblages* of women who sustained, inspired, and collaborated with one another, forming vibrant *planes of consistency* that championed the transformative potential of knowledge and solidarity. These *assemblages* transcended geographic and disciplinary boundaries, creating dynamic fields of mutual empowerment that sustained intellectual and political flows. For Kovalevskaya and others in her milieu, such connections were generative configurations, enabling access to education, pathways for professional growth, and *rhizomatic* spaces for revolutionary becoming. Far from static or purely social, these systems were living, responsive networks, constantly reconstituting themselves to navigate shifting terrains of possibility and constraint.

Across this study, women mathematicians' lives were embedded in intricate webs of collaboration, inspiration, and interdependence. These *assemblages*—rooted in familial bonds, intellectual partnerships, and political affiliations—emphasise that their pursuits were never solitary. Through shared study, correspondence, and collective advocacy, they cultivated conditions for growth, discovery, and resilience. These connections enabled the flow of resources, ideas, and encouragement, creating multiplicities that sustained their intellectual pursuits while challenging the limitations imposed upon them.

The relational narratives of these women mathematicians highlight not only their individual contributions but also the transformative capacities of their *assemblages*. These collective efforts generated diffractions that reconfigured the intellectual and political landscapes, broadening the scope of who could participate in the creation of mathematical knowledge. The *assemblages* that connected women across generations and geographies remind us that the history of mathematics is not a solitary pursuit but a process of interconnection and shared becomings. These networks amplified their voices and expanded the possibilities of their work, embodying the power of collaboration in reshaping knowledge production.

Moreover, these *assemblages* were not only spaces of solidarity but platforms for imagining and enacting alternative futures. As Kovalevskaya expressed, her destiny intertwined science and justice, asserting that the pursuit of knowledge could serve as an act of resistance. The work of Kovalevskaya and her contemporaries reveals how women's networks were vital to transforming the intellectual and political environments of their time, offering a testament to the collective potential of *assemblages* in reimagining the future. In doing so, they all took leaps into the unknown, engaging with challenging mathematical and philosophical problems and questions of their times, as I will discuss next.

Leaps into the unknown

On 27 November 1960, the French artist Yves Klein unveiled one of his most provocative works, *Leap into the Void*.[10] The image is a staged photograph of Klein himself, suspended mid-air in a graceful yet unsettling pose, as if in the act of leaping from a rooftop into the empty space below. At first glance, it appears to capture a moment of defiance, a literal plunge into the unknown. But the photograph is not what it seems. Carefully constructed and edited, it challenges the viewer's perception of reality and possibility, blurring the line between the actual and the virtual, the material and the imagined.[11]

For Klein, the *Leap into the Void* was not an act of reckless abandon, nor a metaphor for existential courage. It was a gesture towards the uncharted, a breaking away from the constraints of gravity, of representation, of the given. It was a moment of pure potential, where the familiar was suspended, and the unknown—a void full of virtualities—became tangible. Inna Semetsky (2010) has drawn on the philosophies of Charles Sanders Peirce, John Dewey, and Gilles Deleuze to explore how intuition, taken as a leap into the unknown, plays a crucial role in the learning process and the creation of new knowledge. As she notes, the leap, the moment of breakthrough, is what initiates a *line of flight*—a movement that disrupts the stability of being. In Deleuze's thought, this disruption unsettles the existing order yet it is along this very trajectory that events unfold and processes of becoming emerge (Semetsky 2010, 234). This leap is both a risk and a creative act, a shift in intensity where something new emerges. It is not guided by a structure or an overarching goal; it is an encounter with the virtual, the potential, and the yet-to-be-articulated.

Mathematics, too, can be seen as a series of such leaps, where thought ventures into the void of the unproven, the unresolved, the yet-to-be. The women mathematicians in this study took their own leaps—not as acts of defiance against structures, but as participations in the generative flux of mathematical becoming. Like Klein's staged photograph, their work blurred distinctions: between personal and intellectual, between creativity and rigour, and between the known and the possible. Each theorem, each correspondence, each new idea was a step into a void, a moment where the known world gave way to the infinite potential of the unknown.

In this light, their mathematical practices were not linear or predictable but were instead acts of *becoming-mathematical*—a continuous deterritorialisation of established fields and norms. Their work did not operate within the confines of fixed identities or predefined roles but instead unfolded in the liminal spaces of creativity and inquiry, where new connections and *assemblages* formed. Their work exemplifies a dual leap: grappling with the mathematical unknown while simultaneously navigating the cultural and institutional barriers that sought to define their roles and limit their ambitions. As we have seen throughout this book, they took intellectual and personal risks, tackling abstract problems and pioneering innovative ideas, often without the institutional support or acknowledgement afforded to their male counterparts. For these women, mathematics was not only a discipline but also a space of resistance and redefinition—a site where they could challenge existing frameworks and forge new paths.

Consider Du Châtelet's ambitious project to construct a unified philosophical framework that synthesised and critically engaged with the ideas of Descartes, Newton, and Leibniz; Germain's groundbreaking work on elasticity theory, tackling a problem deemed insoluble by Lagrange; Lovelace's visionary insights into the computational potential of the Engines; Kovalevskaya's elegant resolution of the historical problem of a rigid body's rotation; and the bold contributions of Agnesi and Somerville in introducing differential calculus to mathematical literature within the hesitant intellectual climates of their respective nations. Each of these efforts represents a leap into the unknown, expanding the horizons of mathematical and philosophical thought. These leaps involved making intuitive guesses or hypotheses that went beyond existing knowledge and understanding. Indeed, women mathematicians often relied on their intuition to make breakthroughs in their work, trusting their inner sense of direction and understanding. Their ability to think differently and embrace intuition allowed them to contribute unique perspectives and solutions to mathematical problems.

As Semetsky (2010) has argued, intuition and abduction are essential for the learning process, enabling individuals to make connections and create new knowledge. For Dewey, Semetsky notes, all thinking and learning involve a process of moving from the known to the unknown—a leap that reaches beyond the present towards what is absent. This act, akin to the notion of a *line of flight* (Deleuze and Guattari 1988), is an inevitable jump into uncertainty, grounded in the conditions that enable it. Dewey emphasises the importance of attending to these conditions, as they shape the transformative process of understanding and discovery: 'The very inevitableness of the jump, the leap to something unknown, only emphasizes the necessity of attentionto the conditions under which it occurs' (Dewey 1991, cited in Semetsky 2010, 234).

In thus excavating the conditions of possibility for such leaps to occur, throughout the book, I have shown that women mathematicians' work was not a triumph over barriers in the light of their 'exceptional abilities', but an *event*, a process of creating new possibilities, new ways of thinking and doing, that exceeded the rigid boundaries of their historical contexts. These women became nodes of intensification in mathematical *assemblages*, reshaping what mathematics could do and how it could unfold.

By venturing into the unknown, these women mathematicians did more than contribute to the field; they reconfigured what it meant to be a mathematician. Their engagement with mathematics was not a mastery of problems but a way of participating in the continuous flux of mathematical becoming. Each theorem, correspondence, or proof was a moment of opening—*a line of flight* that connected them to the broader *assemblages* of knowledge, history, and imagination. These leaps into the unknown were not acts of will but expressions of immanence, where the potential of mathematics and the becoming of the mathematician coalesced into something new, something unexpected, and something yet to come.

And what about the present? The persistent under-representation of women in mathematics—the genealogical question that sparked the research underpinning this book—remains as pressing as ever. This inquiry into the histories and trajectories of women mathematicians began, perhaps quite unconsciously, during my long-standing participation as an observer in a mathematics colloquium held in a secluded chalet in the French mountains. Over the years, this colloquium has become a space of recurring familiarity, a gathering of mathematicians who come together each year to ski in the mornings, hold sessions in the afternoons, and share conversations over meals and late-night drinks. Since my first visit in 2005, the venue, the rhythms, and even the ratio of women to men—roughly 1 to 6—have remained unchanged.

Despite this disparity, the colloquium has always included women mathematicians, not only from France but from across the globe. Over these two decades, I have had the privilege of speaking with many of these women—about their careers, their struggles, their aspirations, and their disappointments. These candid discussions may well have planted the seeds for the project underpinning this book: a feminist genealogy of *automathographies*, exploring the conditions under which women have navigated the complex terrain of mathematics across time.

At the 2025 colloquium, twelve women were among the participants, representing a diverse range of experiences. Some were established professors, others were early-career researchers or PhD students. Their reflections echoed the enduring challenges faced by women in the field. Over dinner one evening, an associate professor from a prominent European university remarked, 'It's not easy to be a woman mathematician', after hearing about my research. On the final evening, a senior woman professor confided that she had confronted one of the conference organisers, questioning why, in the five days of sessions, only one woman had been assigned as a session chair.

Do things change? Have they ever? Will they? These questions linger, unanswered yet urgent. What I have come to understand, however, is that the genealogical inquiry that runs through this research is not about finding definitive resolutions. It is about creating conditions for thinking differently, for unsettling the present by tracing its entanglements with the past. Michel Foucault once wrote that the void is not a deficiency or a gap to be filled, but 'the unfolding of a space in which it is once more possible to think' (1970, 342).

Perhaps the insights drawn from this book and its underpinning research can open such spaces—triggering new questions, new provocations, and new leaps into the unknown. The struggles and stories of women mathematicians, past and present, are not merely accounts of marginalisation but acts of intellectual creation and transformation. By engaging with their histories, we might begin to reimagine the possibilities of what mathematics could be and who might inhabit its spaces, now and in the future.

Notes

1 The concept of the *carnivalesque* appears in Mikhail Bakhtin's *Rabelais and His World* (1984 [1965]). Derived from medieval and Renaissance carnival traditions, it describes a temporary suspension of hierarchies and norms, where social structures are inverted, and laughter, parody, and bodily imagery create a subversive and liberating space for critique and renewal.
2 I refer here, of course, to the lyrics of 'The Impossible Dream', the best-known tune from the 1965 Broadway musical *Man of La Mancha.*
3 Oxford Bodleian Libraries, Archive of the Noel Byron and Lovelace Families, Dep. Lovelace Byron. In the text and thereafter, LB followed by the folder and folio number.
4 Nickname for Sofia Vladimirovna Kovaleskaya.
5 After Fufa's death in 1952, several recollections of her life with her mother were found and published, and they are now in the Archive of the Russian Academy of Sciences. See Kochina 1985, 318.
6 I use first names in this section to avoid confusion between the two sisters.
7 See Carrer and Petrucci 2010, Britton 1981, and De Jongh 1979.
8 In Deleuze and Guattari's conceptual vocabulary (1988), rhizomatic connections refer to non-linear, decentralised, and interwoven networks of relationships and ideas, inspired by the botanical concept of a rhizome—a root system that spreads horizontally and creates multiple interconnected pathways. Unlike hierarchical structures, rhizomatic

connections emphasise fluidity, multiplicity, and collaboration, allowing for diverse and dynamic exchanges across different domains. See Tamboukou 2004 for an explication of the concept and a discussion of its application in educational contexts.

9 Sofia Kovalevskaya to Mittag-Leffler, Library of the Mittag-Leffler Institute [IML/SK-ML/L1].

10 https://images.app.goo.gl/HbmiB4zpPTPNN5Nu6

11 See Symons 2015 for a detailed discussion of this photograph in light of Deleuze's idea of the leap.

References

Agnesi, Maria Gaetana. 1801 [1748]. *Analytical Institutions for the Use of the Italian Youth*. Translated by John Colson. London: Taylor and Wilks.

Anzoletti, Maria Luisa. 1900. *Maria Gaetana Agnesi*. Milan: Cogliati.

Bakhtin, Mikhail. 1984 [1965]. *Rabelais and His World*. Translated by Hélène Iswolsky. Bloomington: Indiana University Press.

Britton, Carolyn. 1981. 'The Life and Keyboard works of Maria Teresa Agnesi, Milan (1720–1795)'. In *Rodgers and Hammerstein Archives of Recorded Sound, The New York Public Library. 'European Women Musicians of the Classical Period: Paper Presentations'*. New York: Public Library Digital Collections. Accessed 7 January 2025. https://digitalcollections.nypl.org/items/4e513f30-b83a-0133-7492-60f81dd2b63c

Carrer, Pinuccia, and Barbara Petrucci. 2010. *Donna Teresa Agnesi Compositrice Illustre (1720–1795)*. Genoa: edizioni San Marco dei Giustiniani.

De Jongh, Carolyn. 1979. 'The Life and Keyboard Works of Maria Teresa Agnesi'. D.M.A. thesis, University of Minnesota.

Deleuze, Gilles, and Felix Guattari. 1988. *A Thousand Plateaus: Capitalism and Schizophrenia*. Translated by Brian Massumi. London: The Athlone Press.

Deleuze Gilles and Parnet Clare. 2002. *Dialogues II*, translated by Hugh Tomlinson and Barbara Habberjam. London: Continuum.

Dewey, John. 1991. *How We Think*. Buffalo: Prometheus Books.

Du Châtelet, Emilie. 1740. *Instutitions de Physique* [Foundations of Physics]. Paris: Prault.

Du Châtelet, Émilie. 1961. *Discours sur le bonheur, édition critique et commente par Robert Mauzi* [Discourse on Happiness, Critical Edition and Commentary by Robert Mauzi]. Paris: Les Belles Lettres.

Du Châtelet, Émilie. 2009. *Selected Philosophical and Scientific Writings*. Edited and with an Introduction by Judith Zinsser, translated by Isabelle Bour and J. Zinsser. Chicago and London: The University of Chicago Press.

Eichner, C. J. 2004. *Surmounting the Barricades: Women in the Paris Commune*. Bloomington: Indiana University Press.

Foucault, Michel. 1970. *The Order of Things: An Archaeology of the Human Sciences*. London: Tavistock.

Koblitz, Ann Hibner. 1988. *Science, Women and Revolution in Russia*. New York: Routledge.

Koblitz, Ann Hibner. 1989. 'Career and Home Life in the 1880s: The Choices of Mathematician Sofia Kovalevskaia'. In *Uneasy Careers and Intimate Lives: Women in Science, 1789, 1979*, edited by Pnina G. Abiram and Dorina Outram, 172–90. New Brunswick, NJ: Rutgers University Press.

Koblitz, A., H. 1993 [1983]. *A Convergence of Lives: Sofia Kovalevskaia: Scientist, Writer, Revolutionary*. New Brunswick and London: Rutgers University Press.

Kochina, Pelageya. 1985. *Love and Mathematics: Sofia Kovalevskaya*. Translated by Michael Burov. Mir Publishers.

Kölving, Ulla, et Andrew Brown, eds. 2018. *La Correspondance d' Émilie Du Châtelet*. 2 vols. Ferney-Voltaire: Centre international d'étude du xviiie siècle.
Kovaleskaya, Sofia. 1951. *Воспоминания и письма* [Memories and Letters]. Edited and annotated by Solomon Yakovlevich Shtraikh. Moscow: USSR Academy of Sciences. [Second edition 1961]
Kovaleskaia, Sofia Vladimirovna (Fufa). 1951. 'Memories of My Mother' [Воспоминан ця о материъ]. In Kovaleskaya, Sofia, *Memories and Letters*, 360-368. Moscow: USSR Academy of Sciences.
Kovalevskaya, Sofia. 1978. *A Russian Childhood*. Translated by Beatrice Stillman. New York: Springer-Verlag.
Krukovskaya, Anna, V. 1864a. 'The Dream'. *Epokha*, No. 8.
Krukovskaya, Anna. V. 1864b. 'Mikhail'. *Epokha*, No. 9.
Leffler, Anna Carlotta. 1895. 'Memoir'. In *Sonya Kovalevsky. Biography and Autobiography*, translated by Louise Von Cossel, 1–166. London: Walter Scott.
Lermontova, Iulia, V. 1951. 'Memories of Sofia Kovalevskaya' [Воспоминанця О Софье Ковадевской']. In Kovaleskaya, Sofia. *Memories and Letters*, 357-387. Moscow: USSR Academy of Sciences.
Lewis, Judith, S. 1995. 'Princess of Parallelograms and her daughter. Math and Gender in the Nineteenth Century English Aristocracy'. *Women's Studies International Forum* 18 (4): 387–94.
Martini, Carlotta. 2017. *Zwei Frauenleben für die Wissenschaft im 18. Jahrhundert: Eine vergleichende Fallstudie zu Émilie Du Châtelet und Maria Gaetana Agnesi* [Two Women's Lives for Eighteenth-Century Science: A Comparative Case Study of Émilie Du Châtelet and Maria Gaetana Agnesi]. Bearbeitet und herausgegeben von Gudrun Wolfschmidt (Hamburg:ntredition Nuncius Hamburgensis; Band 43).
Mazzotti, Massimo. 2007. *The World of Maria Gaetana Agnesi, Mathematician of God*. Baltimore: Johns Hopkins University Press.
McMillan, Dorothy. 2001. 'Brief Biographies'. In *Queen of Science: Personal Recollections of Mary Somerville*, edited by D. McMillan, 371–455. Edinburgh: Canongate.
Moore, Doris Langley. 1977. *Ada. Countess of Lovelace: Byron's Legitimate Daughter*. New York: Harper & Row.
Musielak, Dora. 2020. *Sophie Germain: Revolutionary Mathematician*. 2nd ed. Cham: Springer.
Pugh, Evelyn L. 1980. 'John Stuart Mill and the Women's Question in Parliament, 1865–1868.' The Historian 42 (3): 399–418.
Pugh, Evelyn L. 1980. 'John Stuart Mill and the Women's Question in Parliament, 1865–1868.' *The Historian* 42 (3): 399–418.
Semetsky, Inna. 2010. 'Eagerness for experience: Dewey and Deleuze on the Problematic of Thinking and Learning.' In *John Dewey and Continental Philosophy*, edited by Paul Fairfield, 233–65. Carbondale: Southern Illinois University Press.
Somerville, Mary. 1872. Oxford Bodleian Libraries, Dep. c. 355, MSAU-2, fols. 72–363.
Somerville, Mary. 2001 [1873]. *Queen of Science: Personal Recollections of Mary Somerville*. Edited by Dorothy McMillan. Edinburgh: Canongate.
Stein, Dorothy. 1985. *Ada, a Life and a Legacy*. Cambridge, MA: The MIT Press.
Stenhouse, Brigitte. 2021a. 'Mary Somerville: Being and Becoming a Mathematician'. PhD thesis The Open University. https://doi.org/10.21954/ou.ro.0001346f
Stenhouse, Brigitte. 2021b. 'Mister Mary Somerville: Husband and Secretary'. *The Mathematical Intelligencer* 43 (1): 7–18.
Stupuy, Jean-Léon-Hippolyte. 1896 [1879]. 'Étude sur la Vie et les œuvres de Sophie Germain. [Study on the Life and Works of Sophie Germain]'. In *Œuvres philosophiques de Sophie Germain, suivies de pensées et de lettres inédites,* 1–73. Paris: Firmin-Didot.
Symons, Stéphane. 2015. 'Leap into the Surface: Photography, Repetition and Recollection'. In *The Marriage of Aesthetics and Ethics*, edited by S. Symons, 220–33. Leiden: Brill.

Tamboukou, Maria. 2004. 'Nomadic Trails in the Unfolding of the Self. *Spacesofidentity. Net* 4 (2). https://doi.org/10.25071/1496-6778.8009

Tamboukou, Maria. 2025. 'Exceptional Women in Science Education? Émilie du Châtelet and Maria Gaetana Agnesi'. *Paedagogica Historica: International Journal of the History of Education* 61 (2): 207–27. https://doi.org/10.1080/00309230.2023.2238621

Tamboukou. Maria. 2023a. 'Ever Yours, Mathematically': Women's Letters and the Mathematical Imagination'. *Gender and Education* 35 (8): 742–57.

Tamboukou, Maria. 2023b. 'Feeling Happiness, Feeling Science: Diffractive Readings of Émilie Du Châtelet's and Sophie Germain's Philosophical Writings'. *Journal of the History of Women Philosophers and Scientists* 2: 68–88.

Tamsin E. Lorraine. 2000. 'Becoming-Imperceptible As a Mode of Self-Presentation: A Feminist Model Drawn from a Deleuzian Line of Flight'. *Resistance, Flight, Creation: Feminist Enactments of French Philosophy*: 179–94. https://works.swarthmore.edu/fac-philosophy/93

Taylor, Brian, W. 1998. 'Annabella, Lady Noel-Byron: A Study of Lady Byron on Education'. *History of Education Quarterly* 38 (4): 430–55.

Zinsser, Judith, P. 2006. *Daring Genius of the Enlightenment, Émilie du Châtelet.* New York: Penguin.

Index

Note: ‘n’ denotes footnote numbers

academies 6, 51, 155, 234; Bologna Academy of Sciences 6, 43, 125, 139, 148–149; French Institute 98, 102, 167–170, 174; Paris Academy of Sciences 9, 11–12, 85, 101–102, 125, 181, 189, 235; Royal Academy of Sciences 9, 85; Royal Society in London 106, 301
acoustics 12, 101
affects 3, 22, 39–40, 45, 126, 164, 194–195, 199, 207, 215–217, 254–255, 295
Agnesi enigma 11; *Analytical Institutions* 9, 11, 92, 96, 149, 155, 189, 234, 243; oration 10, 164, 312; *Philosophical Propositions* 11, 195, 223–224, 232, 234–237, 239–246, 302
agonism 4, 22, 159–162, 166, 174, 179, 187, 191; agonistic learning 166; agonistic spirit 167, 174; *see also* Arendt
Algarotti, F. 149–150, 155, 225, 242, 244–245
amelodic 278
Anzoletti, L. 11, 47, 165–166, 188n4, 240, 302
archives 21, 30, 33, 35–36, 38, 40–41, 45, 47, 51, 54, 61–62, 69, 71, 106, 150, 164, 295; archival assemblages 40; archival sensibility 31, 48, 61; Bodleian Libraries in Oxford 25, 34, 54, 61, 70, 111, 307; the British Library 61, 70, 155n6; meta-archives 62, 67; Moreniana Biblioteca in Florence 49, 70; The Paris manuscripts 230–231; physical archives 69; Veneranda Biblioteca Ambrosiana in Milan 45–46, 48–49, 70, 72, 92, 165–166, 195, 294; visual archives 164
assemblages 22–24, 28, 36, 38–40, 52, 83, 186, 207, 214, 270, 280, 294–296, 298–303, 311–312, 315–320
Arendt, H. 4, 22, 62, 159–162, 171–172, 174, 176, 180, 182, 196, 218, 222; *see also* agonism
astronomy 13, 14, 19, 20, 99, 110, 141–142, 152, 173–174, 239, 256–258, 303, 315
Audin, M. 30, 65–66, 74n62
autobiography 2, 13, 15, 25, 54–55, 57, 63, 104–105, 111, 141–143, 150, 154, 171, 173, 181, 196, 200, 275–277, 304, 307, 311, 315; automathographies 2–3, 20, 35, 82, 111, 153, 203, 294, 321
awards 13, 102, 125–126, 170–172, 183, 288; medals 167, 171–172, 174, 181, 188; Prix Bordin 18, 288

Babbage, C. 14, 16, 60–61, 110–114, 116–118, 141–145, 154–155, 167, 179, 198, 259, 297, 301–302, 305; Analytical Engine 116, 142–145, 179, 198; Difference Engine 113, 118, 144
Bachelard, G. 163, 270–279, 281, 283, 285–290, 291n3
Bakhtin, M. 214–215, 217, 298, 321n1
Barad, K. 21, 35, 126, 214, 245, 271
Bassi, L. 30, 125, 148, 155n5, 242
becomings 19, 31, 186, 218, 246, 270, 296, 298, 303, 315, 318; becoming imperceptible 47, 299
Bergson, H. 210–211, 219n10, 277–282, 286–287, 289–290, 293n1
Berlin 17–18, 49, 64, 118, 120–124, 183, 314
Bernoullis 85, 87–88, 139, 143, 167, 213, 229

Bölling, R. 65, 82, 119, 129n34, 180
Bologna University 9, 11, 93–94, 148, 244
Bonnel, R. 85, 88–91, 127n3
Brading, K. 225–227
Brown, A. 42–43, 71n2
Brown, H. 134, 137, 139, 145–147, 149, 152–153, 154n1
Burntisland 13, 57, 73n50
Byron, A. 15, 112–113; Lady Byron 16, 58, 60–61, 305; Lord Byron 16, 58, 260, 297, 305

calculus 9, 17, 93–94, 105, 111, 119, 128n26, 131, 141, 148–149, 168, 173–174, 176, 178, 180, 302, 319; differential calculus 17, 94, 128n26, 141, 148–149, 173–174, 176, 319; priority dispute 128n26, 141, 149
Cavarero, A. 31, 64, 68, 161, 176
Cavazza, M. 6, 10, 92, 127n10
Centre for the History of Women Philosophers and Scientists 42
Château de Cirey 8–9, 42–44, 72n19, 87–88, 90, 131–136, 148, 161, 194, 204, 308
Cicero 7, 159, 208; ciceronian rhetoric 234
Clairaut, A. 85, 125n3, 139–141, 213, 216
Cobbe, F. P. 55–56, 315–316
Coleman, S. D. 151, 280, 282–284, 290n1
Coleridge, S. T 261–262; *see also* romanticism
Comte, A. 49, 246, 252–254, 264n13
conversazioni 6, 38, 92, 164; *see also* salons
correspondence 4, 13, 15, 22, 41, 43–44, 50–52, 60–61, 65–66, 82–92, 100–107, 109–120, 122, 125–127, 141, 143–148, 153, 159, 164, 169–171, 173–175, 177–179, 181–182, 195, 197, 200, 205, 210–216, 229, 243, 297, 300–305, 307–309, 315, 318–319; amorous 86, 195, 211, 216; polemical 90, 147, 187; scientific 22, 43, 60, 82, 85, 89–91, 105, 125–127, 159, 297; *see also* epistolarity
Cupillari, A. 47

Del Centina, A. 25n15, 50–51, 73n32; and Fiocca 52, 74n38, 127n12, 128n13
Deleuze, G. 37, 185–186, 219n12, 319, 322n11; and Guattari, F. 36, 38–40, 67, 83, 186, 189n25, 270–271, 274, 277, 290, 295–296, 299, 303, 320, 321n8
De Mairan, J-J. 85, 147–148, 163–164, 229
De Morgan, A. 16, 61, 116–117, 176–179, 302, 306
Derrida, J. 48, 275
De Saint-Lambert, J-F. 9, 194, 204, 210–213, 216
Descartes, R 8, 127n8, 207, 226, 231, 237, 239, 244, 253, 319; Cartesian 90, 93–94, 127n8, 148, 188n2, 207, 225–226, 228, 236, 238–239, 241, 243–244, 251; cogito 238–239
Detlefsen, K. 5, 161, 230
diffractions 20–21, 318; diffractive readings 70, 216
digitisation 51, 70; contextual loss 51
Du Châtelet, Gabrielle-Pauline 8, 308–309; Dissertation on the Nature and Propagation of Fire 9, 16, 89; Foundations of Physics [*Institutions de Physique*] 9, 42, 88–90, 125, 139, 146–147, 211, 223–226, 230–231, 308; *Mathematical Principles of Natural Philosophy* [*Principia*] 9, 14, 90, 138–141, 146, 148, 168, 194, 227
durée 210–211, 219n10, 271, 278–279, 283, 285–286, 290, 291n4

echo 24, 53–54, 261, 269–271, 273–276, 278, 285, 288, 290
edifying literature 217
Edinburgh Review 113
elasticity theory 13, 50, 223, 300, 319
Encyclopédie 139–140, 154n3, 184
Enlightenment 4, 6, 8–10, 41, 46, 134, 139, 141, 204–205, 207–209, 226, 228, 230–231, 233, 237–238, 242, 258–259, 295–297, 313, 317; Catholic Enlightenment 10, 242
entanglements 22, 24, 35–36, 54, 67–68, 82, 100, 126, 185, 194, 199, 213, 216, 219, 270–271, 274–275, 277, 291, 296, 298, 321
epistolarity 82–83; dispute-letter 89; epistolarium 44, 82–83, 85, 125, 211; epistolary rhythm 176; epistolary sensibility 22, 83–84, 125–126; epistolary waves 71; letter-gazette 89, laboratory letter 90–91, 163
Époque Émilienne 8, 140, 264n5
Euler, L. 85, 127n3, 162, 164, 167, 173, 183, 188n2
event 8, 11, 23, 30, 62, 67–68, 120, 159, 183–188, 210, 319–320

exceptional woman 22, 184–187
existential choices 194–195, 219

Farge, A. 67–68, 70–71
feminism 1, 244; feminist genealogies 22, 82
fictitious marriage 119, 200, 285, 302, 313
Findlen, P. 10, 30, 134, 165, 234–235, 242–243, 245
fonds 51
Foucault, M. 3, 30–36, 38–40, 44, 46, 67–68, 184–185, 321; dispositif 35–40; scarcity principle 32–35, 44, 46
Fourier, J. 73n34, 102, 104, 188n7, 300, 307
French Revolution 12, 42, 72, 304
Frend, W. 16, 112, 175, 261, 305
Frenkel, E. 248
Frisi, A. F. 9–10, 47, 165, 242; *Elogio Storico di Donna Maria Gaetana Agnesi* 47

Gauss, K. 11–12, 51–52, 73n34, 84, 97–102, 127n12, 172, 188n7, 300, 302; *Disquisitiones Arithmeticae* 11–12, 97
genealogy 31–32, 35, 63, 82, 153, 184–185, 253, 291n8, 294, 296, 321; *see also* feminist genealogies
geographies of science and mathematics 4
Germain, M. S., *Considérations Générales* 11, 247; *Œuvres philosophiques* 11, 223–224, 247, 264n11
German idealism 252–253
Girton College 55, 316
Göttingen University 17, 51, 97, 124, 316
Gottsched, L. 147
Govoni, P. 1, 24n1, 217
graves 47, 52, 57, 66, 68, 299; English cemetery in Naples 57; Hucknall, Nottinghamshire 17, 58; Norra begravningsplatsen 66; Père-Lachaise 52

Hagengruber, R. 8, 88, 91, 127n6, 140, 147, 154n4, 163, 230, 264n5
Halmos, P. 2–3, 20
happiness 7, 18, 23, 196, 203–210, 224, 243, 283–284, 316; *Discourse on Happiness* 159, 195, 210
Hardy, G. H. 250
Hegel, G. F. F. 252–253; Hegelian dialectics 252
Heidelberg 17, 64, 124, 280, 313, 316
Herschel, J. 56, 105–111, 128n27, 141; Herschel, C. 315
history of mathematics 23–24, 54, 306, 318
Honig, B. 161
humanism 4

imagination 15–16, 18, 24, 35, 59, 68, 71, 135, 144–145, 185–186, 199–200, 211, 224, 231, 249–250, 252, 255–263, 277–278, 280–281, 285, 288–289, 291, 320; mathematical imagination 250; scientific imagination 198
instant 271–274, 276–279, 283
intellectual freedom 17, 194, 317
intra-actions 86, 126; *see also* Barad
Italian states 6, 95, 134, 155n5, 234, 241

Jaclard, A. V. 120, 181, 313; and the Paris commune 17, 313–314
Jacquier, F. 85, 148, 213, 308
Jesuits 234–235

Kant, I. 164, 250–253, 264n10; transcendental idealism 250, 253, 264n10
Keller, F. E. 1, 32, 218
Key, E. 63–64, 74n63, 199–201
Koblitz, A. H. 17, 277, 297, 310, 314
Kochina, P. 3, 64–65, 82, 120–121, 129n34, 179, 199–201, 269, 281–282, 310, 313, 316–317, 321n5
Kovalevskaya, S.; *Nihilist Girl* 151, 269, 287, 298; *A Russian Childhood* 63; *The Sisters Rajevski* 63, 150; *Struggle for Happiness* 283
Kovalevskaya, S. V (Fufa) 17, 64, 74n64, 309–310, 321n5
Kovalevskii, M. 18, 181–182, 200, 285, 287; Kovalevskii, V. 17, 313
Krukovskaya, A. 18, 313; Krukovskaya, S. 17

Lacoue-Labarthe, P. 215, 271, 274–275, 277, 289–290
Lagrange, J-L. 12, 97, 167–169, 173, 183, 307, 319
Laplace, P-S. 15, 99, 105–106, 110, 141–142, 153–154, 167–168, 171, 173, 176, 197, 301
Latin 5, 7, 10, 49, 66, 72, 139–140, 165, 167, 173, 184, 188n2, 208, 234, 240, 308
leap into the void 318

Leffler, C. 64, 150–151, 181, 281, 314, 317
Legendre, A-M. 97, 102, 104, 115, 128n19, 167–170, 188n7, 300
Leibniz, G. 87, 89–91, 93, 127n8, 128n26, 141, 146–147, 149, 208, 225–226, 228–232, 235, 241, 244–245, 319; principle of sufficient reason 229, 264n4n5
Lermontova, I. 74n64, 120, 200, 309–310, 316
Lherbette, J-A. 246–247, 314
Libri, G. 11–13, 49–51, 73n30n32, 97, 103–104, 196, 247, 300
lines of flight 183, 186, 188–189, 189n25, 246, 296, 299, 303, 314
London 14, 16–17, 51, 54, 57, 64, 109, 111, 116, 171, 176–177
love and mathematics 3, 23, 194, 202, 218
Lovelace, A., *Sketch of the Analytical Engine* 145; *What is imagination?* 15, 259

Mandeville, B. 9, 42, 133–138; *The Fable of the Bees* 9, 42, 133–134, 136–138, 308
Masotti, A. 45, 47, 240
mathematical sciences 20, 30, 47, 49, 82–83, 99, 103, 170, 183–185, 242, 246, 248, 252; mathematics as poetical science 15, 199, 259, 261–263
Maupertuis, P., L. 8, 43, 85–91, 159, 161–163, 243, 302
Mazzotti, M. 1, 10, 35, 46–48, 92, 97, 148–149, 188n4, 195, 234–237, 239, 242, 245, 306, 311–313
memory 9, 21, 30, 32–35, 47, 49–50, 52–53, 59, 62, 66, 68, 145, 196, 199, 205, 211, 271, 277, 280–281, 285, 288, 307; archival memory 34; counter-memory 3, 31–32, 35, 50; cultural memory 33, 53, 71; feminist memory 52; memorial practices 53; memory assemblages 39; memory *dispositif* 35, 37; memory machines 21, 30, 41, 47, 53, 67–68, 71; memory studies 3, 21, 33, 35–37, 39, 67, 71; memory vortex 37; memory wars 58, 61; memory work 31, 34–42, 51, 58, 60, 67; mnemonic technologies 33–35, 54, 59–60, 62; sites of memory [*lieux de mémoire*] 32, 34, 44, 52, 57, 67; *see also* Nora, Rigney; working memory 33–34
Menabrea, L. 15, 62, 116, 118, 142, 145, 179, 197, 306
Mendelson, M. 62, 64, 74n62, 150, 180
metaphysics 8, 23, 42, 87, 90–91, 146–147, 163, 196, 226, 228, 230, 238, 247, 253, 290
Middle Ages 251
Milan 9–10, 45–47, 49, 70, 92–93, 95, 148, 156n9, 164, 234–235, 242, 244–245, 294, 298, 311–312
Mill, J. S. 316
Minonzio, F. 232, 234–235, 237, 241–243
minor gentry 172, 296, 300–301
Mittag-Leffler, G. 17–18, 66, 180–183, 202, 281, 303, 317; *Acta Mathematica* 18
Moscow 17, 124, 281, 309
Musielak, D. 12, 51–52, 97–98, 102–103, 167, 247, 307
mythographies 18, 66, 200, 210

Napoleonic wars 12, 98
narratives 19, 21, 23–24, 31, 59, 67, 83, 86, 88, 122, 176, 185, 194, 200, 202–203, 217–219, 225, 235, 256, 270–271, 289, 294, 298, 305, 318; auto/biographical narratives 64, 194, 200; epistolary narratives 80, 83, 88, 216; extravagant narratives 86; narrative rhythmanalysis 269–275, 284, 287–288, 290; relational narratives 176, 318
Newton, I. 8–10, 14, 74, 85, 87–88, 90, 93, 105, 126n26, 138–142, 146–150, 153–154, 161, 163, 168, 176, 192, 211–213, 225–232, 235, 239, 241, 243–245, 251, 264n2, 319; *Principia* 9, 14, 90, 138–141, 146, 148, 168, 194, 227, 264n3
New York 44, 150
Nora, P. 32, 34, 44

occasionalism 238
œuvre à faire 176, 225, 231–232, 279–280
ontology 235, 238, 253
other voice 4, 64

Palibino 17, 150–151, 276–278, 313
parallax 256–257
Paris 6, 8, 12–14, 17–18, 44, 50–51, 53, 72, 74, 85–86, 97–98, 102–104, 120, 125, 141–142, 146, 150, 161–162, 167, 171, 176, 180–183, 194, 212–213, 216, 230–231, 246, 300, 304, 308, 313–314; Eiffel Tower 49
pathos 214–215, 217; discourse of pathos 214; epistolary pathos 214–215
Patterson, E. 15, 55, 110–111, 257, 308

personal and scientific 82, 87, 98, 101, 107, 109, 112, 118, 126, 218, 302; *see also* epistolarity
philosophy 3, 6–7, 11, 20–21, 23–24, 36, 40, 49, 87, 91–92, 126, 133–134, 136, 139–140, 145, 150, 165, 185–186, 195, 199, 213, 216, 223–226, 228–233, 235, 237, 240–244, 246, 257, 263, 274, 290, 294–295, 297, 302, 315; educational philosophy 177; mathematics and philosophy 223, 294; moral philosophy 205; natural philosophy 4, 164, 226, 229, 239; philosophy of knowledge 23, 225; philosophy of rhythm 271, 291; philosophy of science 91, 140, 228, 231, 288; philosophy of time 219n10, 273; philosophy of translation 133; poetical philosophy 199; political philosophy 160; positive philosophy 49, 248, 252, 253; process philosophy 67, 253, 271; rationalist philosophy 237; speculative philosophy 226; Stoic philosophy 264n9; *see also* Whitehead
Pinheiro dos Santos, L.A. 279–281; *see also* rhythmanalysis
Piscopia, E. 165
plane of consistency 188, 295–296, 301, 314; *see also* Deleuze
poetry 3–4, 6, 19–20, 23, 123, 135, 137, 151–152, 199, 202, 249–250, 259, 262, 269–270, 280–281, 288, 291, 305; poetical philosophy 199; poetical science 15, 199, 259, 261–263; *The Rainbow* 59
portraits 19, 44–46, 53, 57, 60, 64, 72n23, 74n51; pen-portraits 7, 18, 20
process epistemologies 223–224, 246
Project Vox 188n3

Rampinelli, R. 92–96, 98, 302
rhizomatic connections 303, 315, 321n8
rhythm 23–24, 45, 52–53, 59, 67, 69–71, 152, 166, 176–177, 211, 215, 269–290, 320; epistolary rhythm 176; poetic rhythm 211, 280–281; polyrhythmic reality 273, 278, 283; refrain 12, 65, 270–271, 273–275, 277–278, 290, 304; rhythmanalysis 24, 270, 280–281; rhythm assemblage 270–271, 274; *see also* narrative rhythmanalysis
Riccati family 82, 92–96, 98, 103, 232, 302
Rigney, A. 32–35, 37, 44, 46–47, 53, 55, 59–60, 62, 64, 67, 71; convergence of memories 33, 62; memory transfers 33, 62; recursivity in remembrance 32–34, 47, 53, 62; recycling of models of remembrance 33, 62; selectivity of recall 62; vortex of memory 34, 37, 44, 62
Riviera 18, 285–288
Roero, C. S. 92, 95–96, 127n11
Rossi, E. 234–235, 239

salons 6, 8, 38, 206, 297; *see also* conversazzioni
Schiebinger, L. 6, 32, 187
scholarly persona 4–5, 32, 45, 161
science and mathematics 4, 6, 11, 16, 23, 32, 38, 54, 84, 91, 105, 109, 111, 114, 125, 139, 149, 153, 164, 184–186, 188, 209, 244, 312; public science 5, 161, 166, 296
Scotland 13–14
situated knowledges 40
sociomental topography of the past 32, 50
Söderqvist, T. 195, 217–219
Somerville, M. *Mechanism of the Heavens* 15, 105, 108, 141, 173–174, 197; *On the Connexion of the Physical Sciences* 56, 256–258; *Personal Recollections* 13–15, 53–57, 112, 151, 197, 311; Somerville daughters 32, 54–56, 310–311; William 14, 141, 300–301
space/time 48, 66, 251, 276
spectacle 10–11, 47, 92, 126, 126n10, 165, 195, 206, 298–299, 311
Spinoza, B. 207
Stanley, L. 43, 82–83, 122, 125–126, 211
Steedman, C. 71
Stenhouse, B. 15, 54, 56, 105–106, 109–110, 128n26, 172–174, 188n12, 300–301
Stockholm 17–18, 30, 62, 66, 119, 150, 179, 182–183, 281, 309, 317
Stupuy, H. 11–12, 49–50, 247, 252, 254, 264n1, 299

Tamboukou, M. 18, 22–23, 25, 28–29, 31, 34–36, 40–41, 48, 51–52, 66–68, 70–71, 74n67, 82–83, 125, 161, 166, 188n11, 189n25, 196, 200, 216, 269–270, 290, 295
Toole, B. A. 25n19, 74n54, 128n32, 175, 261
traces 21, 29–31, 33, 35, 39, 45–46, 66–67, 69, 71, 86, 91, 95, 102, 126, 144, 147, 195, 200, 218, 280
transdisciplinarity 20

translation 5, 9, 15, 22, 24, 33–35, 42, 46, 49, 51–53, 62, 64–65, 84, 90, 105–106, 108, 110, 116, 118, 127n9, 129n34, 133–135, 137–154, 179, 188nn3, 7, 8, 194, 204, 223, 277, 295, 306; cultural turn in translation 134, 153; cultures of translation 133–134; the dynamics of translation 22, 133, 154; political labour of translation 150; the politics of translation 133, 152, 153
transpositions 20–21
Turin 97, 142, 294
Turing, A. 61

uniqueness 146, 160, 211, 218, 272, 274, 296, 303

Valéry, P. 275, 288–290
Victorian era 112
vis-viva 87, 127n8
Voltaire 7–8, 41–42, 44, 87, 89

Wade, I. O. 24n12, 42, 204, 219n9
Wallace, W. 105–106, 172–174, 188n12; Wallace J. 105–106, 172, 174
waves 69; epistolary waves 71
Weierstrass, K. 17, 64–66, 74n65, 82, 85, 118–125, 129n34, 180, 182–183, 202, 281, 302
Whitehead, A. N. 67, 185, 204, 219n6, 253–255, 264n14
Whitehead, B. 205, 207, 209, 212
Wolff, C. 85, 147, 226, 229–231

Zen, S. 232, 242, 245–246
Zhenyi, W. 7, 19–20, 134, 152–154, 195, 294–295; *Collection of the Defeng Pavilion* 19–20, 152
Zinsser, J. 5, 7–8, 24n4n12, 72n21, 85–87, 135, 137–140, 148, 161, 163–164, 219n6, 226–227, 303–304, 308

FSC
www.fsc.org
MIX
Papier | Fördert gute Waldnutzung
FSC® C083411

Kolibri 360 GmbH
Ferdinand-Jühlke-Straße 7
99095 Erfurt, Deutschland
produktsicherheit@kolibri360.de